THE GENETICS OF BEHAVIOUR

NORTH-HOLLAND RESEARCH MONOGRAPHS
FRONTIERS OF BIOLOGY
VOLUME 38

Under the General Editorship of
A. NEUBERGER
London
and
E. L. TATUM
New York

**NORTH-HOLLAND PUBLISHING COMPANY – AMSTERDAM · OXFORD
AMERICAN ELSEVIER PUBLISHING COMPANY, INC. – NEW YORK**

THE GENETICS OF BEHAVIOUR

Edited by

J. H. F. van Abeelen

*Genetics Laboratory, Catholic University,
Nijmegen, The Netherlands*

1974

NORTH-HOLLAND PUBLISHING COMPANY – AMSTERDAM · OXFORD
AMERICAN ELSEVIER PUBLISHING COMPANY, INC. – NEW YORK

© *North-Holland Publishing Company* – *1974*

All rights reserved. No part of this publication may be reproduced, stored in a retrieval system, or transmitted, in any form by any means, electronic, mechanical photocopying, recording or otherwise, without the prior permission of the copyright owner.

Library of Congress Catalog Card Nr: 74-80118

ISBN North-Holland for this series 0 7204 7100 1
ISBN North-Holland for this volume 0 7204 7137 0
ISBN American Elsevier. 0 444 10670 7

PUBLISHERS:
NORTH-HOLLAND PUBLISHING COMPANY – AMSTERDAM
NORTH-HOLLAND PUBLISHING COMPANY, LTD. – OXFORD

Sole distributors for the U.S.A. and Canada:
AMERICAN ELSEVIER PUBLISHING COMPANY, INC.
52 VANDERBILT AVENUE, NEW YORK, N.Y. 10017

PRINTED IN THE NETHERLANDS

Contents

General preface XV

Foreword XVII

Preface XXI

Part I. Biometrical genetic and evolutionary aspects

    Chapter 1.   How to analyse the inheritance of behaviour in animals – the biometrical approach   1

1.1. Introduction   1
    1.1.1. Fundamentals   1
    1.1.2. Classical and biometrical approaches   3
1.2. The biometrical model   5
    1.2.1. Polygenic system   5
    1.2.2. Determinants of the phenotype   7
    1.2.3. Limitations   9
1.3. Scaling   10
    1.3.1. General considerations   10
    1.3.2. Scaling tests   11
        1.3.2.1. Parents, F_1, F_2, and backcrosses   12
        1.3.2.2. Diallel crosses   17
        1.3.2.3. Triple-test cross   18
    1.3.3. Alternative procedures   19
        1.3.3.1. Adequate scale   19
        1.3.3.2. Non-allelic interaction   23
1.4. Maternal effects and sex-linkage   24
1.5. Non-inbred subjects   28
1.6. Genotype-environment interaction   29

1.7. Heritability estimates: their uses and limitations	33
1.8. The problem of heterosis	34
1.9. Estimates of the number of genes	36
1.10. Elaborations of the model	37
1.11. Conclusions	38
References	39

Chapter 2. What genetical architecture can tell us about the natural selection of behavioural traits — 43

2.1. Introduction	43
2.2. Initial separation: phenotype and genotype	44
2.3. Further subdivisions: the genetical architecture	45
2.4. Inferences to natural selection	46
2.4.1. Kinds of selection	46
2.4.2. Genetical consequences of selection	47
2.4.3. Experimental evidence	49
2.4.4. Behavioural application	50
2.5. The environmental dimension	52
2.5.1. Genotype and environment	52
2.5.2. Micro-environmental variation	53
2.5.2.1. Stability	53
2.5.2.2. Heterosis	54
2.5.3. Macro-environmental variation	55
2.5.3.1. An example of its application to learning theory	56
2.6. Conclusions	59
References	61

Chapter 3. A diallel study of exploratory behaviour and learning performances in mice — 65

3.1. Introduction	65
3.2. Methods	65
3.3. Findings	69
3.3.1. Exploratory behaviour	69
3.3.1.1. Apparatus and procedure	69
3.3.1.2. Results and discussion	70
3.3.1.3. Summary	73
3.3.2. Passive avoidance	74
3.3.2.1. Apparatus and procedure	74
3.3.2.2. Results and discussion	74
3.3.3. Runway	76
3.3.3.1. Apparatus and procedure	76

3.3.3.2. Results and discussion	76
3.3.4. T-maze	81
3.3.4.1. Apparatus and procedure	81
3.3.4.2. Results and discussion	83
3.4. General discussion and summary	86
References	88

Chapter 4. Applications of biometrical genetics to human behaviour   91

4.1. Statistical and genetical models	92
4.2. Assumptions underlying the model	94
4.2.1. Genetic expectations	96
4.3. Applications	98
4.3.1. General intelligence	98
4.3.2. Educational attainments	108
4.3.3. Schizophrenia	111
4.4. Conclusion	116
References	116

Chapter 5. The genetic basis of evolutionary changes in behaviour patterns   119

5.1. Introduction	119
5.2. Hybrid behaviour	121
5.2.1. Methodological limitations	121
5.2.2. Demonstration of the genetic basis of behavioural differences	122
5.2.3. The genetic analysis of behavioural differences	128
5.3. Discussion	133
5.3.1. What does hybrid behaviour say about the problem of genetic and environmental determination of behaviour traits?	133
5.3.2. The polygenic basis of evolutionary changes	134
5.4. Summary	137
References	137

Chapter 6. A behavioural variant in the three-spined stickleback   141

6.1. Introduction	141
6.1.2. Breeding cycle	141
6.1.3. Courtship	142
6.1.4. Creeping through	143
6.1.5. 'Double' creeping through	144
6.1.6. Morphological features and their geographical distribution	145
6.2. Material and methods	147
6.3. Results	148
6.4. Discussion	162
References	165

Chapter 7. The role of sexual selection in the maintenance of the genetical   167
heterogeneity of Drosophila populations and its genetic basis

7.1. Introduction   167
7.2. Methodology   169
7.3. Sexual selection in large experimental populations   171
7.4. Measurement of the vigour of males and the receptivity of females   172
7.5. Behavioural differences between 'v' and 'cn' males   174
7.6. The genetical basis of the receptivity of females and the vigour of males   175
7.7. Advantage of inversion heterozygotes in sexual selection   177
7.8. Selective advantage of hybrids from crosses between wild-type stocks   178
7.9. Single-gene effects in sexual selection   179
7.10. Selection for quantitative aspects of sexual behaviour   180
7.11. Conclusions   181
References   183

Chapter 8.   Variability in the aggressive behaviour of Mus musculus   185
domesticus, its possible role in population structure

8.1. Introduction   185
8.2. The 'Kooipolder' population   187
8.3. Dominant, subordinate, and tolerant males   190
8.4. A possible role of tolerance in population dynamics   193
8.5. Dominance and reproduction   194
8.6. Attack latency and reproduction   195
8.7. Genetics of aggressive behaviour   197
References   198

Part II.   Phenogenetic and regulatory aspects

Chapter 9.   Activity and sexual behaviour in Drosophila melanogaster   201

9.1. Introduction   201
9.2. Selection for changes in locomotor activity   205
9.3. Biochemical substrate of activity   209
9.4. Courtship and mating behaviour   210
9.5. Genetic variation in courtship and mating speed   216
9.6. Visual stimuli   222
9.7. Auditory stimuli   225
9.8. Physiological factors influencing courtship and mating   230
9.9. Activity in relation to courtship and mating behaviour   243
9.10. Incipient sexual isolation   248
9.11. Conclusions   252
References   253

Chapter 10. Genes affecting behaviour and the inner ear in the mouse — 259

- 10.1. Introduction — 259
- 10.2. Mutants with shaker-waltzer syndrome — 259
- 10.3. Behaviour — 260
- 10.4. Pathology — 262
- 10.5. Causal relationships — 264
- 10.6. Mutants with uncomplicated deafness — 266
- 10.7. Audiogenic seizures — 267
- 10.8. Wider considerations — 269
- References — 271

Chapter 11. Personality development in XO-Turner's syndrome — 273

- 11.1. Introduction — 273
- 11.2. Turner's syndrome — 275
- 11.3. Cognitive development in Turner's syndrome — 275
- 11.4. Personality development — 278
- 11.5. Gender role and gender identity — 281
- 11.6. Concluding remarks — 287
- References — 288

Chapter 12. Emotional responsiveness and interspecific aggressiveness in the rat: interactions between genetic and experiential determinants — 291

- 12.1. Introduction — 291
- 12.2. The concept of 'emotional and social responsiveness' — 292
- 12.3. Genetic and experiential determination of aggressive behaviour — 294
- 12.4. Mouse-killing behaviour in the rat and its underlying brain mechanisms — 296
    - 12.4.1. Appetitively motivated attack-behaviour — 297
        - 12.4.1.1. Processes and mechanisms involved in the release of appetitively motivated attack behaviour — 298
    - 12.4.2. Aversive motivations and their behavioural consequences — 302
        - 12.4.2.1. Interference with spontaneous attack-behaviour in the experienced killer-rat — 302
        - 12.4.2.2. Elicitation of an affective' kind of mouse-killing behaviour in the non-killing rat — 304
        - 12.4.2.3. Changes in the emotional responsiveness of the rat — 305
        - 12.4.2.4. Interference with ontogenetic shaping of social-emotional adaptations — 312
- References — 314

Chapter 13. Genetic determination of aggressive behaviour — 321

13.1. Measurement of aggression — 322
13.2. Strain differences — 324
13.3. Selective breeding for aggressiveness — 328
13.4. Variation in the physiological mechanisms of aggressive behaviour — 332
13.5. Interaction of genotype and environment in the development of aggressiveness — 336
13.6. Conclusion — 342
13.7. Summary — 343
References — 343

Chapter 14. Genotype and the cholinergic control of exploratory behaviour in mice — 347

14.1. Introduction — 347
14.2. Methods — 350
    14.2.1. Breeding conditions — 350
    14.2.2. Observation techniques — 350
    14.2.3. Determination of enzyme activity — 352
    14.2.4. Drug administration — 352
    14.2.5. Statistical evaluation — 353
14.3. Results and discussion — 354
    14.3.1. Inbred strains and hybrids — 354
        14.3.1.1. Behavioural differences — 354
        14.3.1.2. Acetylcholinesterase-activities — 357
        14.3.1.3. Responses to drugs — 358
    14.3.2. Selected lines — 364
        14.3.2.1. Behavioural differences — 364
        14.3.2.2. Acetylcholinesterase-activities — 367
        14.3.2.3. Responses to drug treatment — 368
14.4. Summary and prospects — 371
References — 372

Chapter 15. Genetic factors in the control of drug effects on the behaviour of mice — 375

15.1. Heterogeneous stocks — 378
15.2. Inbred strains — 379
15.3. Recombinant inbred strains — 388
References — 392

Chapter 16. Neurochemical correlates of behaviour in inbred strains of mice — 397

16.1. Introduction — 397

16.2. Behavioural characteristics of inbred strains of mice	398
16.3. Serotonin, norepinephrine, gamma-aminobutyric acid, and behaviour in mouse strains	399
16.4. The cholinergic system and behaviour in mouse strains	401
16.5. Multiple neurochemical correlates of behaviour	402
16.5.1. Choline acetyltransferase and acetylcholinesterase	402
16.5.2. Norepinephrine levels and turnover	406
16.5.3. Dopamine levels and turnover	408
16.5.4. Serotonin levels and turnover	409
16.6. Conclusions	409
References	411

Chapter 17. Behavioural effects of psychoactive drugs influencing the metabolism of brain monoamines in mice of different strains   417

17.1. Introduction	417
17.2. Methods	418
17.3. Results	419
17.3.1. Rearing and locomotion	419
17.3.2. Effects of d,1-amphetamine sulfate	422
17.3.3. Effects of anticholinergics	425
17.3.4. Imipramine-like antidepressants	426
17.3.5. Antagonism to reserpine-like benzquinolizine Ro 4-1284	427
17.3.6. Effects of reserpine and its derivatives	428
17.4. Conclusion	429
References	430

Subject index   *433*

List of contributors

J. H. F. van Abeelen, *Genetisch Laboratorium, Catholic University, Nijmegen, The Netherlands*

F. J. Bekker, *Psychologisch Instituut, State University, Leiden, The Netherlands*

E. Boesiger, *Laboratoire de Génétique Expérimentale des Populations, Montpellier, France*

P. L. Broadhurst, *Department of Psychology, University of Birmingham, England*

B. Burnet, *Department of Genetics, University of Sheffield, England*

J. Busser, *Biological Centre, State University, Groningen, The Netherlands*

J. P. Chaurand, *Centre de Neurochimie du CNRS, Laboratoire de Neurophysiologie, Strasbourg, France*

K. J. Connolly, *Department of Psychology, University of Sheffield, England*

M. S. Deol, *Department of Animal Genetics, University College London, England*

A. Ebel, *Centre de Neurochimie du CNRS and Institut de Chimie Biologique, Faculté de Médecine, Strasbourg, France*

F. Eclancher, *Centre de Neurochimie du CNRS, Laboratoire de Neurophysiologie, Strasbourg, France*

D. Franck, *Zoologisches Institut und Zoologisches Museum der Universität Hamburg, Germany*

D. W. Fulker, *Department of Psychology, University of Birmingham, England*

M. 't Hart, *Zoölogisch Laboratorium, State University, Leiden, The Netherlands*

J. L. Jinks, *Department of Genetics, University of Birmingham, England*

P. Karli, *Centre de Neurochimie du CNRS, Laboratoire de Neurophysiologie, Strasbourg, France*

E. Kempf, *Centre de Neurochimie du CNRS and Institut de Chimie Biologique, Faculté de Médecine, Strasbourg, France*

J. M. L. Kerbusch, *Psychologisch Instituut,Catholic University, Nijmegen, The Netherlands*

K. M. J. Lagerspetz, *Department of Psychology, The Swedish University of Turku, Finland*

K. Y. H. Lagerspetz, *Zoophysiological Laboratory, Department of Zoology, University of Turku, Finland*

I. P. Lapin, *Laboratory of Psychopharmacology, Bekhterev Psychoneurological Research Institute, Leningrad, U.S.S.R.*

G. Mack, *Centre de Neurochimie du CNRS and Institut de Chimie Biologique, Faculté de Médecine, Strasbourg, France*

P. Mandel, *Centre de Neurochimie du CNRS, Strasbourg, France*

A. Oliverio, *Laboratorio di Psicobiologia e Psicofarmacologia CNR, Rome, Italy*

G. A. van Oortmerssen, *Genetisch Instituut, State University, Groningen, The Netherlands*

P. Schmitt, *Centre de Neurochimie du CNRS, Laboratoire de Neurophysiologie, Strasbourg, France*

P. Sevenster, *Zoölogisch Laboratorium, State University, Leiden, The Netherlands*

M. Vergnes, *Centre de Neurochimie du CNRS, Laboratoire de Neurophysiologie, Strasbourg, France*

A. Zweep, *Biological Centre, State University, Groningen, The Netherlands*

General preface

The aim of the publication of this series of monographs, known under the collective title of '*Frontiers of Biology*', is to present coherent and up-to-date views of the fundamental concepts which dominate modern biology.

Biology in its widest sense has made very great advances during the past decade, and the rate of progress has been steadily accelerating. Undoubtedly important factors in this acceleration have been the effective use by biologists of new techniques, including electron microscopy, isotopic labels, and a great variety of physical and chemical techniques, especially those with varying degrees of automation. In addition, scientists with partly physical or chemical backgrounds have become interested in the great variety of problems presented by living organisms. Most significant, however, increasing interest in and understanding of the biology of the cell, especially in regard to the molecular events involved in genetic phenomena and in metabolism and its control, have led to the recognition of patterns common to all forms of life from bacteria to man. These factors and unifying concepts have led to a situation in which the sharp boundaries between the various classical biological disciplines are rapidly disappearing.

Thus, while scientists are becoming increasingly specialized in their techniques, to an increasing extent they need an intellectual and conceptual approach on a wide and non-specialized basis. It is with these considerations and needs in mind that this series of monographs, '*Frontiers of Biology*' has been conceived.

The advances in various areas of biology, including microbiology, biochemistry, genetics, cytology, and cell structure and function in general will be presented by authors who have themselves contributed significantly to these developments. They will have, in this series, the opportunity of bringing together, from diverse sources, theories and experimental data, and of integrating theses into a more general conceptual framework. It is unavoidable, and probably even desirable, that the special bias of the individual authors will become evident in their contributions. Scope will also be given for presentation of new and challenging ideas and hypotheses for which complete evidence is at present lacking. However, the main emphasis will be on fairly complete and objective presentation of the more important and more rapidly advancing aspects of biology. The level will be advanced, directed primarily to the needs of the graduate student and research worker.

Most monographs in this series will be in the range of 200–300 pages, but on occasion a collective work of major importance may be included somewhat exceeding this figure. The intent of the publishers is to bring out these books promptly and in fairly quick succession.

It is on the basis of all these various considerations that we welcome the opportunity of supporting the publication of the series '*Frontiers of Biology*' by North-Holland Publishing Company.

E. L. TATUM
A. NEUBERGER, General Editors

Foreword

Genetics is generally thought of as beginning with the rediscovery of Mendel's paper in which the relation between segregation of developed phenotypes and segregation of genes in gametogenesis was established. More and more detailed study of this relationship, especially after the discovery of linkage and proof of the chromosome theory, together with attempts to explain the causal sequence connecting the genotype of the fertilised egg with the observable phenotype, have led to genetics invading, pervading, and hybridising with disciplines with other labels, a process that continues with increasing momentum. It does not seem so long ago that one was asked whether one was a plant geneticist or an animal geneticist; but nowadays we have molecular genetics, biochemical genetics, physiological genetics, immunogenetics, developmental genetics, population genetics, biometrical genetics, behaviour genetics, and pharmacogenetics, quite apart from genetics qualified with a great range of taxonomic adjectives.

Of these 'branches' of genetics, biometrical genetics and behaviour genetics are in fact the elder, for both stem from Galton. But the genetics of behavioural variables was slow to develop and it is only relatively recently that it has approached the explosive stage, considerable impetus having been given by the publication of Fuller and Thompson's book in 1960.

The reasons why this development has been slower than that in other areas are many. Extreme environmentalism among psychologists and behaviourists has been one cause, but the main problem has been the intrinsic difficulty that arises from the complexity of behavioural phenotypes, the difficulty of defining them in useful ways, and the involvement of environmental factors difficult to specify. Continuous variation with complex genotype-environment interactions must be a major component of behavioural genetics, so that full development of the subject had to wait upon techniques that could only be worked out in other branches of genetics.

Thus it is that much behaviour genetics was for a long time largely confined to demonstrations of the heritability of behavioural variables, either through the simpler biometrical approaches or through selection experiments, and to the study of the effects of a few major gene differences with behavioural consequences.

More numerous studies and more sophisticated techniques now hold hope of rapid progress, and, hence, we are at a stage when it is desirable to give especial consideration to the techniques that might help towards particular ends. The genetics of behavioural variables is of course of interest for itself, and with respect to man of social interest. But there are three specific targets on which we should keep our sights.

First, genetical understanding is necessary if we are to make conclusions about past evolution, to make predictions about likely future trends that result from any correlations between behaviour variables and fitness, and to understand the population structure of species where this is affected by behaviour. This aim may be aided by the study of polymorphisms and of major mutants, and generally by investigation of courtship behaviour and mating choice as in Boesiger's, and Connolly and Burnet's papers. But in its general aspects it necessitates the techniques of biometrical genetics as exemplified particularly in the paper of Broadhurst and Jinks.

Second, behaviour variables are complex variables, and one of the cardinal problems of behaviour studies is to know how to describe and measure such variables in terms of biologically meaningful components, how in fact to dissect such characters. Study of the ways in which genetic segregation at specific loci affects components of the variables provides one way of looking for such components. This aim can be aided by comparative study of different strains, by study of major mutants, and by biometrical studies, especially if the development and physiology of the behavioural variants are examined as for example in the papers by van Abeelen, Lapin, and Karli.

Third, the study of such segregations may also be a way, perhaps the only way, of determining which biochemical and physiological differences are causally connected with the behaviour differences, thus providing a valuable tool for the study of the biochemistry, physiology, and development of such variables. Further, given gene differences in genetically well-controlled stocks, it should be possible to study their interaction with specific environmental variables and hence aid the specific understanding of the environmental factors involved in the genotype-environment interaction that must be so important. Here I would stress that in the longer run, for character dissection, for physiological and for biochemical purposes, we must also seek to get at the effects of single-locus segregations whose effects are within the normal range. This requires the techniques that have sometimes allowed us, with non-behavioural continuous variables, to abstract, and study separately, single segregating units, thus supplementing the approach of biometrical genetics. The paper by Oliverio, which uses a technical approach essentially similar to that which has been used to locate polygenes in *Drosophila* and wheat, provides an example of this last approach.

I am grateful for the opportunity of welcoming a book in which the editor has collected together papers exemplifying all these approaches. I am sure that it will prove a great stimulus to future research on all these lines.

J. M. Thoday
University of Cambridge, England

Dedicated to:
JOHN L. FULLER, *Explorer and Founder*

Preface

Like other phenotypes, behaviour is accessible to genetic analysis and manipulation. It is apparent that genetic procedures can contribute much to a deeper understanding of the variability and the causation of behaviour. Two main approaches to the behavioural phenotype may be distinguished: firstly, biometrical analysis and the placing of observed variations into an evolutionary perspective and, secondly, the study of behavioural development and of underlying phenogenetic mechanisms. Both facets can be found in Jerram L. Brown's (1969) definition: '... instinct is the collection of mechanisms through which the effects of evolution on behaviour are mediated'. This volume is accordingly divided into two parts which cover, respectively, the quantitative-genetic and phylogenetic aspects (Chapters 1 through 8) and the gene-physiological and ontogenetic aspects (Chapters 9 through 17). The species most frequently used for investigations into the genetic foundations of behaviour are fruit flies, some fish species, Norway rats, house mice, and man; each of these categories is represented in the book. The past few decades have witnessed the development of the inter-discipline of behaviour genetics in the U.S.A. and, shortly thereafter, in Europe. This volume intends to provide an overview of contemporary behaviour-genetic research from European laboratories. It should be useful to researchers and interested students alike.

J. H. F. van Abeelen

PART I

Biometrical genetic and evolutionary aspects

CHAPTER 1

How to analyse the inheritance of behaviour in animals - the biometrical approach*

J. L. Jinks and P. L. Broadhurst

1.1. Introduction

1.1.1. Fundamentals

"The idea of a nature-nurture controversy is simply meaningless to a well-trained biologist". So wrote Ghiselin (1969, p. 209), expressing a view which will, we hope, need little emphasis to the reader of this book but which has frequently had to be stressed in behavioural writings hitherto. The truism that behaviour is never wholly environmentally determined and never wholly genetically determined but is always the outcome of varying degrees of joint determination of both is one which, we take it, is so widely accepted nowadays as to need no further elaboration. But the dependent proposition that the *analysis* of behaviour into those two major classes of determinants and their sub-divisions can only be accomplished by recognising and accepting the limitations and the potentialities of the fundamental genetic equation that

$$\text{Phenotype (P)} = \text{Genotype (G)} + \text{Environment (E)} \qquad (1)$$

cannot be said to have received comparably wide acceptance. And this despite such expression of views as that of the distinguished entomologist Wilson (1971), who writes: "Few behavioral biologists ask any longer whether the responses are instinctive or learned, this having proved to be

---

* Some of the work reported in this chapter was supported by grants from the (British-) Medical Research Council. Its preparation was furthered by P.L.B.'s Resident Fellowship (1973–74) at N.I.A.S. (Nederlands Instituut voor Voortgezet Wetenschappelijk Onderzoek op het Gebied van de Mens- en Maatschappijwetenschappen).

an inefficient classification. Instead a series of questions with a more operational basis are posed. For example, exactly how stereotyped is a given behavioral act? If it varies among individuals belonging to the same species, we next enquire how much of the variation is genetic. Under ideal conditions the genetic component of the variance can be precisely measured, with the fraction it constitutes being labeled the heritability. The remainder of the variance by definition is acquired by interaction with the environment, the effects of which are manifested as deviations in organogenesis, as physiological change in established organs, or as learning in the strict sense" (p. 197).

Thus it is that the phenotype is all that we can ever directly observe. Recognition that simply altering the environment markedly from the normal, as in, for example, the isolation experiment (Lorenz, 1966), can never lead us to valid deductions about the genotype has been slow to come. In such a situation, of course, we merely observe a new, different and perhaps extreme phenotype, a result of possibly complex genotype-environment interaction, perhaps involving part of the genotype not normally concerned with phenotypic determination within the normal environmental range. Support for this view comes from instances which have been detected of the "buffering" by maternal genotypes of the genotypic response of offspring to stress (Fulker, 1970) to be discussed in detail later (§ 1.4). Such analyses, and, indeed, any others which seek to examine the extent to which genotype and environment play a part in determining the resultant phenotype and which seek, moreover, to go beyond this simple objective and to unravel the often subtle complexities of their interaction cannot avoid involving the genetic process itself as an experimental variable. That is to say, nothing can be deduced from studying the phenotype of one generation by itself, and indeed, little can be deduced from comparing the phenotypes of several generations or subspecies, or even species unless they stand in direct, family, i.e. genetical, relationship to one another. The basis for this assertion is not far to seek: only in the process of reproduction does the genetical mechanism come into action and can the effects of meiosis and chromosomal inheritance come to be observed. This is not to say, of course, that comparisons of sexes, of varieties or of strains, given that they have been reared in such a way that the effects of environmental variation can be discounted, cannot contribute to our knowledge, or that comparisons of species cannot contribute to taxonomic clarification and support inferences about evolutionary distinctions, but in both such cases these comparisons can be suggestive only. What can be deduced in the first case is a presumption of the strength of genetic determination, and in the second case evidence, often of a very precise kind, which bears on theories of, for example, adaptive radiation and/or convergent evolution. But for secure, quantitative estimates of the genetic mechanisms involved – the "genetic architecture" – and its interaction with the environment there is at this time no possible substitute for breeding studies.

It should be stressed that the adoption of the fundamentally biological view of eq. (1) of the problem of behavioural inheritance as one of analysing phenotypic differences does not imply any view as to the relative importance of either heredity or environment in the outcome of the analysis. While it is true that geneticists have tended to be more interested in the hereditary aspects and psychologists, most prominently among behavioural workers, in the environmental, the approach which we shall advocate is, in principle, entirely neutral in respect of the so-called nature-nurture controversy referred to above. If either major class of determinant is specially preponderant in producing the phenotype as measured, this fact will be revealed rather precisely with little room for speculative argument. The inferences relating to, for example, the evolution of behaviour which can be drawn from such a knowledge of the genetic architecture and the genotype-environment interaction of a behavioural phenotype are, of course, another matter and will be explored separately in ch. 2.

1.1.2. Classical and biometrical approaches

The purpose of this present chapter, therefore, is to lay before the interested worker who seeks to analyse animal behaviour and its inheritance a methodology, procedure and analysis of proven capacity for the task he may set himself. This is biometrical genetics (Mather and Jinks, 1971). Its limitations, which usually relate to the material at hand and the assiduity of the investigator rather than being inherent in the method, will be noted as we proceed (see especially §§ 1.7 and 1.9), and the advantages of using techniques securely based in the fundamental principles of Mendelian genetics will also, we hope, become apparent to the interested reader.

However, before embarking on this expository task we must first dwell a little on the nature of behaviour as a phenotype, or, to be more exact, behaviour as one of the multitudinous aspects of the phenotype. The first consideration which strikes the observer is the absence of relatively clear-cut "types" in the behaviour of animals or, indeed, of humans for that matter.

In this connection, it may be mentioned in passing that the biometrical genetical approach is equally applicable to the analysis of human behavioural phenotypes, given certain adaptations consequent on the nature of the population and the breeding structure and the inadmissibility of manipulating it for experimental purposes in the way readily done with animals, especially in the laboratory. These aspects of the application of biometrical genetical to behaviour are dealt with in ch. 4 by Fulker and will not concern us further here. If clear-cut types, that is differential patterns of behaviour, can be shown to exist, as is the case among, for example, some species of insects, then analysis of their inheritance can proceed along lines made familiar in the classical Mendelian analyses of morphological characters. Rothenbuhler's (1964) dissection of the hygienic behaviour of honey bees (*Apis mellifera* L.) as dependent on the action of two major genes is an example which is cited with deserving frequency. But such cases are rare, especially in the animal kingdom, and we are usually faced with variation which is gradual and continuous, that is, quantitative in kind and which, moreover, is sensitive to environmental manipulation to an extent which the actions of genes of major effect are often not. The reason for this state of affairs is not far to seek: discontinuities due to major genes are relatively rare. The more usual situation is one in which many genes, each having an effect which is minor but supplementary to one another, act together on the resultant phenotype. That techniques for the analysis of genes of major effect – the classical Mendelian analyses – came to be developed first is not surprising for such cases naturally came to light by their prominence.

But classical analyses based on one-, two-, or few-gene models are not able to cope with the complexity of quantitative variation and the relative lack of success of the attempts to make them do so delayed the progress of psychogenetics for a considerable period. Meanwhile, within genetics itself, the problem was only resolved when the brilliant insight of Fisher (1918), which lay largely unrecognised for three decades, came to be elaborated and assimilated in the 1950's as a viable method for the analysis of quantitative variation, largely as a result of the efforts of Mather (1949) and his colleagues at Birmingham (Mather and Jinks, 1971) in applying it to the analysis of plant phenotypes. Fisher had shown how the classical Mendelian approach to particulate inheritance, as exemplified in the work of pioneers such as Bateson and Morgan, could be integrated with the biometrical approach of those such as Galton and Pearson whose concern was with quantitative variation. His was an essentially statistical approach, based strictly on

Mendelian models, but assuming, *inter alia*, the operation of a large number of genes of additive effect, thereby including the operation of one or few genes of major effect as a special case. Thus it is that, even if the fundamental assumption of biometrical genetics, polygenic determination, turns out to be erroneous, the method will still be applicable: simple genetic determination will inevitably be revealed. An example illustrative of this unusual eventuality already exists in the behavioural literature in the work of Wilcock (1969), who reanalysed data on avoidance conditioning in mice (Collins, 1964) to the effect that: "Collins' investigation shares with that of Rothenbuhler (1964) the distinction of having provided evidence for effects of single genes on normal behavior without studying mutants and without looking for such effects. What is particularly interesting about the reanalysis is the demonstration that if the genetic substrate is simple, biometrical analysis will not disguise the fact. No psychogeneticist would knowingly perform a complex analysis if it were clear that the underlying genetic structure were simple, but where the situation is initially not clear it is reassuring to find that a biometrical analysis will reveal the underlying simplicity" (p. 24).

1.2. The biometrical model

1.2.1. Polygenic system

So we can see the essential feature of the biometrical approach to be the idea that the variation is governed by multiple factors, both genetic and environmental. The genetic contribution is regulated by numerous genes which are inherited in a Mendelian way and which have effects similar to one another, supplementing each other, and small as compared with the total variation due to all causes, both heritable and non-heritable. In this way smooth, continuous variation of the phenotype results from discontinuous, quantal variations of the genotype and of the environment. In general the effects of single genetic factors will be too small to be followed individually, and the same will often be true of many "micro"-environmental factors also (§ 1.2.2).

An immediate consequence of a multiple gene control is that the balanced effect of many genes, some having a positive, increasing effect on the measure, and others having a negative, decreasing effect, determines the genotypic contribution to the phenotype. Hence various combinations of genes may make the same contribution. Indeed, the more genes included in the polygenic system governing a particular trait, the greater the number of possible combinations of genes that will, on balance, make precisely the same contribution to the phenotype. Further, this multiple control results in the apparent paradox that genotypes which produce identical phenotypes may differ among themselves by as many genes as do genotypes which produce widely different phenotypes. Thus the presence of a uniform environment is no guarantee that phenotypic identity means genetic identity. Neither do we assume that degree of phenotypic divergence means genotypic divergence in any direct sense. Genetic analysis alone can tell us whether or not phenotypic similarity is due to genetic similarity or to a similar balance achieved by different gene combinations.

Also following from multiple gene control is the consequence that the relationship between an F_1 (the first filial cross) and the pair of inbred strains from which it is derived ceases to give the indication of the dominance relationships in the way it does when the genes involved are few in number. Thus the metrical position of an F_1 which is halfway between its parents (P_1 and P_2) on the scale on which the phenotype is measured need not as a consequence imply the absence of dominance. Such a metrical position could be compatible with complete dominance at all gene loci with the increaser allele being dominant at half of them and the decreaser dominant at the other half, in a word, *ambidirectional* dominance. Similarly, if an F_1 falls outside the parental range in performance, this does not necessarily indicate a high level of dominance or even overdominance (§ 1.8). On the contrary, such a finding is compatible with a low level of unidirectional dominance of alleles which are distributed more or less equally as between both parental strains. Because of this absence of a simple relationship between an F_1's metrical position and dominance, it is customary to refer to the deviation of the F_1 from the mid-parental value as *potence* rather than dominance to avoid ambiguity.

Thus only a fraction of the total genetical variation between inbred strains and between them and the F_1's to which they give rise is visible at any one time. This situation follows from the neutralising effect of increasing and decreasing alleles and of dominant increasers and dominant decreasers. Much variation is hidden in the potential state by internal balancing of this kind, but is can be released by outcrossing and segregation into a free state from which it can be fixed by inbreeding and selection (Mather, 1953, 1973).

Indeed, the analogy has been pushed further: genetic variation is like an iceberg – the greater part of it, the potential variation, is invisible below the surface, and a change of, shall we say? temperature in the shape of genetic recombination or of massive environmental intervention, melts the ice and frees hitherto potential effects.

1.2.2. Determinants of the phenotype

We can proceed, therefore, to a method which demands no preconceptions and admits of the possibility of environmental determination ranging from the trivial to the considerable and of genetic determination from the simple to the most complex.

Each of these major determinants of the phenotype may be subdivided in numerous ways: in summary, these take the following main directions. First, let us consider the environmental effects which, as has been proposed (Fulker et al., 1972), can be considered in terms of micro- and macro-environmental effects. The former are those which can be attributed to minor, incidental variations in conditions of husbandry or early (or other) life experience over which there has been no conscious control and which are largely inseparable from any situation involving behavioural measurement. However, they present two special features which need careful consideration, especially in work with mammals. The first of these is the essential distinction between variation occurring *within* families as opposed to that occurring *between* families. In the biometrical model, which we must now begin to consider in greater detail, this distinction is recognised by a subdivision of the environmental variance, E (eq. 1), where

E_1 = variation within families, and
E_2 = variation between families.

This fundamental distinction, not always explicitly recognised, even by geneticists, for example, Parsons (1967b), finds a place in any appropriate biometrical model, as we shall see later, either as an effect in its own right, or in interaction with other components. In animal experimentation, the distinction is often encountered in the form of a difference within and between litters, either from the same or from different parents.

The second group of special micro-environmental effects which it is important to delineate are those subsumed under the term *maternal effects*. They comprise a range of possible causations, including effects of the nutrients and other possible influences contained in the egg (extra-chromosomal variation), pre-natal hormonal effects, pre- and post-natal nutrition, including lactation, effects attributable to maternal behaviour prior to weaning and so on. All of these can be complicated by maternal age and parity. Maternal effects can give rise to the phenomenon termed correlated environments which refers to the fact that environmental differences influencing members of different families (or litters) tend to be larger than those between members of the same one, i.e. $E_2 > E_1$. Random variation, uncontrolled and often inseparable from behavioural measurement, may also be regarded as micro-environmental in origin.

Macro-environmental effects are those which follow from some treatment or condition explicitly imposed as an environmental variable, not only early in the animal's experience but also later in the course of experimentation, including such disturbances of the organism's internal environment as is occasioned by food deprivation, shock intensity, or by drugs or hormones, all of which differences can be incorporated in the biometrical analysis (see § 1.6 for an illustrative example).

The genetic contribution can similarly be subdivided into three major influences on phenotypic variation: additive, dominance and interactive genetic effects. The first kind of heritable variation, the additive or "fixable", arises from differences between genes in homozygous condition. For the case of the single locus with two alleles this is the metrical difference between the homozygotes AA and aa, and is designated by the symbol d_a. The second heritable variation to be recognised is dominance variation which arises from the metrical deviation of the heterozygote Aa – obtained by crossing AA with aa – from the mid-parental mean of these two parental homozygotes. Because it is heterozygous it is symbolised h_a and is referred to as "non-fixable" since it is subject to segregation. Finally, there is variation which is non-allelic in character, so called to distinguish it from allelic, that is dominance, variation. This is the variation which arises from the modification of the contribution to the phenotypes of the genes at the A-a locus just considered (AA, Aa, and aa) caused by the presence of allelic differences at another locus we may call B-b. Three types of non-allelic interaction of this kind can be recognised, namely interaction between homozygous combinations at both loci, for example, AA or aa with **BB** or **bb**, and symbolised i_{ab}, that between a homozygous combination at one locus and a heterozygous one at the other, for example, Aa with **BB** or **bb** and Bb with AA or aa and symbolised j_{ab}, and, finally, that between heterozygous combinations at both loci as in Aa with Bb for which the symbol is l_{ab} (see § 1.3.2.1).

Finally, to complete this brief survey of the biometrical model to be employed, full details of which the reader may find in Mather and Jinks (1971), we must point to a further source of variation. This is genotype-environment interaction, which arises because not all environmentally produced variation is additive in the sense that it affects all individuals in the same way and to the same extent irrespective of their genotype. Some genotypes respond more, and others less, than the average. Thus we have a type of variation which is neither specifically environmental norge netic in origin but results from the interaction between the two. This contribution to variation is usually symbolised g_{ij}, which refers to the interaction of the ith genotype with the jth environment (see § 1.6).

1.2.3. Limitations

Thus we arrive at a point at which the essential characteristics of the biometrical model necessary for the analysis of continuous variation of the kind so often encountered in behavioural work have been specified and we must now proceed to see how the model may be put to work in practice. Two points need to be made here: first, in theory there is no constraint on the complexity of the biometrical model which can be used, and, indeed, models appropriate for situations considerably more complex than those which are usually encountered in practice are available in the genetical literature. But, secondly, because in practice constraints do inevitably arise – limitations dictated by the nature of the subjects available, the facilities for testing them, and the size of the investigation possible – it follows that the comprehensive model, in which estimates are made of all sources of environmental, genetical and interactive contributions to the phenotypic variation, can hardly be achieved in one investigation. So extensive are these limitations on information that the complete model needs considerably more parameters – that is, different kinds of information – than are usually available. Hence we must cut our coat according to our cloth and minimise or even discount some sources of variation in order to concentrate on others. Only in this way can we hope to obtain valid estimates for the more important components of the model as outlined above. Fortunately, the use of inbred strains of animals maintained under controlled environmental conditions and of crosses between them allows us to proceed a considerable way in distinguishing the various contributions, and in what follows it will be assumed such is generally the case, though see §§ 1.3.2.3 and 1.5.

To concentrate in this way on crossbreeding is, it is realised, to discount other approaches to psychogenetics which have been favoured by ethologists and psychologists such as those using comparisons between species, sometimes involving hybridisation, the study of mutations or of major genes and the genetical method alternative to the cross-breeding approach, that of selective breeding. The problems which, in our view, render the first two approaches less than satisfactory, if the purpose is to analyse genetic determination and genotype-environment interaction, have been explored by Broadhurst (1968) and Wilcock (1969) respectively; the limitations of genetic selection in this respect have been touched on by Broadhurst (1969). Despite the importance of this methodology in early experimental work in psychogenetics (for example, Tryon, 1940; Hall, 1938), it is thought that, briefly, selection is not a technique which lends itself at all well to the partitioning of heritable and environmental variation on a biometrical model. Strains successfully selected for extremes of behaviour do, of course, provide valuable material which can then be used in many ways, including for example, cross-breeding experiments but the analysis of the process of selection itself tends to stop short at estimates of heritability (§ 1.7) and does not provide the kind of fine-grain analysis of gene effects we can achieve by strain-crossing techniques.

1.3. Scaling

1.3.1. General considerations

The next consideration of importance in the biometrical approach is the choice of scale on which the behaviour to be analysed is measured – we leave aside here the problems common to all kinds of behavioural measurement as such, but not specifically related to the partitioning of its variation into hereditary and environmental components, as beyond the scope of this chapter. We define a scale as adequate for this purpose as one which maximises the predictive value of the estimates of the parameters obtained from a biometrical genetical analysis. As we have seen, this procedure of necessity means that complex effects such as non-allelic interactions and genotype-environment interactions requiring many parameters for their specification must be minimised, since we are rarely able to allow for them in the models fitted because of paucity of available data. If we cannot allow for them in our analyses, we must ensure that they are unimportant by choosing a scale on which the contributions arising from these effects are minimised. To do so does not involve any new principle. The scale on which we measure behaviour is already arbitrary in that it cannot be based on prior knowledge of how genotypic and environmental causes of individual differences act and interact. Moreover, the transformation of data from one arbitrary scale to another to improve the validity and precision of tests of significance is an accepted procedure of statistical analysis. A scalar change to remove interactions and hence improve the predictive power of a biometrical analysis is a logical and valuable extension of this procedure.

That scalar changes can achieve the desired result, has often been demonstrated, both with actual data and with theoretical models. Suitable changes can minimise the effects of interactions between gene loci and between the genotype and the environment. The number of components we have to consider when partitioning the phenotypic variation can then be limited to four, namely, additive, dominance, environmental, and maternal (including extrachromosomal) components. On the other hand, we can expect to meet certain types of interactions not easily removed by scalar changes without at the same time obliterating the phenotypic differences we seek to analyse. In consequence we must determine empirically what is the best approach in each case.

1.3.2. Scaling tests

As our definition of an adequate scale is one on which interactions are absent or are minimal, we require scaling tests to decide whether or not the criterion is satisfied for any scale, that is, we must be able to detect interactions if they are present in the data. But tests to decide the adequacy of the original transformed scale can be applied only to crosses between inbred strains and the generations that can be derived from them by sib-mating or back-crossing. Let us consider three cases in which we have measured (a) the two inbred parental strains and the F_1, F_2, and backcross generations derived from them; (b) several inbred lines crossed in all possible combinations to give a diallel set of crosses; and (c) each of a random collection of individuals from any source crossed to two contrasting inbred strains and to their F_1.

1.3.2.1. Parents, F_1, F_2, and backcrosses. The various methods of detecting non-allelic interactions can be illustrated by reference to the expected generation means based, for purposes of illustration, on the full model which makes provision for interactions (Broadhurst and Jinks, 1961, as modified by Mather and Jinks, 1971). These expectations are given in terms of six parameters: m, [d], [h], [i], [j], and [l], which are the sum or balances of the effects over many genes of the kinds of gene action and interaction shown in Table 1.1.

TABLE 1.1

Expected generation means in terms of the parameters additive [d] and dominance [h] components and the interactions between pairs of homozygous genes [i], pairs of heterozygous genes [l], and between pairs of genes, one each being in the homozygous and the other the heterozygous state [j] (see text).

Generation[1] Description	Designation	Parameters[2] m	[d]	[h]	[i]	[j]	[l]
Mean of larger[3] inbred parental strain	$\bar{P}_1 =$	1	+1		+1		
Mean of smaller[3] inbred parental strain	$\bar{P}_2 =$	1	−1		+1		
Mean of cross between P_1 and P_2	$\bar{F}_1 =$	1		+1			+1
Mean of cross between sibs of F_1	$\bar{F}_2 =$	1		+½			+¼
Mean of backcross of F_1 to P_1	$\bar{B}_1 =$	1	+½	+½	+¼	+¼	+¼
Mean of backcross of F_1 to P_2	$\bar{B}_2 =$	1	−½	+½	+¼	−¼	+¼

[1] Additional generation means can be derived similarly, but are omitted here as unlikely to be of use in behavioural work using mammals as subjects.

[2] To be read as indicating, for example, that
$$\bar{P}_1 = m + [d] + [i], \quad \text{and}$$
$$\bar{B}_2 = m - \tfrac{1}{2}[d] + \tfrac{1}{2}[h] + \tfrac{1}{4}[i] - \tfrac{1}{4}[j] + \tfrac{1}{4}[l]$$

[3] That is to say, having a higher or lower value, respectively, on the metric on which the performance is measured.

We can proceed to detect non-allelic interactions by deriving from the table relationships between the generation means which hold – within the sampling errors of these means – only if interactions are absent, that is, if [i] = [j] = [l] = 0. If we assume the absence of non-allelic interactions the following relationship between some of the various generation means holds:

$$\bar{F}_2 = \tfrac{1}{2}\bar{F}_1 + \tfrac{1}{4}\bar{P}_1 + \tfrac{1}{4}\bar{P}_2 \qquad (2)$$

from which we can generate the expectation that

$$\bar{P}_1 + \bar{P}_2 + 2\bar{F}_1 - 4\bar{F}_2 = C \qquad (3)$$

where C = 0, within the sampling error of C which is

$$V_{\bar{P}_1}+V_{\bar{P}_2}+4V_{\bar{F}_1}+16V_{\bar{F}_2} \qquad (4)$$

for the sum of degrees of freedom of the four variances of the means (V = sampling errors - SE^2) in this expression. A list of similar relationships, applicable to mammals, which also hold but only in the absence of non-allelic interactions, is given in Table 1.2, thus providing a statistical test of the correctness or otherwise of the assumption of non-allelic interaction, which can therefore be detected by means of these tests.

TABLE 1.2

Scaling tests: relationships between generations means which hold on the assumption of the absence of non-allelic interaction, the failure of which provides scaling tests for detecting the presence of such interactions.

Scaling tests	Relationship
A	$\bar{P}_1+\bar{F}_1-2\bar{B}_1 = 0$
B	$\bar{P}_2+\bar{F}_1-2\bar{B}_2 = 0$
C	$\bar{P}_1+\bar{P}_2+2\bar{F}_1-4\bar{F}_2 = 0$
D	$\bar{B}_1+\bar{B}_2-2\bar{F}_2 = 0$

Another way of detecting non-allelic interactions is to estimate the interactive components [i], [j], and [l] directly as in Table 1.3. The standard errors of these components can then be obtained and tests of their significance applied by the customary methods. Again, in the absence of non-allelic interactions, the estimates of [i], [j], and [l] will not differ significantly from zero.

TABLE 1.3

Estimates of the interaction parameters and their sampling errors.

Parameter	Estimate of parameter	Sampling Error
[i]	$2\bar{B}_1+2\bar{B}_2-4\bar{F}_2$	$4V_{\bar{B}_1}+4V_{\bar{B}_2}+16V_{\bar{F}_2}$
[j]	$\bar{P}_2-\bar{P}_1+2\bar{B}_1-2\bar{B}_2$	$V_{\bar{P}_2}+V_{\bar{P}_1}+V4_{\bar{B}_1}+4V_{\bar{B}_2}$
[l]	$\bar{P}_1+\bar{P}_2+2\bar{F}_1+4\bar{F}_2-4\bar{B}_1-4\bar{B}_2$	$V_{\bar{P}_1}+V_{\bar{P}_2}+4V_{\bar{F}_1}+16V_{\bar{F}_2}+16V_{\bar{B}_1}+16V_{\bar{B}_2}$

Yet a third test for non-allelic interaction is the joint scaling test (Cavalli, 1952; Mather and Jinks, 1971). In this test, weighted least squares values for m, [d], and [h] are estimated from the generation means, once more assuming the absence of non-allelic interactions. The weights used are the reciprocal of the squared standard errors of the generation means, that is $1/V\bar{P}_1$, $1/V\bar{P}_2$, etc.. The weighted squared deviations of the expected and observed generation means are then a χ^2 with $(n-3)$ degrees of freedom, where n is the number of observed generations means. If the χ^2 is not significant the m, [d], [h] model can be judged to be adequate and hence

$$[i] = [j] = [l] = 0$$

Application of these tests to behavioural data can be illustrated by some findings relating to mouse activity. McClearn (1961) observed the number of 5-inch square subdivisions which 50-day-old mice entered when placed for 3 minutes on a 30-inch square floor inside a cabinet, with parallel barriers about 5 inches high on alternate square borders. The subjects were drawn from two inbred strains, one black and one albino (C57BL/Crgl and A/Crgl, respectively) and the F_1, F_2, and backcrosses (B_1, B_2) between them. Table 1.4 shows the application of the scaling tests of Table 1.2 to McClearn's data, both in its original form and after a square-root transformation. This transformation McClearn found to be the most suitable of the various ones he tried, and we can see from the final column of the table the extent to which it reduces the significance of the deviations from zero and hence minimises the effects of the non-allelic interactions.

TABLE 1.4

The application of the A, B, C, and D scaling tests applied to activity in mice (McClearn, 1961) on the original and on a transformed (square root) scale. Units for this Table and Tables 1.5–1.7 and 1.9 are the number of squares entered (see text).

	Test	Deviation	Standard Error	Significance
		Original Scale		
A	$\bar{P}_1 + \bar{F}_1 - 2\bar{B}_1$	12.40	17.78	n.s.
B	$\bar{P}_2 + \bar{F}_1 - 2\bar{B}_2$	53.00	13.38	$P < 0.001$
C	$\bar{P}_1 + \bar{P}_2 + 2\bar{F}_1 - 4\bar{F}_2$	84.00	36.06	$P < 0.05$
D	$\bar{B}_1 + \bar{B}_2 - 2\bar{F}_2$	9.30	17.60	n.s.
		Square-root transformation		
A	$\bar{P}_1 + \bar{F}_1 - 2\bar{B}_1$	0.40	0.88	n.s.
B	$\bar{P}_2 + \bar{F}_1 - 2\bar{B}_2$	3.10	1.06	$P < 0.01$
C	$\bar{P}_1 + \bar{P}_2 + 2\bar{F}_1 - 4\bar{F}_2$	3.70	2.30	n.s.
D	$\bar{B}_1 + \bar{B}_2 - 2\bar{F}_2$	0.10	0.91	n.s.

The alternative approach of Table 1.3, when applied to these same data, gives the results to be found in Table 1.5, which shows the interaction parameters along with those for the mean, and additive and dominance components also. As can be seen, only [j] reaches marginal significance on the original scale. Finally, the joint scaling tests yields the results shown in Table 1.6 when applied to McClearn's data. It will be seen that there is a

TABLE 1.5

Estimates of the parameters in the full interaction model for activity in mice (McClearn, 1961) on the original scale (see text and Table 1.4).

Parameter	Estimate	Standard Error	Significance
$\hat{m}$	58.8	35.35	n.s.
$[\hat{d}]$	67.6	3.21	$P < 0.001$
$[\hat{h}]$	−8.6	81.33	n.s.
$[\hat{i}]$	18.6	35.21	n.s.
$[\hat{j}]$	40.6	18.39	$P < 0.05$
$[\hat{l}]$	46.8	49.88	n.s.

highly significant deviation from the additive-dominance model on the original scale, which confirms the need for the square-root transformation used by McClearn in assessing the nature of gene effects determining locomotor activity in the strains of mice he used. For example, quite clear

TABLE 1.6

Cavalli's joint scaling test applied to activity in mice (McClearn, 1961) on the original scale (see text and Table 1.4).

	Generation Means			
Generation	Observed	Expected on additive-dominance model	Estimates of parameters in additive-dominance model	Goodness of fit of model
P_1	145.0	145.68		
P_2	9.8	8.75		
F_1	97.0	72.89	$\hat{m} = 77.216$	
F_2	66.2	75.05	$[\hat{d}] = 68.463$	$\chi^2_{(3)} = 17.58$
B_1	114.8	109.28	$[\hat{h}] = -4.330$	$P < 0.001$
B_2	26.9	40.82		

evidence of dominance for alleles governing higher, rather than lower, activity in this situation was demonstrated by McClearn.

In the absence of genotype-environment interaction, the contribution of the environment to phenotypic variation will be definition be the same for all the genotypes concerned since the inbred lines and the F_1 cross between them are completely uniform genetically, comprising two homozygous strains and one (uniformly) heterozygous one, respectively. We can measure the magnitude of environmental contribution in these generations since it alone is responsible for all the observable variation within them. This follows from the fact that, if the parental and F_1 generations are equally exposed to the environmental causes of variation, as they will be in an adequately designed experiment, the variation within each parental strain and within the F_1 must be equal, in the absence of genotype-environmental interactions. If such is not the case, genotype-environment interaction is indicated.

McClearn's data from the same experiment (1961) also provide an example of this effect and illustrate the extent to which it can be countered by transforming the scale on which the raw data are gathered. The values shown in Table 1.7 are drawn from his Table 1. As he notes, "... it may be seen that the parental-strain variances have been essentially equated by the transformation although the F_1 variance is still greater than the parent-strain variances".

TABLE 1.7

The transformation of scale to reduce genotype-environment interaction applied to activity in mice (McClearn, 1961; see text and Table 1.4).

Generation	n	Variance on	
		Original Scale	Square-root transformation
P_1	19	723.7	1.3
P_2	18	58.4	1.6
F_1	21	1660.6	5.4

It is worth noting that where, as in this case, the highest variance (that of the F_1 generation) is associated with an intermediate mean it may be extremely difficult, if not impossible, to find a transformation which will equalise the variance. The square root transformation used by McClearn, while not entirely satisfactory, is probably as good as can be found.

1.3.2.2. Diallel crosses. A set of crosses using n strains, each crossed with every other yielding an n^2 table of means, is known as a diallel cross, to which technique Broadhurst (1967) provides an introduction. From a diallel of this kind it is possible to detect both non-allelic interactions and genotype-environmental interactions using data from the n parental strains and their F_1 generations only. Here the test for non-allelic interaction depends on the fact that the n^2 parental and F_1 families can be grouped into what are termed arrays, an array being a set of n families (parents and F_1) that have one parent in common and hence the members of which are half-siblings. Two statistics can be obtained from each array, namely, V_r (the variance of family means in the array), and W_r (the covariance of the family means in an array on their non-common parent). In the absence of non-allelic interactions, $(W_r - V_r)$ is a constant over all arrays. There are, however, two difficulties associated with this test.

The first is a purely statistical problem involving the testing of the constancy of $(W_r - V_r)$ over the n arrays. Two methods have been used, both of which lack statistical perfection. One is to test the variation of $(W_r - V_r)$ over the arrays against its own error obtained from replicated estimates of $(W_r - V_r)$ for each array. The other is to test the regression of W_r on V_r which is expected to be a linear regression with unit slope, if $(W_r - V_r)$ is constant. The linearity of the regression and its deviation from unity can be tested by the usual statistical techniques. The second problem arises from the fact that the failure of another assumption which is unique to diallel analysis, namely that distribution of the genes among the inbred parental strains is random, also leads to the failure of the W_r/V_r relationship. While failures from this latter cause have rarely been met with in practice, it is safer to regard the W_r/V_r relationship as a test for failures of both assumptions jointly rather than as a specific test for non-allelic interactions only.

Since every one of the n^2 families in a diallel set of crosses is either a parent or an F_1, all variation within the families reared will be purely environmental in origin. Given a properly designed experiment, therefore, we can immediately detect the presence of genotype-environmental interactions by considering the heterogeneity of the within-family variances.

Analyses of emotional reactivity in the rat will serve as an example of these techniques. Each subject is exposed for two minutes per day on four successive days in an arena or open field to noise and light stimulation of controlled intensity. The experience is mildly stressful and rats respond by defecating, ambulating around, and other indices of emotional arousal.

Most weight, however, has been placed on the first two, and it was these measures which were analysed in a replicated diallel cross of six strains of rats (Broadhurst, 1959, 1960). Application of the test of the variation of $(W_r - V_r)$ over the six arrays in the diallel table gave non-significant results for both the defecation and the ambulation scores. Similarly, the regression of W_r on V_r did not depart from linearity in either case. These indications, therefore, did not point to any significant non-allelic interactions in the data for this cross-breeding experiment. With respect to genotype-environment interaction, the situation was not, however, so satisfactory. While the ambulation scores showed no significant inhomogeneity of the within-family variances, the defecation scores did. A square-root transformation reduced this inhomogeneity to non-significant proportions, without affecting the linearity of the W_r, V_r regression, and the analysis proceeded on this basis (Broadhurst and Jinks, 1966).

1.3.2.3. Triple-test cross. A test for non-allelic interactions more widely applicable has been described by Kearsey and Jinks (1968) and Jinks et al. (1969). This test is an extension of Comstock and Robinson's F_2 and backcross design (1952). It involves crossing any individual, whether inbred or not, or any homogeneous strain, to three testers. Two are contrasting inbred lines, the third is the F_1 between them. If we refer to the means of the three families thus produced from any one set of crosses as $\bar{L}_1$, $\bar{L}_2$, and $\bar{L}_3$, respectively, then under all circumstances $\bar{L}_1 + \bar{L}_2 - 2\bar{L}_3 = 0$, in the absence of non-allelic interactions, irrespective of the genotype of the individual or homogeneous strain used as a common parent in all three families. No example of this relatively new approach as applied to behavioural phenotypes is as yet available in the literature but its efficiency has been demonstrated for morphological traits in insects and in plants (Jinks et al., 1969; Jinks and Perkins, 1970), and a partial reanalysis of some of Broadhurst's (1960) data on emotionality in rats has been attempted along these lines (Fulker, 1972).

The method is currently being applied in our laboratory to the study of escape-avoidance conditioning in rats. The results of this triple-test cross analysis confirmed the findings of a diallel cross analysis of this same phenotype (see Fulker et al., 1972; also ch. 2, § 2.5.3.1). That is, a moderate amount of ambidirectional dominance was demonstrated for the total number of avoidances over the whole sequence of 80 trials, which, on further analysis of this overall score by blocks of five trials, emerges as dominance for low avoidance initially, but changing to dominance for high avoidance later. The narrow heritability for number of avoidances being low (in the region of 10–15 %), the additive and dominance components do not reach striking levels of significance (V. Owen, personal communication). These outcomes indicate that the method has a place in psychogenetic analysis, but in cases where heritability is low the method is inevitably less powerful than the diallel cross of inbred strains in which different genotypes are more abundantly represented.

1.3.3. Alternative procedures

At this point we can consider two alternative situations which can arise. If the scaling criteria are satisfied on the original or on a transformed scale, our simple model of gene and environmental action is adequate. On the other hand, no adequate scale may be found to fit the simple model and hence we must proceed either to an approximate analysis making assumptions about interaction that are probably unjustified, or to generate and fit more complex models that allow for the interaction.

1.3.3.1. Adequate scale. Let us therefore first consider model fitting on an adequate scale, using first-degree statistics. Where the scale is adequate, weighted least squares estimates of the additive [d] and dominance [h] components of the generation means are obtained as part of the joint scaling test along with their standard errors in the way described in § 1.3.2.1 above (see Tables 1.5 and 1.6 also). These estimates are sufficient, along with m, to predict the means for any generation that can be derived from an initial cross between two inbred strains. For the reasons noted earlier (§ 1.2.1), the ratio [h]/[d] is not a dominance ratio but a potence ratio whose value depends as much on the distribution of genes between the two parents and the direction of dominance as it does on the level of dominance as such.

Second-degree statistics can also be used as, firstly, in the case of crosses between pairs of inbred lines. On an adequate scale there are only three sources of variation (if we ignore maternal effects, etc.), namely, $D = Sd^2$, $H = Sh^2$, and E, the additive environmental variation. The expected contribution of these three components of variance to the commoner generations encountered in animal breeding studies are given in Table 1.8. The square-root ratio $\sqrt{(H/D)}$ measures the true dominance level. Once again we can illustrate the use of these formulae by reference to McClearn's (1961) experiment on activity in mice. Table 1.9 shows the components of variation, as calculated both on the original and the transformed, square-root scale. It will be noted that the value for H is shown as zero; this is because the estimates as calculated proved negative, manifestly impossible for a second-degree statistic. As a consequence the $\sqrt{(H/D)}$ ratio measuring true dominance cannot be calculated, and no differentiation can be made between

TABLE 1.8

The contribution of the additive (D), dominance (H), and environmental (E) components of the variation in the generations specified, as measured on an adequate scale.

Generation and Component of Variation		Expectation
Description	Designation	
Variation within a family of an inbred line	V_{P_1}, V_{P_2}, etc.	E_1
Variation between family means of an inbred line	$V_{\bar{P}_1}$, $V_{\bar{P}_2}$, etc.	E_2
Variation within an F_1 family	V_{F_1}	E_1
Variation between family means of an F_1	$V_{\bar{F}_1}$	E_2
Variation within an F_2 family	V_{1F_2}	$\frac{1}{2}D + \frac{1}{4}H + E_1$
Variation within a B_1 family	V_{1B_1} } summed	$\frac{1}{2}D + \frac{1}{2}H + 2E_1$
Variation within a B_2 family	V_{1B_2}	

the "narrow" and "broad" heritability ratios (see § 1.7 below). Those given in Table 1.9, however, are somewhat lower than the values for broad heritability McClearn gives for the two scales (60% and 69%, respectively), which are based on rather less extensive analyses than those presented here.

TABLE 1.9

Estimates of the components of variation assuming an additive, dominance model for activity in mice (Mc Clearn, 1961) with dominance taken as zero (see text and Table 1.4).

	Estimates of Component	
Component	Original Scale	Square-root transformation
D	972.925	6.925
H	0	0
E_1	1025.825	3.425
Heritability-"Broad"	32%	50%
$ht_b = \dfrac{\frac{1}{2}D + \frac{1}{4}H}{\frac{1}{2}D + \frac{1}{4}H + E_1}$		

The present analysis is admittedly broadly unsatisfactory but is nevertheless included to illustrate the difficulties occasionally encountered in biometrical genetical analysis. It is clear that in this case no scalar transformation can be effective in removing the large non-allelic interactions and genotype-environmental interactions which give rise to the difficulty in fitting the model. We must therefore conclude from our analyses that an additive, dominance model is inappropriate for these data, but that more data – for example, from second backcross and sib-mated F_2 generations – are needed to determine the correct model to use and to investigate more fully the nature of the undoubtedly complex gene effects involved.

Secondly, we can consider second-degree statistics in the case of diallel sets of crosses derived in the way described above (§ 3.2.2). Given an adequate scale, four genetic parameters are required to account for the heritable component of variation in all the statistics obtainable from a diallel set of crosses. These parameters are:

$$\begin{aligned} D &= S4uvd^2 \\ H_1 &= S4uvh^2 \\ H_2 &= S16u^2v^2h^2 \\ F &= S8uv(u-v)dh \end{aligned} \quad (6)$$

where u is the frequency of increaser, and v the frequency of decreaser alleles. These frequencies are simply relatable to the corresponding parameters, D_R and H_R, for randomly mating populations as used in Table 1.11 below. When $u = v = \frac{1}{2}$, as it does in a simple cross between a pair of inbred lines, D, H_1, H_2, and F, which refers to the balance between dominant and recessive alleles in the parental population (Mather and Jinks, 1971), reduce to the D and H described earlier in this section, as does the D_R and H_R just mentioned.

TABLE 1.10

The contribution of D, H_1, H_2, F, and E_2 to the statistics obtained from a full diallel set of crosses, including reciprocals, between n lines measured on an adequate scale (see text).

Description	Statistic Designation	Expectation
Variation between inbred parents	V_p	$D + E_2$
Mean variation within an array	$\bar{V}_r$	$\frac{1}{4}D + \frac{1}{4}H - \frac{1}{4}F + E_2(n+1)/2n$
Mean covariation within an array	$\bar{W}_r$	$\frac{1}{2}D - \frac{1}{4}F + E_2$
Variation between array means	$\bar{V}_r$	$\frac{1}{2}D + \frac{1}{4}H_1 - \frac{1}{4}H_2 - \frac{1}{4}F + E_2(n+1)/2n^2$

The contributions of these parameters to the variances and covariances which we can obtain from a diallel set of crosses are given in Table 1.10, and an example of the application of these expectations can be found in

TABLE 1.11

Estimates of the random mating components of variation from a diallel cross for emotionality in rats (Broadhurst and Jinks, 1966). Units are, for defecation, the number of fecal boluses deposited per day, and, for ambulation, the number of the floor units traversed per day (see text).

	D_R	H_R	E_2†	ht_n (Heritability- "Narrow")
Day		Defecation (square-root transformation)		
1	0.24**	0.09	0.20	41%
2	0.13	−0.05	0.27	20%
3	0.06	−0.04	0.27	13%
4	0.07	0.05	0.17	16%
		Ambulation		
1	3.4***	0.2	1.0	62%
2	3.1***	1.4*	2.5	30%
3	4.6***	1.8*	2.3	45%
4	4.1***	3.7*	2.5	38%

* significantly different from zero at $P < 0.05$
** significantly different from zero at $P < 0.01$
*** significantly different from zero at $P < 0.001$
† It is not possible to assess the significance of the E_2 component in the same way as the other components since the inter-block difference itself defines the error variance.

the diallel cross of rats for emotionality referred to earlier (§ 1.3.2.2). Table 1.11 shows the components of variation calculated, in this case, for each of the four successive days on which the open-field test was administered. This technique allows a sensitive estimate to be made of the interaction of genotype with changes in environment resulting from the growing experience of the situation by the subject – clearly a learning phenomenon attributable to habituation. Taking other calculations not shown here into consideration, the authors (Broadhurst and Jinks, 1966) concluded that these data show that, for defecation, the test increases over time the expression of genes responsible for low defecation scores, and, for ambulation, genes responsible for intermediate scoring emerge as dominant. The evolutionary implications of these findings have been assessed elsewhere (Broadhurst, 1968; see also ch. 2 by Broadhurst and Jinks).

On an adequate scale we can also use the W_r/V_r relationship discussed earlier to tell us more about the dominance properties of the genes. Some examples may be briefly cited: the mean value of $W_r - V_r$ over all arrays = $\frac{1}{4}(D - H_1)$ after a correction for environmental components has been applied, hence we can test whether D is greater than, equal to, or less than H_1, and the arrays with the smallest values of W_r and V_r contain most dominant alleles, similarly those with the largest value contain most recessive alleles, and by examining the ranking of the arrays for their mean performance and their proportion of dominant to recessive alleles, we can see whether or not dominance is associated with a high, low, or intermediate performance, and hence we can infer the direction of dominance.

In the absence of epistasis, the triple-test cross provides a simple yet powerful method for detecting and estimating D and H. The variance of $\bar{L}_1 + \bar{L}_2$ or of $\bar{L}_1 + \bar{L}_2 + \bar{L}_3$ (see § 1.3.2.3) provides estimates of D, and the variance of $\bar{L}_1 - \bar{L}_2$ an estimate of H. These estimates are independent and direct tests of their significances are given by an analysis of variance (Jinks and Perkins, 1970). The utility and sensitivity of this breeding design for detecting and estimating additive and dominance components of variation has, once again, already been demonstrated for plant breeding material.

1.3.3.2. Non-allelic interaction. Turning to the problems which arise if we are attempting model-fitting in the presence of demonstrated non-allelic interactions, we begin with the case involving first-degree statistics. Now, to fit the full interaction model to the generation means we must have data from at least six different generations. The simplest set of these comprises the two parents, P_1 and P_2, and the F_1, F_2, and backcrosses of the F_1 to each parent, B_1 and B_2. From these we can obtain a perfect fit solution of the six parameters m, [d], [h], [i], [j], and [l], though to test the adequacy of this model we require more than six generations. If fewer than six generations are available we can estimate only various combinations of the parameters as described by Broadhurst and Jinks (1961).

We can obtain an estimate of the potence ratio from the relative values of [h] and [d], in addition to classifying the predominant type of interaction present from a consideration of the relative signs of [h] and [l], (Jinks and Jones, 1958; Mather and Jinks, 1971). Thus if [h] and [l] have the same sign (and are both significant), the interaction is mainly of a complementary type, that is, the pairs of interacting genes have a greater effect when associated than is expected from their individual effects when separate, a kind of synergistic action, while if they have opposite signs action is mainly of a duplicate type, that is, the pairs of interacting genes have a smaller effect when associated than is expected. This distinction is important when considering the genetical basis of heterosis and the genetical architecture of a character (see § 1.8). Examples of this approach are not common in the behavioural literature, though many cases of complementary and duplicate interaction can be found in plant genetics. This is probably because psychologists have based their experiments on strains of laboratory animals usually well separated, for example, by selection, in respect of the aspects of the phenotype which is desired to investigate (§ 1.2.3). Thus, in such cases, the tendency will always be for genes of increasing and decreasing effect to be well assorted in the various strains so that heterotic effects are at a minimum.

While second-degree statistics have not as yet been used to estimate the contribution of non-allelic interactions to the variances and covariances obtained from generations derived from crosses between inbred lines, they have been used to detect and classify interactions. In particular the W_r/V_r relationship has been used to distinguish between interactions which are mainly of the complementary type from those that are mainly of the duplicate type (Mather and Jinks, 1971).

1.4. Maternal effects and sex-linkage

So far we have only considered models in which F_1, F_2, and backcross generations have the same expectations irrespective of whether P_1 or P_2 served as male parent in the initial cross or of whether the F_1 or the inbred parent served as the male parent of the backcross. When such crosses are made in both directions, two circumstances can arise which may lead to a failure of models which assume that such reciprocal crosses have identical phenotypes, namely, maternal effects and sex-linkage. In such cases differences between reciprocal crosses may be detected.

Let us consider these effects in turn, dealing first with maternal effects (§ 1.2.2) in crosses between pairs of inbred lines. If we make the two possible reciprocal crosses between strain 1 and 2, there are two F_1's deriving from a mother from strain 1 crossed with a father from 2, and vice versa, that is $1♀ × 2♂$ and $1♂ × 2♀$. If we rear the F_2's by mating siblings of opposite sex

TABLE 1.12

The various ways in which the two backcrosses, B_1 and B_2, to two parental (P) strains, 1 and 2, can be generated reciprocally from the F_1.

	B_1			B_2	
P_1		F_1	P_2		F_1
1♀	×	♂(1♀×2♂)	2♀	×	♂(1♀×2♂)
1♂	×	♀(1♀×2♂)	2♂	×	♀(1♀×2♂)
1♀	×	♂(1♂×2♀)	2♀	×	♂(1♂×2♀)
1♂	×	♀(1♂×2♀)	2♂	×	♀(1♂×2♀)

in each F_1, there will be two F_2's corresponding with the two kinds of F_1 parent they come from, namely ♀(from 1♀×2♂)×♂(1♀×2♂) and ♀(1♂×2♀)×♂(1♂×2♀). Clearly there can be two further F_2's which involve reciprocal crosses between animals from the two *different* reciprocal F_1's noted above though they are infrequently reared. When we come to the backcrosses there are four possibilities for each backcross. Thus B_1 and B_2 can be generated in the various ways shown in Table 1.12. In each generation we can, of course, merely test the means of reciprocal crosses for significant differences in the usual way without any expectation as to the direction or consistency of these differences over generations. Alternatively, we can construct models, in which case the range of complexity is immense. However, all models have a common basis, namely, the assumption that the heritable contribution to the offspring phenotype can be split into two major components. The first is the contribution of the offspring genotype which is designated $[d_o]$, $[h_o]$, $[i_o]$, $[j_o]$, and $[l_o]$ etc., and the second is the maternal genotype which contributes $[d_m]$, $[h_m]$, etc. (see Barnes, 1968).

We have, therefore, five parameters m, $[d_o]$, $[h_o]$, $[d_m]$, and $[h_m]$ on an adequate scale. If none can be found there are various possibilities. There may be interaction among the genes of the offspring genotype, or interaction among the genes of the maternal contribution or both. There is a further possibility not allowed for in the models, namely, of interaction between genes in the offspring genotype and genes in the maternal genotype, that is $[i_{om}]$, $[j_{om}]$, and $[l_{om}]$. However, we can only proceed on the assumption that such complexities of gene action and interaction on behaviour are unimportant until such time as we are obliged to accept otherwise.

An example of this approach may be found in Fulker's (1970) partial reanalysis of Joffe's data (1965; see also 1969) on the open-field ambulation of adult rats whose mothers had been stressed prior to mating. The subjects were rats of two strains which have been selectively bred for high and low emotional elimination in the open-field test. They are known as the Maudsley strains (Broadhurst, 1960, 1962a; Eysenck and Broadhurst, 1964), and the high defecating strain is the Maudsley emotionally reactive, (MR), and the low, the non-reactive, (MNR). Joffe also used the two F_1 reciprocal crosses

TABLE 1.13

Estimates of components of the maternal effect on ambulation of rats whose mothers had been stressed prior to mating (Joffe, 1965). Units are number of floor units traversed per day. (From Fulker, 1970.)

	Means		
Generation	Controls	Stressed	
P_1 (MNR)	44.97	35.33	
P_2 (MR)	27.25	36.53	
F_{11} {MNR♀ × MR♂	33.45	37.40	
{MNR♂ × MR♀	28.77	24.95	
Parameter	Estimates		Standard errors
$[\hat{d}_o]$	6.53	−6.83	$[d_o], [h_o] = 3.60$
$[\hat{d}_m]$	2.34	6.23	$[d_m] \quad = 2.55$
$[\hat{h}_o]$	−5.00	−4.75	

between these strains which were at the 23rd generation of selection (S_{23}) at the time. Thus all four possible combinations of the two strains (a 2 × 2 diallel) were bred. All their mothers were subjected to regimens designed to heighten pre-natal anxiety by exposing them to training in escape-avoidance conditioning in a shuttle box in which they had to learn to avoid shock. Table 1.13 shows the results for these parental strains and their F_1 crosses, both for the subjects bred from stressed mothers and the controls whose mothers had not been treated in this way. Thus we have the possibility of observing a genotype-environment interaction, the latter being mediated, in this case, via the maternal effect on the offspring. This maternal effect $[\hat{d}_m]$ is significant only in the case of the subjects whose mothers had been stressed, being more than twice the value of its standard error, and not significantly different from zero for the controls. The additive genetic effect for the offspring $[\hat{d}_o]$ borders on significance for both groups, though the effect of the maternal stress has clearly been to reverse its direction, being positive in the controls but negative in the pups whose mothers had been stressed. Thus it is acting in opposition to the additive genetic effect suggesting a maternal buffering mechanism for moderating the phenotypic expression of the offspring genotype. The dominance or rather potence effect is consistent in magnitude, though in the negative direction for both groups. This finding indicates that genes governing lower rather than higher ambulation in the open field tend to be dominant, a result generally supported by other psychogenetical experiments using these Maudsley strains.

The second effect not hitherto considered is sex-linkage. Sex-linkage expresses itself by a reciprocal difference which is usually confined to progeny of one sex only in each generation reared. Thus in a cross between two inbred strains, sex-linkage leads to reciprocal differences in the F_1 generations within the heterogametic (XY) but not within the homogametic (XX) sex (males and females respectively in mammals). In the F_2's raised by sib-mating within each of the two types of reciprocal F_1's, it is the homogametic sex (females) which shows a difference between reciprocal crosses. Consequently the expectation in Table 1.2 that $\overline{P}_1 + \overline{P}_2 + 2\overline{F}_1 - 4\overline{F}_2 = 0$ no longer holds even in the absence of non-allelic interaction, if the test is made for each sex separately for each reciprocal cross. However, it will hold if we average over reciprocal crosses in each generation. Hence by rearing reciprocal crosses we can unambiguously separate sex-linkage from the effects of non-allelic interaction.

So far we have been considering first-degree statistics, but sex-linkage can also affect the second-degree statistics obtained from segregating generations. For example, a single autosomal gene contributes $\frac{1}{2}d^2 + \frac{1}{4}h^2$ to the variance of the F_2 generation in both reciprocal crosses in both sexes. Each sex-linked

TABLE 1.14

The contribution of a single sex-linked gene to the variation of the F_2 generation according to (a) the direction of F_1 crossing of strains 1 and 2, and (b) the sex of progeny, illustrated for mammalian species.

Cross		Sex of Progeny	F_2 variation
Mother	Father		
$(1♀ \times 2♂) \times (1♀ \times 2♂)$		Female	$\frac{1}{2}(d_x - h_x)^2$
		Male	d_x^2
$(2♀ \times 1♂) \times (2♀ \times 1♂)$		Female	$\frac{1}{4}(d_x + h_x)^2$
		Male	d_x^2

gene (d_x, h_x), on the other hand, makes a different contribution as shown in Table 1.14. Hence we can detect the effect of sex-linked genes on the second-degree statistics by the failure of a model which assumes only autosomal gene contributions, providing, of course, that we have first ruled out other causes of failure by carrying out the appropriate scaling tests in the ways described earlier (§ 1.3.2).

If we wish to detect maternal and sex-linked gene effects in a diallel set of crosses between inbred lines, then we must ensure that the table is complete, that is, each cross must be reared reciprocally. The effect of their contributions to the differences between each pair of reciprocal crosses will, of course, be identical with those just described for a cross between a single pair of inbred strains and their derived generations, and hence we can detect and estimate their contributions in a comparable way. There are, however, a number of standard analyses of variance of diallel sets of crosses which contains items, e.g., the 'c' and 'd' item of the Hayman analysis (1954), which specifically test for maternal effects and their interaction with the offspring genotype, respectively (see Mather and Jinks, 1971). This analysis can readily be extended to test for maternal effects which are confined to the heterogametic sex, that is, those that arise from sex-linkage.

1.5. Non-inbred subjects

The problems posed by analyses without inbred strains are considerable. Where inbred strains are not available and the subjects are unrelated individuals or members of a population in which each individual is prospectively unique, then the individual rather than a strain or population becomes the unit of any mating programme designed to investigate the cause of variation. Such a situation is common in human psychogenetics and, though perhaps as yet has been little encountered in laboratory work with animals, it clearly applies to field investigations. It also applies with some force to possible laboratory work in psychogenetics with primates, among whom the instability of family groupings, coupled with this present restriction relating to the absence of inbred material, poses a considerable problem for the interested worker (Broadhurst, 1962). But it is clear that the unavailability of inbred strains has a number of consequences, both analytical and practical, which we can identify. Among them we simply list the following:
(a) There is no method of testing whether or not non-allelic interactions

and genotype-environment interactions are present on the chosen scale: consequently, there are no criteria for choosing an adequate scale. (b) The complexity of the mating programme is limited by the number of matings in which a single individual can participate. This limitation is usually more severe for females, since the additional problems potentially created by effects of maternal age and parity on the behaviour of progeny reduce the advantages of repeated matings with the same mother. (c) Since each male and female in the population is prospectively unique, reciprocal mating is not available as a means of detecting maternal effects and sex linkage. (d) We cannot obtain estimates of the environmental component of variation from the variation observed within inbred lines and F_1's.

However, there are solutions to these problems, though inevitably the precision of the outcome of the analyses may suffer. Such analyses are largely based on two separate techniques, first on the analysis of variance and second on the study of correlations between parents and offspring. The interested reader is referred to a somewhat more succinct but wider ranging treatment of biometrical genetics, on which the present chapter is based (Broadhurst and Jinks, 1974), which gives details and examples to illustrate these approaches (see also Jinks and Broadhurst, 1965).

Moreover, the possible alternative offered by the triple-test cross (§§ 1.3.2.3 and 1.3.3.1) as a method for investigating the genetic architecture of a non-inbred population, or even of a wild one, should not be overlooked.

1.6. Genotype-environment interaction

In the presence of genotype-environment interaction the expectations for generation means for inbred parents and the generations than can be derived by crosses between them have been given by Jones and Mather (1958), Mather and Jones (1958), Bucio-Alanis (1966), and Bucio-Alanis and Hill (1966), and the models extended to many inbred lines and a diallel set of crosses between them by Perkins and Jinks (1968). If we assume a simple additive and dominance model for the heritable contribution, the phenotype of an inbred line P_i which has been reared in environment e_j can be expressed as

$$\bar{P}_{ij} = \mu + [d]_i + e_j + g_{dij} \qquad (6)$$

where $\mu =$ the mean averaged over i lines and j environments, $[d]_i =$ the additive genetic contribution to the ith line, e_j = the additive environmental contribution of the jth environment, g_{dij} = the genotype-environment interaction of the ith inbred line with the jth environment. As defined

$$S_j \hat{e}_j = 0 \quad \text{and} \quad S_j \hat{g}_{dij} = 0 \tag{7}$$

If we consider the case of a pair of inbred lines, P_1 and P_2, reared in each of j environments, then

$$\mu = S_j (\bar{P}_{1j} + \bar{P}_{2j})/2j \tag{8}$$

$$[d] = S_j (\bar{P}_{1j} - \bar{P}_{2j})/2j \tag{9}$$

$$e_j = (\bar{P}_{1j} + \bar{P}_{2j})/2 - \mu \tag{10}$$

$$g_{dj} = (\bar{P}_{1j} - \bar{P}_{2j})/2 - [d] \tag{11}$$

where $\bar{P}_{1j}$ is the mean of P_1 in environment j, and so on. Similarly the F_1 obtained from a cross between the parental strains when reared in the same range of j environments has the expectation

$$\bar{F}_{ij} = \mu + [h] + e_j + g_{hj} \tag{12}$$

where μ and e_j are defined as above, and $[h]$ = the dominance genetic contribution as before and g_{hj} = the genotype-environment interaction of the F_1 with the jth environment. We can then write

$$[h] = S_j (\bar{F}_{1j} - \tfrac{1}{2}(\bar{P}_{1j} + \bar{P}_{2j}))/j \tag{13}$$

$$g_{hj} = \bar{F}_{1j} - \tfrac{1}{2}(\bar{P}_{1j} + \bar{P}_{2j}) - [h] \tag{14}$$

TABLE 1.15

Expected generation means in the presence of genotype-environment interactions for the usual generations (see text).

Generation Description	Symbol	General Expectation	Special Case of $g_{ij} = \beta_i e_j$
Inbred parents	$\bar{P}_{1j}$	$\mu + [d]_1 + e_j + g_{d1j}$	$\mu + [d]_1 + (1 + \beta_{d1}) e_j$
F_1	$\bar{F}_{1j}$	$\mu + [h]_1 + e_j + g_{h1j}$	$\mu + [h]_1 + (1 + \beta_{h1}) e_j$
F_2	$\bar{F}_{2j}$	$\mu + \tfrac{1}{2}[h]_1 + e_j + \tfrac{1}{2} g_{h1j}$	$\mu + \tfrac{1}{2}[h]_1 + (1 + \tfrac{1}{2}\beta_{h1}) e_j$
B_1	$\bar{B}_{1j}$	$\mu + \tfrac{1}{2}[d]_1 + \tfrac{1}{2}[h]_1 + e_j + \tfrac{1}{2} g_{d1j} + \tfrac{1}{2} g_{h1j}$	$\mu + \tfrac{1}{2}[d]_1 + \tfrac{1}{2}[h]_1 + (1 + \tfrac{1}{2}\beta_{d1} + \tfrac{1}{2}\beta_{h1}) e_j$
B_2	$\bar{B}_{2j}$	$\mu - \tfrac{1}{2}[d]_1 + \tfrac{1}{2}[h]_1 + e_j - \tfrac{1}{2} g_{d1j} + \tfrac{1}{2} g_{h1j}$	$\mu - \tfrac{1}{2}[d]_1 + \tfrac{1}{2}[h]_1 + (1 - \tfrac{1}{2}\beta_{d1} + \tfrac{1}{2}\beta_{h1}) e_j$

The expectations of other generations of general interest in this connection are summarised in Table 1.15. These expectations can readily be extended in three ways: first, by adding g_{dij} and g_{hij} to the expectations of Table 1.1, assuming no genotype-environment interaction, with the same sign and coefficient as $[d]_i$ and $[h]_i$, respectively; second, by including non-allelic interactions; and third, by including interactions of these non-allelic interactions themselves with the environment by introducing g terms corresponding to [i], [j], and [l]. If we estimate the genetic components from the observed generation means in any one environment, j, using the formulae in § 3.2.1, the estimates are biassed by the interactions with the environment as follows: for environment j the estimates of [d] is in fact $[d]+g_{dj}$, the estimates of [h] becomes $[h]+g_{hj}$ and so on. Thus we can see that the estimates of [d] and [h] will differ from one environment to another to an extent dependent on the magnitude and sign of the g_j for each environment. Further, it has been empirically established (Bucio-Alanis, 1966; Perkins and Jinks, 1968) that the estimates of the g_{dj}'s and g_{hj}'s are often a linear function of the corresponding e_j's, that is

$$g_{ij} = \beta_i e_j + \delta_{ij} \qquad (15)$$

where β_i is the linear regression of g_{ij} on e_j for the ith line and δ_{ij} is the deviation from the linear regression line for the ith line. We can therefore write expectations

$$\overline{P}_{ij} = \mu + [d]_i + (1+\beta_i) e_j + \delta_{ij} \qquad (16)$$

and so on for the other generations.

Where this model fits, that is to say the values of δ can be shown to be zero, each strain can be characterised by two parameters, $[d]_i$ which is its relative mean performance over all environments and β_i, the rate of change in its behaviour with unit change in the environment. If the sign of β_i is positive, we can say that the strain has greater than average sensitivity to the environment among the strains studied, that is a greater rate of change, whereas if β_i is negative it has a below-average sensitivity – a lower rate of change.

We illustrate with a reanalysis of some data presented by Gray, Levine and Broadhurst (1965), again using the Maudsley strains (MR and MNR) and the F_1 between them. Gray and his co-workers were interested in the effect of injections of gonadal hormones given in infancy on adult behaviour in the open field, since such treatments are known to affect both sexual behaviour directly, as well as sex differences in such indices as emotional elimination. The treatments comprised the injection of both male and female hormones (androgen and estrogen) to both sexes at five days of age, and included a group given a placebo (oil) and a control group given nothing. Table 1.16 shows the estimates of the various components of the means and

TABLE 1.16

Estimates of the genotype, environmental, and genotype environment interaction components of the emotional elimination of rats exposed in infancy to different environmental (hormonal) treatments (Gray, Levine and Broadhurst, 1965). Units are number of fecal boluses deposited per day.

		Environmental treatments (j)			
	Control	Estrogen	Androgen	Placebo	Overall
		Females			
$\hat{\mu}$					1.981
$[\hat{d}]$					1.969
$\hat{e}_j$	−0.794	−0.306	0.506	0.594	0
$\hat{g}_{dj}$	−0.781	−0.294	0.494	0.581	0
β					0.979*
$\bar{P}_{MR_j} = 3.950 + 1.979\ e_j$					
$\bar{P}_{MNR_j} = 3.950 + 0.021\ e_j$					
		Males			
$\hat{\mu}$					2.153
$[\hat{d}]$					2.103
$\hat{e}_j$	−0.316	−0.091	−0.003	0.409	0
$\hat{g}_{dj}$	−0.365	−0.065	−0.028	0.459	0
β					1.124*
$\bar{P}_{MR_j} = 0.426 + 2.124\ e_j$					
$\bar{P}_{MNR_j} = 0.426 - 0.124\ e_j$					

* Both β's are significantly different from zero but not from one another.

their interactions, as described above in this section, for the defecation scores observed in the adult subjects, and given separately for the two sexes. As may be seen, the additive environmental components ($\hat{e}_j$) attributable to the three treatments are larger than their control values in both sexes. Furthermore, the genotype-environment interaction components ($\hat{g}_{ij}$) are linearly related to the corresponding environmental components with highly significant regression slopes (β values). Hence, the performances of the two strains under the different treatments can be described by the simple relationships given in Table 1.16. These relationships show that the MR strain is more sensitive to environmental effects than the MNR strain, in that it has a larger coefficient of e_j. Since, as noted earlier, the $\hat{e}_j$ and $\hat{g}_{dj}$ values must sum to zero the contribution of these two components to the total variation can be obtained as the sum of their squared values. In the present case the additive environmental and genotype-environment components make virtually identical contributions to the total variation. A comparable analysis of the less complete F_1 data (omitting estrogen), provided by Gray, Levine and Broadhurst, shows that the F_1 is intermediate in its sensitivity to the remaining environmental effects, having a value that is not significantly different from the average β for the two parental strains.

Similar models can be developed for second-degree statistics. Thus in any environment e_j our estimate of $D = Sd^2$ is in fact $Sd^2 + Sg_{dj}^2$, while $H = Sh^2 + Sg_{hj}^2$ and so on for the diallel and random-mating forms of D and H (see §§ 1.3.2.2 and 1.3.3.1). Hence, estimates of these components of variation will differ from one environment to another in the presence of genotype-environment interactions. While Mather and Jones (1958) have indicated how these methods can be extended to random-mating populations, the only practical application of this approach to non-inbred populations has been not to animal but to human populations. Jinks and Fulker (1970) have shown that the special properties of twins can be used to detect genotype-environmental interactions for various behavioural measures.

1.7. Heritability estimates: their uses and limitations

The results of our biometrical analyses are usually summarised in the form of heritabilities. These are of two kinds, "narrow" and "broad". In general the "narrow" form is the proportion of the total phenotypic variation that is additive in origin, while the broad form is the proportion of the total that is genetic in origin. In terms of the simple additive, dominance model the "narrow" form is

$$ht_n = \tfrac{1}{2}D_R/(\tfrac{1}{2}D_R + \tfrac{1}{4}H_R + E_1 + E_2) \tag{15}$$

and the "broad" form is

$$ht_b = (\tfrac{1}{2}D_R + \tfrac{1}{4}H_R)/(\tfrac{1}{2}D_R + \tfrac{1}{4}H_R + E_1 + E_2) \tag{16}$$

for a random mating population. Heritabilities, however, are more than a convenient means of portraying the proportion of the variation that can be ascribed to various causes. In fact, they have a predictive value for short-term selection experiments. It can be shown that over a few generations of selection the expected response is the narrow heritability of the population on which selection is being practised multiplied by the selection differential, which is the deviation of the selected sample mean from the generation mean (Falconer, 1960). Thus the expected response is proportional to the heritability for the same selection differential; the higher the heritability the greater the response and vice versa.

An estimate of heritability strictly applies only to the circumstances in which it is obtained, although it is frequently used in a looser, more general sense. Thus the values of D and H will be characteristic of the inbred population under investigation. They will not apply to another population or to the same population if it is changed by selection. Equally the values of E_1 and E_2 will be characteristic of the environment in which the subjects have been reared, and they will not be appropriate if the environment changes. Hence, heritability is a property of the population and not of the traits. It is, therefore, incorrect to refer to *the* heritability of a trait as if it had been established for all time for all circumstances.

1.8. *The problem of heterosis*

Heterosis is defined by the extent to which an F_1 produced by crossing two inbred strains exceeds the parent which is better, that is, fitter in an adaptive sense. It is an important phenomenon in commercial breeding where the superior performance of the F_1 is often exploited to give greater yields. It also has evolutionary significance and hence is of interest to the behavioural worker (see ch. 2, § 5.2.2).

The expected magnitude of the heterosis to be expected has been given by Jinks and Jones (1958). If the heterosis results from the F_1 exceeding the higher-scoring parent P_1, it is defined as:

$$\overline{F}_1 - \overline{P}_1 = \text{Heterosis} = ([h]+[l])-([d]+[i]) = \text{positive value.} \quad (18)$$

If, on the other hand, it results from the F_1 having a lower score than the lower-scoring parent P_2 heterosis is defined as:

$$\overline{F}_1 - \overline{P}_2 = ([h]+[l])-(-[d]+[i]) = \text{negative value.} \quad (19)$$

If the appropriate tests show that non-allelic interactions are absent, then these expectations reduce to:

$$\overline{F}_1 - \overline{P}_1 = [h]-[d] \quad (20)$$

and
$$\overline{F}_1 - \overline{P}_2 = [h] + [d] \qquad (21)$$

This is equivalent to noting that in either case there is heterosis when [h] > [d], that is directional dominance, either positive or negative, is exceeding the additive component. This circumstance can arise in one of two ways, first where there is overdominance, and second where there is dispersion of dominant alleles.

These are two alternative interpretations which differ in ways that have important theoretical and practical consequences. They can be illustrated by a two-gene model. The composition of the four possible inbred strains in a two-gene model are AABB, AAbb, aaBB, and aabb, where AA and BB are the increaser alleles adding $+d_a$ and $+d_b$ to the mean, and aa and bb are the decreaser alleles subtracting $-d_a$ and $-d_b$ from the mean, respectively.

The two crosses of interest are those having parents differing in respect of both gene differences. These are: (1) AABB × aabb which gives an F_1 AaBb, and (2) AAbb × aaBB which also gives AaBb.

But in cross 1, where the genes in the parents are associated, that is to say, increaser with increaser and decreaser with decreaser,
$$\overline{F}_1 - \overline{P}_1 = (h_a + h_b) - (d_a + d_b) \qquad (22)$$
and
$$\overline{F}_1 - \overline{P}_2 = (h_a + h_b) + (d_a + d_b). \qquad (23)$$

For heterosis to be observed in this cross, $\pm h_a$ must be greater than d_a and/or $\pm h_b$ must be greater than d_b. That is, the heterozygous state must be superior to either homozygous state at one or both loci (i.e., Aa > AA > aa or AA > aa > Aa, etc.). Hence if the genes in the parents are associated there must be overdominance for heterosis to occur.

However, in cross 2 where the genes are dispersed in the parents, that is, increaser with decreaser,
$$\overline{F}_1 - \overline{P}_1 = (h_a + h_b) - (d_a - d_b) \qquad (24)$$
and
$$\overline{F}_1 - \overline{P}_2 = (h_a + h_b) - (-d_a + d_b) \qquad (25)$$

It follows, therefore that heterosis may arise when $h_a \leqslant d_a$ and $h_b \leqslant d_b$

when the genes are dispersed in the parents, provided h_a and h_b have the same sign. In this way we see that heterosis can, therefore, arise without the heterozygous state at any locus being superior to the homozygous states at the same locus.

When the appropriate tests show that non-allelic interactions are present, the expectations for heterosis given above mean that many combinations of additive, dominance, and interactive effects can lead to heterosis. However, heterosis will clearly be most evident when [h] and [l] have the same sign, [d] is small and [i] is small or negative. These are in fact the conditions in which we would have a complementary type of interaction between the genes which are dispersed between the two parents (see § 1.3.2.2).

If we wish to investigate the cause of heterosis, we must obviously rear, as a minimum, the parents P_1, P_2, their F_1, F_2, and their backcrosses, B_1 and B_2. These six generations will provide sufficient statistics to decide whether or not non-allelic interactions are contributing to the generation means and to estimate all the components of heterosis if interactions are present. Although parents and F_1's alone have been used to investigate heterosis (Bruell, 1964; Parsons, 1967a), they are clearly insufficient, unless they constitute and are analysed as a complete diallel, to determine its cause; they can only be used to detect its presence.

1.9. Estimates of the number of genes

Methods of estimating the number of genes – or, rather, effective genetic factors – involved in determining a phenotype are among the least satisfactory procedures in biometrical genetics. The reason is that the methods depend on a number of unlikely assumptions, the failure of which invariably leads to an underestimate of the number of genes involved. These assumptions are, first, that each polygene makes an exactly identical contribution to the phenotype, i.e., $d_a = d_b = \ldots = d_k$ for all k genes controlling the character, and second, that all the genes of like effect are associated in one parent, that is, all increasing genes yielding positive d's are present in one inbred strain, P_1, and all decreasing genes yielding negative d's in another, P_2. This assumption is likely to hold only in pairs of strains after having been subject to extreme selection (§ 1.2.3). If these assumptions hold, then

$$\tfrac{1}{2}(\overline{P}_1 - \overline{P}_2) = [d] = kd \qquad (26)$$

We would estimate $D = Sd^2 = kd^2$ from the segregating generations (F_2, B_1, B_2, etc.) derived from crossing these strains P_1 and P_2, and D would then equal kd^2. Hence we can arrive at an estimate of the number of genes (or effective factors, that is, groups of genes linked together) for which these strains P_1 and P_2 differ as follows:

$$(\tfrac{1}{2}(\overline{P}_1 - \overline{P}_2))^2/D = k^2 d^2/kd^2 = k \qquad (27)$$

If the first assumption fails then the number we estimate by this formula is k', which is always smaller than k and bears the following relationship to it:

$$k' = k/(1+\sigma^2_d) \qquad (28)$$

where σ^2_d is the variance in the magnitude of the additive effects of the genes.

Similarly if the second assumption fails then what we estimate is k'', which is also always smaller than k, where $k'' = r_d^2 k$ and r_d ranges from 1 to 0 and is a correlation expressive of the degree of association of the genes in the parental strains, being unity for complete association and zero for complete dispersion, which occurs when half the increaser genes and half the decreaser genes are present in each parental strain.

In addition to these two special assumptions, this estimate is subject to all the other assumptions that must be met in order to estimate D satisfactorily from the variation in the segregating generations (§ 1.3). However, it is clear that failure of the special assumptions must always lead to an *under*estimate of k, since k' and k'' are both less than k. Other methods of estimating k, each subject to their own limitations, have been described by Cooke and Mather (1962). In general their complexity renders them beyond the scope of most animal breeding experiments of interest in psychogenetics.

1.10. Elaborations of the model

Mather (1949) and Mather and Jinks (1971) have shown how models can be constructed which allow for the effects of linkage between the genes controlling a metrical trait and these have been elaborated by others to cover all situations that are likely to arise in an animal breeding programme (e.g., Van der Veen, 1959; Jones, 1960). These models have suggested methods for detecting the presence of linkage, determining its phase, that is coupling or repulsion, and estimating the number of linked genes. In practice, however, these require an experimental breeding programme beyond the scope of any laboratory at the present time except for a species such as the fruit fly *Drosophila melanogaster* (Cooke and Mather, 1962). This is also true of the procedures described by Thoday and his colleagues (see Thoday, 1966, for reviews) for determining the number of genes controlling a metrical trait and locating their positions on the linkage map relative to one another and to major-gene loci.

1.11. Conclusions

It has only been possible to hint at the implications which can be drawn from the possibilities inherent in the application of the biometrical approach in discussing the examples of behavioural analyses cited in this chapter. In several cases no behavioural examples are as yet available, though plant data sometimes exist. This situation reflects the tendency, long apparent in this area, for theory to outstrip practice, so that analytical techniques and methods lie fallow while researchers come gradually to learn of their existence and tentatively to try them out. A recent review (Broadhurst et al., 1974) indicates that work using these techniques is now increasing.

One aspect of the methods of analysis described in this chapter as applicable to behavioural data is their generality. It cannot be sufficiently stressed that they can be applied to each and every aspect of the phenotype of the organism it is desired to study, given that the feature concerned is amenable to mensuration. Adequacy of the scale on which that measurement is made can be examined, both in ways familiar to the psychometric psychologist and in respect of the assumptions required by the analysis it is intended to apply. The genetic relationships among the populations or individuals involved, if we are not using inbred material, are to some extent irrelevant, since the consequences stemming from the nature of the relationships are now largely understood and appropriate allowances can be made for them in the analyses we have outlined. It is sufficient that the extent to which a greater degree of imprecision of outcome can be accepted as a consequence of the use of less-than-optimal data should be understood.

Essentially what we are stressing is the freedom the recognition of the potentialities of the biometrical analysis of behavioural phenotypes gives to the worker interested in the impact hereditary and environmental variables may have on the behaviour being measured. The generality of the analyses available allows their effect to be evaluated in as great a detail and with as great a complexity of their mutual interaction as the excellence of the data and the persistence of the experimenter warrant.

The use to which the adoption of these relatively new methods of analysis has as yet led is limited. Examples given in this chapter have mainly been derived from experimental work with laboratory animals of inbred strains. Application to non-inbred populations has hardly begun, though the immediate future promises advances here, too. It is, therefore, perhaps premature to suggest the especial utility of biometrical genetics in behavioural biology, but there is at least one general line which can already be discerned. This relates to the illumination findings in this area may bring to questions regarding the evolution of behaviour, and we explore this theme more fully in ch. 2.

References

Barnes, B. W. (1964): Maternal control of heterosis for yield in *Drosphila melanogaster*. *Heredity*, **23**, 563.

Broadhurst, P. L. (1959): Application of biometrical genetics to behaviour in rats. *Nature (Lond.)*, **184**, 1517.

Broadhurst, P. L. (1960): Experiments in psychogenetics: Applications of biometrical genetics to the inheritance of behaviour. In: Experiments in Personality. Vol. 1. Psychogenetics and Psychopharmacology. H. J. Eysenck, ed. (Routledge and Kegan Paul, London) pp. 1–102.

Broadhurst, P. L. (1962a): A note on further progress in a psychogenetic selection experiment. *Psychol. Rep.*, **10**, 65.

Broadhurst, P. L. (1962b): Note on some possibilities for primate psychogenetics. *Lab. Primate Newsletter*, **1** (4), 1.

Broadhurst, P. L. (1967): An introduction to the diallel cross. In: Behavior-genetic Analysis. J. Hirsch, ed. (McGraw-Hill, New York) pp. 287–304.

Broadhurst, P. L. (1968): Experimental approaches to the evolution of behaviour. In: Genetic and Environmental Influences on Behaviour. Eugenics Soc. Symposia, Vol. 4. Thoday, J. M. and A. S. Parkes, eds. (Oliver and Boyd, Edinburgh) pp. 15–36.
(Also: 1972, Approche expérimentale à l'étude de l'évolution du comportement. In: Problemes de Méthodes en Psychologie Comparée: Congrès et Colloques de l'Université de Liège, Septembre, 1969. Richelle, M. and J.-Cl. Ruwer, eds. (Paris, Masson) pp. 95–114.)

Broadhurst, P. L. (1969): Psychogenetics of emotionality in the rat. *Ann. N.Y. Acad. Sci.*, **159**, 806.

Broadhurst, P. L., D. W. Fulker and J. Wilcock (1974): Behavioral genetics. *Ann. Rev. Psychol.*, **25**, 389.

Broadhurst, P. L. and J. L. Jinks (1961): Biometrical genetics and behavior: Reanalysis of published data. *Psychol. Bull.*, **58**, 337.

Broadhurst, P. L. and J. L. Jinks (1966): Stability and change in the inheritance of behaviour: A further analysis of statistics from a diallel cross. *Proc. Roy. Soc.*, B, **165**, 450.

Broadhurst, P. L. and J. L. Jinks (1974): Psychological genetics, from the study of animal behavior. In: Handbook of Modern Personality Theory. Cattell, R. B. and R. M. Dreger, eds. (Prentice-Hall, New Jersey).

Bruell, J. H. (1964): Inheritance of behavioral and physiological characters of mice and the problem of heterosis. *Amer. Zoologist*, **4**, 125.

Bucio-Alanis, L. (1966): Environmental and genotype-environmental components of variability: I. Inbred lines. *Heredity*, **21**, 387.

Bucio-Alanis, L. and J. Hill, (1966): Environmental and genotype-environmental components of variability: II. Heterozygotes. *Heredity*, **21**, 399.

Cavalli, L. L. (1950): An analysis of linkage in quantitative inheritance. In: Quantitative Inheritance. Reeve, E. C. R. and C. H. Waddington, eds. (H.M.S.O., London) pp. 135–144.

Collins, R. L. (1964): Inheritance of avoidance conditioning in mice: A diallel study. *Science*, **143**, 1188.

Comstock, R. E. and H. F. Robinson (19W): Estimation of average dominance of genes. In: Heterosis. Gowen, J. W., ed. (State College Press, Ames, Iowa) pp. 494–516.

Cooke, P. and K. Mather (1962): Estimating the components of continuous variation: II. Genetical. *Heredity*, **17**, 211.

Eysenck, H. J. and P. L. Broadhurst (1964): Experiments with animals: Introduction. In: Experiments in Motivation. H. J. Eysenck, ed. (Pergamon Press, Oxford) pp. 285–291.

Falconer, D. S. (1960): Introduction to Quantitative Genetics. (Oliver and Boyd, Edinburgh).

Fisher, R. A. (1918): The correlation between relatives on the supposition of Mendelian inheritance. *Trans. Roy. Soc. Edin.*, **52**, 399.

Fulker, D. W. (1970): Maternal buffering of rodent genotypic responses to complex genotype-environment interaction. *Behav. Genet.*, **1**, 119.

Fulker, D. W. (1972): Applications of a simplified triple-test cross. *Behav. Genet.*, **2**, 185.

Fulker, D. W., J. Wilcock and P. L. Broadhurst (1972): Studies in genotype-environment interaction. I. Methodology and preliminary multivariate analysis of a diallel cross of eight strains of rat. *Behav. Genet.*, **2**, 261.

Ghiselin, M. T. (1969): The Triumph of the Darwinian Method. (University of California Press, Berkeley, Calif.).

Gray, J. A., S. Levine and P. L. Broadhurst (1965): Gonadal hormone injections in infancy and adult emotional behaviour. *Anim. Behav.*, **13**, 33.

Hall, C. S. (1938): The inheritance of emotionality. *Sigma Xi Quart.*, **26**, 17 and 37. (Also: 1954, In: Readings in Child Development. Martin, W. F. and C. B. Stendler, eds. (Harcourt Brace, New York) pp. 59–68.)

Hayman, B. I. (1954): The analysis of variance of diallel crosses. *Biometrics* **10**, 235.

Jinks, J. L. and P. L. Broadhurst (1965): The detection and estimation of heritable differences in behaviour among individuals. *Heredity*, **20**, 97.

Jinks, J. L. and D. W. Fulker (1970): A comparison of the biometrical genetical, MAVA and classical approaches to the analysis of human behavior. *Psychol. Bull.*, **73**, 311.

Jinks, J. L. and R. M. Jones (1958): Estimation of the components of heterosis. *Genetics*, **43**, 223.

Jinks, J. L. and J. M. Perkins (1970): A general method for the detection of additive, dominance and epistatic components of variation: III. F_2 and backcross populations. *Heredity*, **25**, 419.

Jinks, J. L., J. M. Perkins and E. L. Breese (1969): A general method of detecting additive, dominance and epistatic variation for metrical traits: II. Application to inbred lines. *Heredity*, **24**, 45.

Joffe, J. M. (1965): Genotype and prenatal and premating stress interact to affect adult behavior in rats. *Science*, **150**, 1844.

Joffe, J. M. (1969): Prenatal Determinants of Behaviour. (Pergamon Press, Oxford).

Jones, R. M. (1960): Linkage distributions and epistacy in quantitative inheritance. *Heredity*, **15**, 153.

Jones, R. M. and K. Mather (1958): Interaction of genotype and environment in continuous variation: II. Analysis. *Biometrics*, **14**, 489.

Kearsey, M. J. and J. L. Jinks (1968): A general method of detecting additive, dominance and epistatic variation for metrical traits: I. Theory. *Heredity*, **23**, 403.

Lorenz, K. (1966): Evolution and Modification of Behaviour. (Methuen, London).

Mather, K. (1949): Biometrical Genetics: The Study of Continuous Variation. (Methuen, London).

Mather, K. (1953): The genetical structure of populations. *Symp. Soc. exp. Biol.*, **7**, 66.

Mather, K. (1973): Genetical Structure of Populations. (Chapman and Hall, London).

Mather, K. and J. L. Jinks (1971): Biometrical Genetics: The Study of Continuous Variation, 2nd ed. (Chapman and Hall, London).

Mather, K. and R. M. Jones (1958): Interaction of genotype and environment in continuous variation: I. Description. *Biometrics*, **14**, 343.

McClearn, G. E. (1961): Genotype and mouse activity. *J. comp. Physiol. Psychol.*, **54**, 674.

Parsons, P. A. (1967a): Behavioural homeostasis in mice. *Genetica*, **38**, 134.

Parsons, P. A. (1967b): The Gene ic Analysis of Behaviour. (Methuen, London).

Perkins, J. M. and J. L. Jinks (1968): Environmental and genotype-environmental components of variability: III. Multiple lines and crosses. *Heredity*, **23**, 339.

Rothenbuhler, W. C. (1964): Behavior genetics of nest cleaning in honey bees. IV. Responses of F_1 and backcross generations to disease-killed brood. *Amer. Zoologist*, **4**, 111.

Thoday, J. M. (1966): New insights into continuous variation. In: Proc. 3rd Int. Congr. Human Genetics, Univ. Chicago, pp. 339–350.

Tryon, R. C. (1940): Genetic differences in maze-learning ability in rats. In: 39th Year Book Nat. Soc. Stud. Educ., Part. I (Public School Publ. Corp., Bloomington, Ill.) pp. 111–119.

Veen, J. H. van der (1959): The 2×2 genotype-environment table. *Heredity*, **13**, 123.

Wilcock, J. (1969): Gene action and behavior: An evaluation of major gene pleiotropism. *Psychol. Bull.*, **72**, 1.

Wilson, E. O. (1971): The Insect Societies. (Belknap Press, Cambridge, Mass.).

CHAPTER 2

What genetical architecture can tell us about the natural selection of behavioural traits*

P. L. Broadhurst and J. L. Jinks

2.1. Introduction

While psychogenetics – or behaviour genetics in the US – is rapidly achieving the status of a respectable specialisation within psychology and behavioural biology, as witnessed by the frequency of its coverage in the *Annual Review of Psychology* (Broadhurst et al., 1974), there still appear to be misconceptions about its nature, its aims, and the appropriate methodology for achieving those aims. It is the purpose of this chapter, therefore, to attempt to put into contemporary perspective some of the considerations which, it seems to us, continue to be ill-understood in the study of animal behaviour. We do not attempt to cover approaches concerned with biochemical and physiological techniques of analysis of the causation and regulation of behaviour.

In particular we wish to draw attention once again to the possibilities offered by the analytical techniques of quantitative genetics of the biometrical kind (Mather, 1949; Mather and Jinks, 1971) in encompassing the problems in the study of animal behaviour. As long ago as 1951, one of us applied these techniques to behavioural work (Broadhurst 1959, 1960) and in 1961 we published a reanalysis of several sets of animal data (Broadhurst and Jinks, 1961, 1963) in the hope that this further demonstration of the efficacy of those techniques would prove convincing. However, progress

---

* Some of the work reported in this chapter was supported by grants from the (British) Medical Research Council, Its preparation was furthered by P.L.B.'s Resident Fellowship (1973–74) at N.I.A.S. (Nederlands Instituut voor Voortgezet Wetenschappelijk Onderzoek op het Gebied van de Mens- en Maatschappijwetenschappen).

remains disappointingly slow outside a few centres, and it seems worthwhile, therefore, to reiterate briefly the main points of the biometrical methodology and analysis which seem to us to suit it eminently well for behavioural analysis. To develop the argument without tedious ellipsis we must go over some of the ground already covered in ch. 1 by Jinks and Broadhurst to which we shall nevertheless refer from time to time.

Biometrical genetics is entirely general in its application, so that, even though it was developed primarily for the analysis of quantitative characteristics of plants, it is equally applicable to any characteristic or trait showing continuous variation. Since behaviour is typically of this kind, it is not surprising that the methodology is as completely applicable in behavioural science as it has been in other fields such as plant breeding where it has been applied with singular success.

2.2. Initial separation: phenotype and genotype

Geneticists make the fundamental distinction between phenotype and genotype: the phenotype is that which can be observed or measured. It need not be a direct measure, but can be a ratio or a transformed score (ch. 1, § 1.3.1). The phenotype is envisaged as being composed of two elements, the genotype and the environment. Thus we write the equation:

$$\text{Phenotype (P)} = \text{Genotype (G)} + \text{Environment (E)}$$

Traditionally, psychologists have concerned themselves with the environmental determinants of behaviour, neglecting, to some extent, genetic determinants. Emphases have fluctuated in this respect, both over time and in different countries: typically American psychology has been more environmentalist and European, especially British, psychology, more hereditarian in outlook, and, again, the three decades between 1920 and 1950 were generally more environmentalist than those which immediately preceded or followed.

However, it is now generally agreed that in considering any behaviour or behavioural phenotype, as we shall now call what the psychogeneticist measures, it is important not to neglect genetical determinants. But it is less often clearly understood *why* this should be so. In essence the reason is obvious: hereditary determinants are transmitted to the next and subsequent generations in a considerably more direct form than those environmental causes which operate on the organism. That is to say, environment in the shape of culture in the broadest sense is transmitted indirectly: heredity directly via genetic material having a known physico-chemical basis. This seems to us so obvious a truism that we feel diffirent about restating it in these simple terms but in it lies the key not only to the whole rationale for the modern revival of interest in psychogenetics but also to much of the misunderstanding of its importance.

This is not to say that we are concerned to stress heredity to the exclusion of environment nor yet, manifestly, he reverse. What we are concerned to argue is that there should be an adequate recognition of the importance of the genetical component in the determination of individual differences in behavioural phenotypes and that the consequences which may be expected to flow from such recognition should be appreciated.

Clearly the most important outcome of the adoption of this viewpoint is that it becomes necessary to estimate the magnitude of the hereditary contribution to the phenotype in order to predict future behaviour. The operation of environmental effects over generations may be important and may ultimately be predictable, but we are in error if we assume they are, without establishing the fact beforehand. The same is, of course, true in respect of hereditary determinants, but it seems to us that the danger in the latter case is less typical of the trend in contemporary psychology than in the former, though the reverse may perhaps still be so in ethology (see ch. 1, § 1.1.1).

In ethology the importance which may be attached to the biometrical approach is perhaps of another kind. It lies in the possibility afforded for the detailed analyses of the patterns of behaviour which have been identified as of biological significance and in the interpretations which can be drawn from them as regards the evolutionary importance of the behaviour observed. Such interpretations can consequently be given a precision not hitherto characteristic of ethological investigations (Broadhurst, 1968) in which, as in psychology, the fact of quantitative variation has impeded progress towards giving our understanding of the evolution of behaviour – in the wild as well as in the laboratory – the firmer basis in genetics which it needs and which it is one of the aims of this chapter to offer.

2.3. Further subdivisions: the genetical architecture

The resolution of the variation within a population thus has a broadly heuristic value, and the initial separation of the genetical portion of it from the environmental allows a precision of prediction not otherwise possible and, indeed, many analyses do not seek to go beyond this point. However, biometrical genetics allows us to proceed yet further by additional subdivision of the genetical components. Similarly experimental psychology permits sensitive and sophisticated analysis of the environmental, that is, stimulus, components operating within a given situation, though it is beyond the scope of this chapter to explore these latter possibilities, except insofar as they interact with genetical determinants in the way to be discussed later (see § 2.5).

The reasons for wishing to subdivide the genetical contribution into the components recognised by biometrical genetics follow from, and are similar to, those which call for the initial distinction of the total variation into environmental and genetical components. Just as environment and genotype influence successive generations in different ways and by different mechanisms, the two major genetical components, recognised as the additive or "fixable" and the dominance or "non-fixable" components of variation (see ch. 1, § 1.2.2) have characteristically different consequences for future generations. That is to say, their effects are transmitted in different ways, depending on the mating system operating in the parental generation in question.

Furthermore, the relative magnitude of these two components of genetic variation, the additive and dominance variation, gives us insight into what geneticists term the "genetical architecture". This phrase implies an understanding of the effects of genes governing a particular phenotype in a given population at a given time. It derives primarily from plant genetics, and *Drosophila* (fruit fly) genetics, where it is found that it is frequently possible to generalise such genetic effects, both across phenotypes and across populations. Thus genetically similar populations tend to similar genetical control of the same phenotype, and the same population tends to similar genetical structure in control of similar phenotypes.

2.4. *Inferences to natural selection*

2.4.1. *Kinds of selection*

This method of inferential argument has been most extensively developed by Mather (1943, 1953, 1960, 1966, 1973), who, with his colleagues, has demonstrated its validity by a series of supporting investigations. First, Mather (1973) recognises three kinds of selection; stabilising, directional, and disruptive, which may be distinguished by their optimum expression in the phenotype, that is, phenotypes which vary from a particular level or mean in the population are less favoured in an adaptive sense and consequently leave, on average, fewer progeny – in a word, are less fit. In the case of stabilising selection, the favoured level of expression is an intermediate one in terms of the phenotypic range, departures from it being less fit, irrespective of whether they exceed or fall short of it. In directional selection, while there is still a single optimal expression, its value is relatively extreme in terms of the phenotypic scale, so genetic mechanisms in the population tend to be such as to move the population mean unidirectionally towards that value. And in the case of disruptive selection, more than one, and usually two, optima are encountered. It is in such a situation that dimorphisms and polymorphisms may arise, which may be in a stable equilibrium within the population or may lead to breeding isolation and incipient speciation. The commonest and perhaps the most striking example of such a diphormism is the existence of the two sexes. But it is with the first two kinds of selection having single optima of expression that we are at present primarily concerned.

2.4.2. Genetical consequences of selection

Stabilising and directional selection have predictably different consequences for the genetical architecture of those phenotypic characters upon which they have operated in the past. Stabilising selection favours the average phenotype; that is, it will be the balanced genotype (see ch. 1, § 1.2.1) which will be favoured at the expense of the extremes which by definition have a phenotype which is the outcome of imbalance between increasers and decreasers and dominant increasers and dominant decreasers. The diagnostic feature of a trait subjected to stabilising selection will therefore be, firstly, linkage of genes in the repulsion phase – that is, positive (increaser) with negative (decreaser) – and, secondly, dominance features having one of two separate possibilities. These are either, on the one hand, the presence of very little or almost no dominance, so that the dominance ratio (see ch. 1, § 1.3.3.1) is very low, or, on the other hand, if dominance is in fact significantly high it will be of an ambidirectional kind. In general, therefore, such traits will show a large additive component of variation, relative to the size of the dominance component. These genetic mechanisms are those which will maximise the amount of potential variation at the expense of the free variation (see ch. 1, § 1.3.3.1), since any free variation will lead to the occurrence in the population of individuals which deviate from the mean, with the unfavourable consequences for fitness which, as we have seen, flow from such deviation. In passing, it is noteworthy that such a trait, contrary to the superficial impression of lack of variation which may be given, will nevertheless show a sustained response to artificial selection (Mather and Harrison, 1949), especially of a bi-directional kind, which delays fixation (inbreeding), and so allows ample opportunity for recombination to break the repulsion linkages hitherto holding increasers and decreasers together in balance. In this way recombinants having predominantly increasing or decreasing alleles will arise and express their effect on the phenotype (see ch. 1, § 1.2.1).

Directional selection, on the other hand, is distinguished by favouring phenotypes towards one extreme of the distribution, and where this situation has occurred as a consequence of natural selection we typically encounter a genetic architecture with the following features. Firstly, directional dominance will be found such that if selection has been, for example, for an increase in the expression of the trait – a higher mean score or level – then the increasing alleles will be the dominant ones, and vice versa. Therefore, and in contradistinction to a trait under the influence of stabilising selection, we would expect a high level of dominance variation to be displayed, relative, of course, to the level of additive variation which may well be small. Nor would we expect to find a reservoir of potential variation of an additive kind. As a second diagnostic feature of directional selection we can point to a type of epistasis known as duplicate gene interaction. This is a form of non-allelic interaction which, as the name implies, differs from allelic or dominance interaction (ch. 1, § 1.3.3.2), and hence is an epistatic interaction. The occurrence of duplicate genes would reinforce the impression of behaviour under directional selection, since it leads to more different genotypes each having the optimal phenotypic effect. The consequence is that a greater proportion of possible genotypes attain the most appropriate level of expression. By providing more pathways whereby the genotype is optimally expressed in the way described, we see a "fail safe" mechanism ensuring that the most important characteristics from the point of view of survival are overdetermined genetically. It is not surprising therefore, that this genetic architecture is the hallmark of the so-called "fitness" characters. Unlike characters subject to stabilising selection, traits directly related to fitness in this way are unlikely to show the same ample response to artificial selection other than that, of course, in a direction *away* from the optimal level of expression since that optimum is maintained under constant pressure from natural selection, with a consequent paucity of additive genetic variation as we have seen.

2.4.3. *Experimental evidence*

These essentially theoretical considerations, which were formulated by Mather, have been supported by a series of experiments using as material the fruit fly. Breese and Mather (1960) studied the genetic architecture governing both viability and the number of sternopleural chaetae (abdominal bristles) in *Drosophila melanogaster*. Bristle number is a character of the fruit fly which is known to show intermediate expression in terms of the range possible when selected artificially for either high or low scores. It numbers about 18–20 in unselected populations. The other character studied, viability, is a fitness character par excellence; clearly, the number of progeny individuals leave is a direct measure of the adaptive success of the genotype they represent, though it should be noted in passing that such may not always be true on an average for the population as a whole and for all generations of it. As would be predicted, Breese and Mather found a genetic architecture for viability characterised by unidirectional dominance (for higher viability) together with duplicate gene interaction. The significance of the former is clear; genes governing high viability show dominance, which, on the average, was complete over those governing low. We must, however, note that the large measure of dominance associated with fitness measures leads to their having a low heritability of the "narrow" kind (see ch. 1, § 1.7) – an apparent paradox soon dispelled when it is recalled that dominance variation is excluded from the numerator in calculating the narrow heritability ratio, leaving only the additive component. The numerator is consequently considerably less than for the broad heritability ratio, or "degree of genetic determination" as it is sometimes called, in which dominance is included. Indeed, this discrepancy between the two ratios may be taken as a mark of the "fitness" character. Moreover, additive variation itself may well be small, so that the narrow heritability estimate will tend to be low for this reason also (see § 2.4.2). The fixable genetical variation has already, as it were, been fixed for fitness characters for which large variation between laboratory strains would naturally not be expected, the *unfit* ones soon having died out during the process of domestication.

The duplicate gene interaction Mather and Breese showed to be characteristic of *Drosophila* viability reinforces the selective advantage of directional dominance, especially by reducing variance (Mather, 1967). We are thus in a position to recognise experimentally a fitness character by the nature of the genetic architecture governing it.

On the other hand, the genetic architecture of the second character (bristle number) which Breese and Mather investigated in the same population was, they confirmed, quite different (see also Mather and Jinks, 1971). The techniques they used enabled them to show that not only different chromosomes but different chromosomal segments also were involved, so there was no question of pleiotropic action – necessarily very unlikely in a polygenic system. While there was a degree of dominance, it was somewhat less than in the case of viability, and less complete, and, moreover, it was ambidirectional.

But how are we to be sure that this intermediate expression has in fact arisen as a result of selective pressures? Clearly there is not for bristle number the same immediately obvious validation for selective advantage of the end product as is the case with viability measures. Kearsey and Barnes (1970; see also Barnes and Kearsey, 1970) provided the next step in supporting Mather's argument by mimicking natural selection in the laboratory. This they achieved by the use of population cages in which wild populations of *Drosophila*, as opposed to ones that had been reared for many years in the laboratory, lived for a considerable period – over six months – consequently involving a number of generations since each matures in approximately two weeks. Samples of eggs were taken at intervals and the emerging flies reared in standardised manner and scored for both bristle number and the viability of their progeny. Kearsey and Barnes showed quite clearly that the intermediate-scoring flies had the greater viability, and moreover, that this relationship was more marked under conditions of increased density of egg laying, that is, where competition was by definition more severe. Thus we can see that a relatively "non-fitness" character can also be subject to natural selection, in this case of a stabilising kind.

2.4.4. Behavioural application

The obvious next question in the mind of the behavioural worker after this somewhat lengthy detour through characters of less than compelling behavioural interest is the extent to which these techniques can be applied to behavioural measures. Again using the fruit fly, Hay (1972) studied the trait of reactivity to disturbance occasioned by mechanical stimulation by using a time-sampling technique. He employed both diallel crosses and the full range of filial and backcross generations (14 in all) readily obtainable with this insect (see ch. 1, § 1.10). He was able to show directional dominance for high activity in this situation, combined with epistasis (non-allelic interaction) of the duplicate gene kind encountered before, pointing strongly to activity as an important fitness trait in this species. Moreover, Hay was also able to relate this genetic architecture to viability by studying the performance under various environmental conditions of the various strains he used in a way which re-established and confirmed the identification of the kind we have previously seen.

As if this were not enough, Angus (1974a) showed that the technique of simulating natural populations used by Barnes and Kearsey (§ 2.4.3) was equally applicable to behaviour. He demonstrated that natural selection for increased spontaneous activity had operated among a crossbred population of *Drosophila* maintained in cages for a period as long as 11 months. These results, taken with his previous finding (Angus, 1974b) of directional dominance and duplicate gene interaction for a higher level of the spontaneous activity, together with his evidence for dominance for a decrease in it in response to shadow stimulation of a kind indicative of predation, further support the inferences from genetical architecture to natural selection we are making.

We have thus reached the further position that we are able to demonstrate the adaptive significance to the species of the quantitative trait which is studied in a laboratory population of it by knowing the genetic architecture and relating this to the level of expression of the trait which is observed. Moreover, it has been established that these procedures apply as much to behavioural traits as to the morphological ones on which they were originally established by Mather and his co-workers.

These techniques have been most extensively applied to mammalian behaviour by Broadhurst and Jinks (1966) in studying the genetic architecture of emotional elimination and of ambulation in the rat, as observed in the open-field test. Several descriptive accounts of these findings have been given elsewhere (see, for example, Broadhurst, 1968, 1969, 1971), so we will only refer to them briefly here. Using data from a 6×6 diallel cross (see ch. 1, § 1.3.3.1 and Table 1.11) we demonstrated directional dominance for a lowering of emotional defecation with increasing experience of the mildly stressful test situation, suggesting a history of directional selection for low expression of this trait. Ambulation, on the other hand, was governed by ambidirectional dominance favouring an intermediate phenotypic score, reflecting stabilising selection for this measure.

Fulker et al. (1972) have also shown how it is possible to determine simultaneously, by the use of multivariate techniques, the adaptive significance of not one but a pattern of levels of several, different behavioural measures, drawn from the open-field test just referred to and from escape-avoidance conditioning also (see § 2.5.3.1). The same breeding design, the diallel cross, was used, but this time with a larger number of rat strains (an 8×8 cross) and the dominance demonstrated was for a combination comprising low defecation and high ambulation in the open-field together with high avoidance, a multivariate pattern strongly suggestive of overall lower emotional responsiveness.

The rationale developed in § 2.4.3 can, of course, be applied to behaviour directly, that is by concentrating on traits having obvious consequences for species survival (Roberts, 1967). An example is seen in Angus's work on the simulation of predation among fruit flies, cited above (1974b), and a start along these lines for rodents has been made by Wilcock (1972), who studied the genetic architecture of escape from water among weanling rats.

2.5. The environmental dimension

The observant reader will have noted that environmental variables are beginning to enter our discussion, since, clearly, if we consider change in test scores over time or in response to a shadow stimulus we are invoking a response to the (test) environment. Indeed, in the work of Hay (1972) cited above, variation in an expressly environmental dimension – the adequacy of the food medium on which the flies were cultured – was introduced as an experimental treatment. We therefore cannot further delay introducing environmental aspects into this discussion as of fundamental importance.

2.5.1. Genotype and environment

As noted earlier, it is not the purpose of this chapter to deal with environmental, that is stimulus, control of behavioural phenotypes, except insofar as it summates with genetic determinants. Inevitably, this is an arbitrary and somewhat artificial distinction to make because it is difficult to conceive of any situation in which pure stimulus control is exercised against a completely non-operative genetic background. That is to say, a broad heritability, or "nature-nurture" ratio, of zero is difficult to imagine for behaviour. By counter, the contrary is also unlikely to occur: some structural phenotypes such as blood group or eye colour – the latter not completely and invariably – may have broad heritabilities of 100 % or approaching the value, but it is difficult to conceive, again, of behavioural phenotypes in this category. Thus *all* behaviour is to some extent the end point of the joint action of genotype and environment, if not their interaction, and to view behaviour as a phenotype (§ 2.2) is to admit as much.

Nevertheless, it is possible to consider situations in which the phenotype measured is subject to genotype-environment interaction in the strict sense, that is, the differential response of different genotypes to the same environment. This phenomenon may be encountered in many situations, but particularly with respect to fluctuations over time. Elsewhere (Broadhurst and Jinks, 1966; see also ch. 1, § 1.2.2), we have characterised two aspects of this interaction over time as stability and change.

2.5.2. Micro-environmental variation

2.5.2.1. Stability. Stability refers to developmental or ontogenetic stability and is essentially micro-environmental in origin (Fulker et al., 1972). Thus it often, though not always, is established during the early individual history of the organism and affects the adult phenotype we observe. The common exceptions arise through imprecisions of behavioural measurement which result in increased phenotypic variances which typically contain components ascribable to test reliability and residual error.

This variation which we subsume under the rubric of stability in the face of micro-environmental variation can itself be analysed by biometrical genetical procedures: thus, as Broadhurst and Jinks (1966) have shown, dominant components of variation are associated with lower variability in behavioural measures, just as is often the case with morphological characteristics. Following the arguments developed here, we can infer an adaptive advantage for stability of development. The implications of this type of finding for behaviour study are of considerable importance. In the case of this phenomenon we seek to investigate and understand the selective pressures which give rise to, not a particular phenotypic level of response, but the extent of the variations about that level which they permit. This variation is itself under genetic control as has been demonstrated, and there are important considerations relating to the degree of inbreeding of the population which affect it. Moreover, the whole question of heterosis is involved here, and this is a concept which has been introduced into psychogenetics in a somewhat unfortunate way, and apparently in ignorance of the lessons to be drawn from the genetical literature on the subject.

2.5.2.2. Heterosis. The facts relating to heterosis are these: at the phenotypic level the phenomenon of heterosis is simple enough and presents itself as an F_1 mean which transgresses the metrical limits defined by those of the inbred parental strains P_1 and P_2. The reasons why the phenomenon occurs are genetically complex (see ch. 1, § 1.8) but they are no more and no less than those which have been discussed above in relation to recognising the results of directional selection (see § 2.4.2), namely high directional dominance with or without supporting epistasis (duplicate genes). Thus we expect and anticipate that heterosis as defined will occur on crossing inbred lines for those characters which have been subjected to directional selection.

Some psychogenetic writers have concentrated on heterosis as an important phenomenon in its own right, and especially as a diagnostic feature of the kind of genetical architecture underlying the behaviour observed. But it is, of course, merely one facet of the way in which traits subjected to selection of this kind manifest their genetic architecture. Moreover, heterosis will occur in one or other of two rather different circumstances. The first of these is overdominance, that is where the dominance ratio exceeds unity. But there is also a second possibility, and this is the case of dispersion of genes (ch. 1, § 1.8). Here, as the name suggests, the increaser and decreaser genes are dispersed between the parental strains so that each contain both increasers and decreasers. In such a case F_1 heterosis can be revealed without there being the distinct heterozygote advantage implied by overdominance. Crossing inbred parents will give an F_1, which is necessarily heterozygous at all loci for which the parents differ, irrespective of the distribution of increasers or decreasers as between the two parents. Hence, provided the dominance deviations at each locus are predominantly of one sign, that is, there is a measure of unidirectional dominance, the F_1 mean will lie outside the parental range.

The whole picture may be complicated by multiple interactions between the members of a polygenic system and, under certain circumstances (see Mather and Jinks, 1971, p. 120 ff.), by linkage, to say nothing of the intervention of environmental considerations of a kind which we have not mentioned previously in this connection. They take the form of genotype-environment interaction which can affect the magnitude of the heterosis (*ibid.* pp. 124–5). However, these are *macro*-environmental sources of interaction with genotype, and it is to these variations which are under the control of the experimenter and which are perhaps more germane to this discussion that we now turn.

2.5.3. Macro-environmental variation

Macro-environmental variation is change due to alterations in the environment which are usually deliberately introduced by the experimenter for the purpose, though it may sometimes be possible to capitalise on the environmental differences inherent in some aspect of the measuring or testing process itself, as we shall see in § 2.5.3.1. But usually macro-environmental manipulation is seen in the form of treatment effects, applied before or during testing. Perhaps prenatal treatment can be regarded as one lying at one extreme of such a temporal continuum, whereas stimulus variables operative within the experiment and even within the organism at the time of testing, such as drug effects, represent the other extreme.

Investigation of such deliberately provoked environmental variation is of relatively recent origin. Early workers in psychogenetics, especially those using selection as the method of choice (for example, Hall, 1938; Tryon, 1940) were concerned to avoid the danger of allowing environmental artefacts to vitiate their findings, so that a necessary but a rather rigid insistence on uniformity of environmental background, against which the better to display the effects of heredity, was the order of the day. In passing, however, it may be noted that geneticists were already concerned with the response to selection under different, systematically varying environments. The work of Falconer (1960b) on selection for body size in mice under different levels of nutrition (see also Falconer, 1960a) provides an example of one as yet unrealized development of selection techniques which might have powerful applications in psychology. But now it is accepted that environmental variation may, so to speak, be accepted in the laboratory as a friend rather than eschewed at all costs as an enemy and incorporated into research designs with the specific intention of studying genotype-environment interaction of the macro-kind. Winston's (1963) was the first published psychogenetic study to adopt this approach and others followed. The interested reader is referred to sources which, inter alia, review this development (Fulker et al., 1972; Broadhurst et al., 1974), but in passing mention should be made of the work of Henderson in this respect. His approach is characterised by a sound grasp of biometrical techniques of design and analysis and an adventurous application of environmental variations having both putatively positive and negative effects on mouse behaviour, such as deprivation and enrichment (see, for example, Henderson, 1972).

2.5.3.1. An example of its application to learning theory. The use to which the results of such investigations can be put will, however, be illustrated by work from our own laboratory by Wilcock and Fulker (1973) which employed macro-environmental variation not systematically imposed, but inherent in the method of testing used. It is chosen for illustration because it is one of the few psychogenetical investigations which have so far appeared which purports to have a relevance beyond the particular, specialist approach and to relate to a theoretical controversy of general psychological interest. This is the two-factor of escape-avoidance conditioning, usually ascribed to Mowrer (1960), and which postulates two processes underlying the development of the kind of behaviour typified in the learning displayed by the rat in a shuttle box. The rat learns to escape the shock to the feet used as the unconditioned stimulus by running to the other, "safe" side, and learns also to avoid the shock by making anticipatory responses of the same kind on hearing the buzzer which serves as the unconditioned stimulus and which precedes the shock, by, in our experimental arrangement, 8 seconds. The two processes underlying the development of the response as described are classical (or Pavlovian) conditioning and instrumental (or operant) conditioning. Mowrer envisaged that, first, the fear responses to shock – emotional and autonomic responses of the kind which might well interfere with efficient learning, and, indeed, give rise to "freezing" – would become conditioned to the buzzer by the process of temporal contiguity typically invoked to explain classical conditioning and hence the buzzer would come to arouse conditioned fear or anxiety, irrespective of whether the rat moved or not. However, escape responses to the shock would also be developing during this stage, so that the rat would also come to experience a diminution in fear on making the crossing-over response, an instrumental response under the control of the rat. This mechanism is also the key to the development of the anticipatory response, in which the rat is reinforced no longer by the cessation of shock but by the reduction of anxiety that is, classically conditioned fear. In recent years this formulation has come under incisive criticism (Herrnstein, 1969), but equally still has its supporters (Rescorla and Solomon, 1967). This digression is intended to set the scene for the findings of Wilcock and Fulker (1973).

They employed a diallel cross design with an environmental dimension (infantile stimulation) with the outcome of which, however, we are not concerned here. Eight strains of laboratory rats were crossed in every possible combination and the offspring, when adult, were subjected to a series of behavioural and other tests, including 30 trials of escape-avoidance conditioning in a Miller-Mowrer shuttle box. Preliminary accounts of the outcome of several of these measures have been published (Fulker et al., 1972; Rick et al., 1971) but it is with one measure only, total number of successful avoidance responses achieved in the 30 trials, that we are concerned here. Standard biometrical analyses of this metric showed no directional dominance for this measure of response acquisition. It was only when attention was focussed on changes over the period of learning that it became clear why this was the case. In the early trials directional dominance was for *low* responding, whereas round about trial 20 a changeover occurred and directional dominance for *high* level of avoidance was revealed: it was not surprising therefore that overall these tendencies cancelled each other out. Wilcock and Fulker repeated the experiment on a smaller scale but with a much larger number of trials (100 in all) and confirmed both that the changeover in directional dominance occurred as before and that it persisted throughout the period studied. Now from the arguments presented before, it will be appreciated that we can deduce the selective pressures which have resulted in these findings regarding directional dominance and attribute an adaptive role to the different levels of responding observed at the two stages. These were the most relevant findings of the study; for other details, the reader may care to consult the original work which has also been discussed elsewhere (Broadhurst, 1971).

What has been demonstrated here is, essentially, the existence of genotype-environment interaction of a macro-environmental kind. The genetic architecture has itself been shown to vary at different stages in the acquisition of a learned response. To put it another way, the genes governing different stages of the same learning process in the same subjects have been shown to be different ones, or, as is logically possible, the same genes acting in a different way at different times. This kind of situation is not unknown in the biometrical genetics of other, often morphological, characters, but is perhaps a novel conception as applied to behavioural traits. It necessitates a rethinking of our approach to psychogenetics and permits a considerable, yet manageable, degree of complexity in our design of experiments and a consequent sophistication in our interpretation of their outcomes which have hitherto been denied us.

In the present case the implications are obvious: if, as seems the case, there is genetic evidence that two different evolutionary processes are at work in conditioned avoidance response acquisition in the rat, and that these processes correspond broadly to the two stages to which the contrasted types of conditioning processes in the two-factor theory apply, then it can be said that this finding offers presumptive support for the existence of the two processes as formulated. This conclusion would appear to gain in weight because the putative evidence comes from so unexpected a source, remote from the usual experimentation designed to test the two-process theory of avoidance conditioning.

It should, moreover, be noted that such interpretive insights are not limited by the precise theoretical formulation discussed, which, as has been mentioned, has been severely criticised. This genetical underpinning can equally apply to other theories which postulate change in behavioural repertoire as crucial to the acquisition of escape-avoidance conditioning. Bolles' (1970) concept of species-specific defence reactions is especially apposite in this connection: it may well be that the sequential discontinuity in genetical architecture we have observed equally supports the notion that one such reaction is succeeded by another. As Wilcock and Fulker put it in the apt words of their summary: "First there appears . . . an adaptive inactivity or 'freezing' behaviour and later . . . an equally adaptive but more active behavior pattern of 'fleeing' " (page 247).

We have concentrated on this study because it illustrates rather clearly both the subtlety possible in the refined biometrical analyses currently available and the important bearing their outcome may have on general behaviour theory. Another application might be to Seligman's hint (1970, pages 409–410) that what he identifies as "prepared" learning should show different genetical determination from "un-" or "contra-prepared" learning. Adoption, therefore, on a wider scale of the techniques of analysis described in ch. 1 combined with the interpretation advanced in the present chapter would no doubt lead to other examples even more striking than the one cited.

2.6. Conclusions

Throughout this chapter we have been stressing the utility of the biometrical approach to animal behaviour described in detail in ch. 1, and, by building on it, seeking to convince the reader of its potentiality in opening new avenues of interpretation, especially with regard to the evolution of behaviour. Presumptuous though it may seem, we consequently point to some research areas which might profitably capitalise on these new-found connections between the analysis by means of the techniques of experimental psychology and the partitioning of behavioural phenotypes by the techniques of biometrical genetics.

The first of these relates to the study of sexual and sex-associated behaviour. Sexual dimorphism is the most obvious example of a genetically determined physical character, a direct outcome of disruptive natural selection, but one which is often forgotten as such. Sexual, that is, reproductive behaviour is usually similarly dimorphic and well-illustrated in many species of typical laboratory animals. Study of the evolutionary implications of different aspects of behaviour comprising species-specific patterns of sexual response should prove rewarding, and might be facilitated by the demonstrated possibility, even in primates, of the macro-environmental manipulation of such behaviour by pheromones as in, for example, Keverne and Michael's work with rhesus monkeys (1971).

Equally accessible to study is sex-associated behaviour – that is, behaviour which is not itself directly related to reproduction but which nevertheless differs characteristically as between sexes. Examples of the application of biometrical techniques to both of these kinds of behaviour are already available in our reanalysis (Broadhurst and Jinks, 1961) of data on direct measures of sexual behaviour in the guinea-pig and in the reanalysis of genotypic-environmental interactive effects of emotional behaviour in the rat presented in ch. 1, § 1.6. This latter also illustrates the extent to which such behaviours can be analysed in terms of macro-environmental manipulation by hormonal treatments as well as leading us to the second of the major areas which, in our view, would well repay study, since the treatments were imposed in infancy and the effect measured in the adult animals.

We refer, of course, to studies of developmental characteristics of all kinds which are especially amenable to laboratory manipulation and study. A start has already been made in this connection, as exemplified in Henderson's work referred to earlier. But the possibilities here are legion, and the profit to be gained from an understanding of the natural selection governing the modes of response discovered may be considerable, not least of all in connection with the possible lessons to be drawn from comparative work to the understanding of human developmental processes, an area of considerable activity in contemporary psychology. This comparative aspect of psychogenetics is one which has not been stressed hitherto in this chapter but which has obvious intuitive appeal.

Of a kind with the preceding, but operative in the shorter term, is the third and final substantive area which we single out for mention as being of contemporary interest and promise. This is what may be broadly characterised as the psycho-pharmaco-genetical, by which polysyllabic term we denote that area of study which concerns itself with differences in response to drugs – viewed as an aspect of the macro-environment – as mediated by the genetic background. Biometrical analysis in this area could well go beyond the mere study of species differences in sensitivity to drugs (Broadhurst, 1964) to the study of dietary and other preferences and of animal analogues of dependence and addiction.

Finally, we would emphasise that, in our view, the methodology proposed is more important than the directions in which we see it as applicable. In this connection the breeding design known as the triple-test cross (ch. 1, § 1.3.2.3) is especially interesting in that it promises to allow by means of laboratory studies the analysis of the genetic architecture of wild populations of unknown, uncontrolled – and uncontrollable – genotype without that circumstance being the complete barrier to this kind of detailed study it has hitherto proved. The technique provides us with a sort of genetical probe by which we can investigate the behavioural phenotypes of a wild population merely by bringing sample individuals – or possibly only their semen if males, or their ovaries for transplantation if females – into the laboratory for crossing with domesticated strains of the same species without the necessity for studying the population as a whole (D. W. Fulker, personal communication). While the limitations and difficulties of the technique are obvious, the freedom it gives is unprecedented and opens new possibilities for psychogenetical analysis.

Finally, there is a corollary to what we have asserted as the general theme of this chapter – that genetical architecture gives us knowledge about natural selection – and it is this: If evidence other than genetic points to certain behaviour having fitness characteristics, then we should expect to find a biometrical analysis of it yielding genetic architecture of an appropriate sort. Thus, if it is hypothesised as being the outcome of natural selection of a directional kind, then we can expect analysis to indicate directional dominance, low narrow heritability, possibly duplicate gene interaction, and so on. If, on the other hand, evidence points towards a history of stabilising selection, then little or ambidirectional dominance, with relatively higher narrow heritability and repulsion linkages can be expected.

Thus the hypothesis can be used not only retrospectively, but also predictively with all the possibilities of corrective emendation and elaboration allowed by the hypothetico-deductive approach. In this sense it is truly heuristic and has some of the characteristics of a strong theory.

References

Angus, J. (1974a): Changes in the behaviour of individual members of a *Drosophila* population maintained by random mating. *Heredity*, **33**, 89.

Angus, J. (1974b): Genetical control of activity, preening and the response to a shadow stimulus in *Drosophila melanogaster*. *Behav. Genet.*, **4**, 317.

Barnes, B. W. and M. J. Kearsey (1970): Variation for metrical characters in *Drosophila* populations: I. Genetic analysis. *Heredity*, **25**, 1.

Bolles, R. C. (1970): Species-specific defense reactions and avoidance learning. *Psychol. Rev.*, **77**, 32.

Breese, E. L. and K. Mather (1960): The organisation of polygenic activity within a chromosome in *Drosophila*. II. Viability. *Heredity*, **14**, 375.

Broadhurst, P. L. (1959): Application of biometrical genetics to behaviour in rats. *Nature (Lond.)*, **184**, 1517.

Broadhurst, P. L. (1960): Experiments in psychogenetics: Applications of biometrical genetics to the inheritance of behaviour. In: Experiments in Personality. Vol. 1. Psychogenetics and Psychopharmacology. Eysenck, H. J., ed. (Routledge and Kegan Paul, London) pp. 1–102.

Broadhurst, P. L. (1964): The hereditary base for the action of drugs on animal behaviour. In: Animal Behaviour and Drug Action. Steinberg, H., A. V. S. de Reuck and J. Knight, eds. (Ciba Foundation Symposium, Churchill, London) pp. 224–236.

Broadhurst, P. L. (1968): Experimental approaches to the evolution of behaviour. In: Genetic and Environmental Influences on Behaviour. Eugenics Soc. Symposia, Vol. 4. Thoday, J. M. and A. S. Parkes, eds. (Oliver and Boyd, Edinburgh) pp. 15–36.

(Also: 1972, Approche expérimentale à l'étude de l'évolution du comportement. In: Problèmes de Méthodes en Psychologie Comparée: Congrès et colloques de l'Université de Liège, September 1969. Richelle, M. and J.-Cl. Ruwer, eds. (Paris, Masson) pp. 95–114.

Broadhurst, P. L. (1969): Psychogenetics of emotionality in the rat. *Ann. N.Y. Acad. Sci.*, **158**, 806.

Broadhurst, P. L. (1971): New lights on behavioural inheritance. *Bull. Brit. psychol. Soc.*, **24**, 1.

Broadhurst, P. L. and J. L. Jinks (1961): Biometrical genetics and behavior. Reanalysis of published data. *Psychol. Bull.*, **58**, 337–362.

Broadhurst, P. L. and J. L. Jinks (1963): The inheritance of mammalian behavior re-examined. *J. Hered.*, **54**, 170.

Broadhurst, P. L. and J. L. Jinks (1966): Stability and change in the inheritance of behaviour in rats: A further analysis of statistics from a diallel cross. *Proc. roy. Soc., B*, **165**, 450.

Broadhurst, P. L., D. W. Fulker and J. Wilcock (1974): Behavioral genetics. *Ann. Rev. Psychol.*, **25**, 389.

Falconer, D. S. (1960a): Introduction to Quantitative Genetics. (Oliver and Boyd, Edinburgh).

Falconer, D. S. (1960b): Selection of mice for growth on high and low planes of nutrition. *Genet. Res.*, **1**, 91.

Fulker, D. W., J. Wilcock and P. L. Broadhurst (1972): Studies in genotype-environment interaction. I. Methodology and preliminary multivariate analysis of a diallel cross of eight strains of rat. *Behav. Genet.*, **2**, 261.

Hall, C. S. (1938): The inheritance of emotionality. *Sigma Xi Quart.*, **26**, 17–27 and 37 (Also 1954, In: Readings in Child Development. Martin, W. F. and C. B. Stendler, eds. (Harcourt Brace, New York) pp. 59–68.)

Hay, D. A. (1972): Genetical and maternal determinants of the activity and preening behaviour of *Drosophila melanogaster* reared in different environments. *Heredity*, **28**, 311.

Henderson, N. D. (1972): Relative effects of early rearing environment and genotype on discrimination learning in house mice. *J. comp. physiol. Psychol.*, **79**, 243.

Herrnstein, R. J. (1969): Method and theory in the study of avoidance. *Psychol. Rev.*, **76**, 46.

Kearsey, M. J. and B. W. Barnes (1970): Variation for metrical characters in *Drosophila* populations. II. Natural selection. *Heredity*, **25**, 11.

Keverne, E. B. and R. P. Michael (1971): Sex-attractant properties of ether extracts of vaginal secretions from rhesus monkeys. *J. Endocrinol.*, **51**, 313.

Mather, K. (1943): Polygenic inheritance and natural selection. *Biol. Revs. Camb. philosoph. Soc.*, **18**, 32.

Mather, K. (1949): Biometrical Genetics: The Study of Continuous Variation. (Methuen, London).

Mather, K. (1953): Genetical control of stability in development. *Heredity*, **7**, 297.

Mather, K. (1960): Evolution in polygenic systems. Int. Colloquium on Evolution and Genetics. (Acad. Naz. dei Licei, Rome) pp. 131–152.

Mather, K. (1966): Variability and selection. *Proc. roy. Soc., B.*, **164**, 328.

Mather, K. (1967): Complementary and duplicate gene interactions in biometrical genetics. *Heredity*, **22**, 97.

Mather, K. (1973): Genetical Structure of Populations. (Chapman and Hall, London).

Mather, K. and B. J. Harrison (1949): The manifold effect of selection. *Heredity*, **3**, 1, 121.

Mather, K. and J. L. Jinks (1971): Biometrical Genetics: The Study of Continuous Variation, 2nd ed. (Chapman and Hall, London).

Mowrer, H. O. (1960): Learning Theory and Behavior. (Wiley, New York).
Rescorla, R. A. and R. L. Solomon (1967): Two-process learning theory: Relationships between Pavlovian conditioning and instrumental learning. *Psychol. Rev.*, **74**, 151.
Rick, J. T., G. Tunnicliff, G. A. Kerkut, D. W. Fulker, J. Wilcock and P. L. Broadhurst (1971): GABA production in brain cortex related to activity and avoidance in eight strains of rats. *Brain Res.*, **32**, 234.
Roberts, R. C. (1967): Some evolutionary implications of behavior. *Can. J. Genet. Cytol.*, **9**, 419.
Seligman, M. E. P. (1970): On the generality of the laws of learning. *Psychol. Rev.*, **77**, 406.
Tryon, R. C. (1940): Genetic differences in maze-learning abilities in rats. In: 39th Yearbook Nat. Soc. Stud. Educ., part 1 (Public School Publ. Corp., Bloomington, Ill.) pp. 111–119.
Wilcock, J. (1972): Water-escape in weanling rats: A link between behaviour and biological fitness. *Anim. Behav.*, **20**, 543.
Wilcock, J. and D. W. Fulker (1973): Avoidance learning in rats: Genetic evidence for two distinct behavioral processes in the shuttle box. *J. comp. physiol. Psychol.*, **82**, 247.
Winston, H. D. (1963): Influence of genotype and infantile trauma on adult learning in the mouse. *J. comp. physiol. Psychol.*, **56**, 630.

CHAPTER 3

A diallel study of exploratory behaviour and learning performances in mice

J. M. L. Kerbusch

3.1. Introduction

The diallel cross constitutes a relatively quick method for surveying the genetic determination of variables and to point to further directions of research. In this way it has been used in recent years by a growing number of psychologists. Although the analysis rests upon an extensive set of assumptions, which have been tabulated by Hayman (1954b), it also provides elaborate checks on the validity of these assumptions and on the additivity-dominance model underlying the analysis. In this way the method may be used to pinpoint characteristics suitable for subsequent research if one wishes to retain the simple model. The diallel method is also especially useful for exploring related characteristics in order to find clues for possible genetic interrelations. Striking examples of this have recently been given by Messeri et al. (1972) and Rick et al. (1971). Within the framework of a larger study undertaken to unravel the relationships between variables of exploratory behaviour and simple learning behaviour, on the one hand, and the activities of some cholinergic enzymes, on the other, a number of behavioural measurements were taken of mice from a 5×5 set of diallel crosses. In the present report, only the genetic analysis of the separate behavioural characteristics are given.

3.2. Methods

The method used was the analysis of a set of diallel crosses; a detailed description may be found in Mather and Jinks (1971). A complete set of crosses consists of the n^2 possible matings between n inbred strains of mice, resulting in the n inbred strains themselves and twice the $\frac{1}{2}n(n-1)$ hybrids (n = 5). On the basis of the work of Mather, Hayman (1954a, b) has developed an analysis of variance to test additive and dominance effects of the genes. An analysis of variance, which conforms largely to the analysis described by Hayman (1954b), is adopted for the data of the diallel cross reported. This rather complicated analysis is to be preferred for psychogenetic experiments since in these studies it will reveal more information, particularly with regard to directional dominance. The model underlying the analysis assumes that genetic variation may be partitioned into additive and dominance components of variation. Genetic and environmental effects are supposed to be additive. In the model, the assumptions are made of absence of interaction across loci and of independent distribution over strains. In Hayman's analysis, four sources of variation are recognized. The a item tests for significance of additive effects, indicating genetic variation among the parental strains. The b item tests for dominance at the relevant loci and this item is further divided into three components. The first, b_1, tests for directional dominance, this test being based on the mean deviation of the F_1 hybrids from the corresponding mid-parental scores. The b_2 item tests whether some strains carry more dominant alleles than others, indicating therefore unequal gene frequencies. The b_3 item tests for dominance deviations not ascribable to b_1 or b_2, that is, unique deviations of F_1's from their mid-parents. The c and d items test for differences between reciprocal crosses. The former is the appropriate source for the average maternal influences of the strains, the latter detecting unique reciprocal effects. The consecutive litters within each cross constitute the randomized blocks in the analysis of variance; the significance of the effects mentioned above are tested against the corresponding interaction of main effect × blocks. An additional factor to test for sex × genotype interactions may be included. Since no such sex differences were present in the diallel tables reported, the factor sex was omitted.

The adequacy of a simple additivity-dominance model may be inferred from second-degree statistics. Most useful is the relationship between the within-array variances (V_r) and the covariance (W_r) of the members of the corresponding array r with the non-recurring parents. If the model is appropriate, there should be a simple relationship between W_r and V_r; $W_r - V_r$ should be constant, i.e. the slope of the regression of W_r on V_r should have the value one. As a test, Mather and Jinks (1971) recommend an analysis of variance of the differences $W_r - V_r$, with blocks as replications, and a joint regression analysis. The absence of a significant $W_r - V_r$ difference suggests that the simple model is appropriate. Significant deviation of the joint regression slope from 1.0 indicates epistatic effects (Hayman, 1954a; Mather and Jinks, 1971). The comparable analysis of variance of the sums $W_r + V_r$ will, in case of a significant strain effect, give additional evidence for non-additive genetic effects. If the simple model holds, the regression of W_r on V_r reveals the rank-order of the strains with respect to the number of dominant alleles controlling the character. On the regression line, the strains carrying the largest number of dominant alleles are situated towards the lower end and those with the fewest towards the top. The intercept indicates complete dominance if it equals zero, partial dominance if it is larger than zero and overdominance if it is negative. In case of absence of dominance, all points should cluster around a single point. On the basis of second-degree statistics it is possible to estimate the genetic components of the model; the reader is referred to Mather and Jinks (1971) for a detailed description of this estimation procedure. From these components the heritability in the narrow sense (h^2) and the heritability in the broad sense, also known as the degree of genetic determination (DGD), may be calculated. Hayman (1954a) also derived an estimate for the lower limit of the number of relevant loci (k).

The conclusions arrived at by the analysis of variance and the W_r, V_r graph may be interpreted in terms of a possible relationship between the genetical architecture and the adaptive significance of the trait studied. Breese and Mather (1960) claim that strong directional dominance, sometimes combined with epistatic interaction, indicates that in evolutionary history the character has been subjected to directional selection, so that some alleles would be favoured. Stabilizing selection would have resulted in a larger additive variation component for the trait and perhaps in some ambidirectional dominance.

The results of the genetic analyses of the data obtained in the present study are summarized in the 20 columns of Tables 2–7. The first seven columns contain the F-ratios for the genetic and environmental effects tested against the corresponding block interactions. The headings of the columns refer to the items of the Hayman analysis. The degrees of freedom are set forth in Table 3.1. Columns 8 and 9 of the tables contain the F-ratios of the analyses of variance for the sums $W_r + V_r$ and the differences $W_r - V_r$. If there are no block effects, their degrees of freedom are 4,5 for exploratory

TABLE 3.1

Item	Effect	Exploratory behaviours block interaction	Other variables block interaction
a	4	4	8
b_1	1	1	2
b_2	4	4	8
b_3	5	5	10
b	10	10	20
c	4	4	8
d	6	6	12

Degrees of freedom for the various effects of the analysis of variance of a diallel table in the case of five inbred strains and two or three litters.

behaviours, and 4,10 for other variables. In case of significant block effects, the F-ratios are for 4,4 and 4,10 degrees of freedom, respectively. Columns 10 through 15 contain the analysis of the regression of W_r on V_r. In column 10 the slope of the joint regression is given; the error belonging to it is printed in column 11. The intercept of the regression line is entered in column 13; if it deviates from zero, the level of significance is indicated. In columns 12 and 14, the t-values are given for, respectively, the deviation of the slope from unity and the variance explained by the joint regression line. The t-values have 7 and 11 degrees of freedom in analyses involving two or three blocks, respectively. The F-ratio for the heterogeneity of regression over blocks is entered in column 15. The F-ratio has 2 and 11 degrees of freedom in the case of three litters; in analyses involving only two litters, it was converted into a t-value with 7 degrees of freedom. In column 16 the dominance order of the strains is given, starting with the strain carrying the largest number of dominant alleles. The next column shows the strains in an increasing rank-order for parental means. The last three columns give the heritability, the degree of genetic determination, and the estimated number of genes involved. The last three values are sometimes missing, due to the estimation procedure of the genetic components revealing negative variances.

The experiment was conducted with a total of 377 mice, 185 male and 192 female. The animals came from three litters produced by each of the 25 possible matings between five inbred strains: viz. C3H/Z ("Z"), C3H/HeJ

("H"), BALB/c ("B"), CBA ("C") and C57BL/6J ("5"). The breeding animals were provided by the Central Animal Laboratory of the Catholic University of Nijmegen, The Netherlands. The three litters from each mating were assigned consecutively to the three blocks used in the analysis of variance. Animals of two of these blocks were observed in an open field at 61 ± 1 days of age. All animals received a passive avoidance training at this age and were tested on the next day. At 63 ± 1 days, the animals received 20 trials of massed escape practice in a runway, followed by T-maze learning the next day.

3.3. Findings

3.3.1. Exploratory behaviour

3.3.1.1. Apparatus and procedure. The open field consisted of a wooden box which measured $50 \times 50 \times 15$ cm; the top and the front wall were made of plexiglass and the floor was divided into 36 equal squares. A single mouse was placed in the box and observation started after 60 sec. The observer was sitting behind a one-way screen in a corridor next to the experimental room. For 17.5 min, the behaviour of the mouse and its location in the field were recorded every seventh second, which was signalled only to the observer by a click. Eleven behavioural categories were used: (1) Jumping, i.e. all four feet off the floor. (2) Walking, i.e. locomotor activity with four legs on the floor. (3) Turning, i.e. walking with two or three legs while at least one hindleg does not move. (4) Rearing, i.e. sitting or standing on the hindlegs, the forepaws free. (5) Leaning, i.e. the same as (4), but the forepaws touch the wall. (6) Sniffing in the air, i.e. sniffing as evident from movements of the vibrissae, without being directed to the floor or the walls. (7) Sniffing at floor or walls. (8) Washing, i.e. all grooming activities with the forepaws directed toward the head. In this category vibrating with the forepaws (cf. van Abeelen, 1965) was included. (9) Grooming the fur, i.e. all other grooming activities, scratching included. (10) Sitting, i.e. resting on four legs. (11) Lying, i.e. resting on the belly and the legs or lying on one side.

In case the above mentioned elements do not exclude each other, the first mentioned category was registered. The last two behavioural categories were not analyzed, since these behaviours were rare. In the present paper only the total frequencies of the behaviours obtained in the 150 samples were analyzed. However, it is possible to make crude time analyses for each behavioural class by dividing the 150 samples into blocks. Two indices of locomotor activity were calculated on the basis of the observed locations in the field at the sampling moments. (a) Movements: if the animal was in a place different from that at the previous click, a "movement" was scored. (b) Distance: from the successively occupied places, an estimate was made of the distance walked by using the city-block metric. (Instead of the Euclidean distance, i.e. the hypotenuse, the sum of the perpendicular sides is used as a distance measure).

Similar behavioural categories and locomotor estimates have been used previously by van Abeelen (1965) and by Vossen (1966).

3.3.1.2. Results and discussion. The results of tests and analyses for the behavioural frequencies are arranged in Table 3.2. For the frequency of *grooming the fur* the model seems to be appropriate since there are no systematic strain differences in $W_r - V_r$. The slope of the regression of W_r on V_r does not depart significantly from 1.0, hence there is no reason to assume epistatic effects. The Hayman analysis of variance reveals no additive effects of genes (a), but pre- or postnatal maternal influences (c) seem to be present. Dominance effects (b) are present according to the analysis of variance although there is no systematic strain difference in $W_r + V_r (F_{4,5} = 0.28; P > 0.80)$; these dominance effects seem to be due to one or more cells showing unsystematic dominance (b_3).

For the frequency of *grooming the head*, the additivity-dominance model seems to be appropriate in so far as there is no strain effect in the $W_r - V_r$ differences, but the slope of the regression of W_r on V_r deviates significantly from unity ($t = 4.16; P < 0.01$), thus epistatic effects are probably present. The c and d items are significant in the analysis of variance, indicating maternal influences and irregular reciprocal differences. Additive genetic variation could not be demonstrated, but dominance seemed to be present, though the latter result is not confirmed by significance of strain differences in $W_r + V_r$.

For *sniffing in the air*, epistatic effects seem to be present since the joint regression slope deviates from 1.0 ($t = 5.73; P < 0.01$). According to the analysis of variance, maternal influences and dominance play a role in the character, though for the latter no definite order of number of dominant alleles can be established due to the epistatic effects. The inbred strains do differ for breeding value.

TABLE 3.2

	Tested against block interaction:											
	a	b_1	b_2	b_3	b	c	d	W_r+V_r	W_r-V_r	b_{1r}
Groom fur	2.22	2.80	1.89	5.64'	3.28'	9.55'	2.32	0.28	0.20	0.81
Groom head	3.15	0.14	2.55	4.24	3.18'	9.25'	6.71'	2.33	2.59	0.15
Sniff air	23.72''	0.59	3.92	7.63'	6.27''	7.55'	2.59	2.17	2.28	0.41
Sniff wall	54.90''	5.05	11.49'	1.47	3.35'	22.10''	4.83'	4.68	0.33	0.92
Lean	35.58''	4.70	5.02	3.50	4.09'	0.51	3.53	1.47	3.63	0.84
Rear	46.09''	0.00	6.97'	1.21	1.95	1.95	6.14'	0.98	0.51	1.25
Turn	21.48''	3.48	1.76	5.94'	3.95'	11.07	20.95''	2.43	0.77	0.45
Jump	33.01''	5.59	1.42	1.49	1.63	10.24'	0.71	1.02	2.18	0.14
Walk	15.10'	35.17	2.38	18.04''	5.40''	15.83'	0.78	8.86'	0.83	0.54
Movement	13.29'	7.48	4.49	5.97'	5.56''	6.27	3.82	1.15	0.85	0.22
Distance	40.79''	14.40	5.87	6.49'	6.37''	2.80	2.92	2.25	0.99	0.53
	error	t_b	interc.	t_{Jr}	t_{hr}	domi-nance	parent means	h^2	DGD	k
Groom fur	0.30	0.62	0.6	2.66'	0.71	CH5ZB	HZCB5	—	—	—
Groom head	0.20	4.16''	0.2	0.76	0.70	B5ZCH	CBH5Z	—	—	—
Sniff air	0.10	5.73''	3.2	4.03''	0.60	5ZHCB	Z5HCB	—	—	—
Sniff wall	0.14	0.60	8.6	6.57''	0.41	Z5HBC	Z5HBC	—	—	—
Lean	0.26	0.63	4.4	3.23''	0.15	ZH5CB	CH5BZ	—	—	—
Rear	0.28	0.86	−12.5	4.38''	0.32	Z5BCH	C5BHZ	0.42	0.97	33
Turn	0.26	2.15'	−0.2	1.77	2.06	Z5BCH	CBH5Z	—	—	—
Jump	0.19	4.49''	0.1	0.71	1.21	HZ5BC	BHC5Z	—	—	—
Walk	0.43	1.08	4.7	1.25	0.56	ZCHB5	BHZC5	—	—	—
Movement	0.08	9.72''	24.8	2.78'	2.61'	Z5CHB	HBC5Z	—	—	—
Distance	0.16	2.87'	−280.3	3.28''	1.06	ZCH5B	HCB5Z	0.34	0.98	107

The genetic analyses of the behavioural frequencies in an open field. ': $P \leq 0.05$; '': $P \leq 0.01$. k is the estimated number of genes involved.

For *sniffing at the floor or wall*, genetic variation among the parental strains (item *a*) was demonstrated. Dominance, maternal influences, and irregular reciprocal differences existed. The line effect of W_r+V_r almost reached significance and the dominance order along the regression line of W_r on V_r, together with the order in the magnitude of parental scores, were very suggestive of directional dominance. The results of the tests on the joint regression line confirm the appropriateness of the model.

Leaning: Although the F-ratio for the line effect of the W_r-V_r differences is quite high, it does not reach significance; consequently there is no reason to doubt the adequacy of the model. Also no epistatic interaction seems to be present since the regression slope equals unity within the limits of the standard error. The analysis of variance very clearly indicates genetic variation among the parental strains and dominance, although the latter result cannot be traced specifically to any one of the three sources.

Rearing: According to the results of the W_r-V_r test and the joint regression analysis, the model is adequate for this character and no epistasis was present. By means of the analysis of variance, additive effects of genes were established, and the strains differed with respect to the number of dominant allleles. The b_1 item, however, by no means reached significance. The latter result indicates that the hypothesis of directional dominance for rearing, as advanced by van Abeelen (1970) cannot be generalized. The estimated DGD is 0.97 and is high compared to van Abeelen's estimate, being 0.77. Also h^2 is quite high, 0.42. This implies the possibility of a rapid selection response as demonstrated by van Abeelen (this volume).

Turning: Although the W_r-V_r test gave no significant F-ratio, the joint regression analysis of the $V_r - W_r$ graph not only revealed a slope deviating from unity, but also a significant heterogeneity of regression for the two blocks. Probably the frequencies of the behaviour were too low to give reliable results. Additive genetic variation, irregular dominance effects, maternal influences, and irregular reciprocal differences were suggested by the analysis of variance.

For *jumping* also very low frequencies were observed. The joint regression analysis did not succeed in explaining a significant part of the variation in V_r and W_r; the slope deviates from 1.0. In the analysis of variance, the presence of maternal effects and additive variation were indicated.

Maternal influences also affect the frequency of *walking*. Dominance (item *b*) plays a role in the inheritance of walking, a finding which is confirmed by the significance of the strain effect for W_r+V_r. Additive genetic variance was found as was the case in Broadhurst's (1960) work with rats.

For both *movements* and distance *activity* scores there is no W_r–V_r line effect, but the joint regression slopes deviate clearly from unity; epistatic interaction is probably present. The analysis of variance reveals additive genetic variation among the inbred strains and dominance at some loci. As in the case of the analysis of walking frequencies, these dominance effects are neither directional (b_1), nor unequally distributed among the various strains (b_2). The main source of non-additive variation is ascribable to b_3, probably indicating unsystematic epistatic effects. The heritability for covered distance could be calculated and equals 0.34, a fairly high percentage, though it does not seem to be in contradiction with previous findings for activity scores.

3.3.1.3. Summary. The purpose of the study was to survey the genetic control of several behavioural categories simultaneously. Using a time-sampling observation method, we obtained nine behavioural measures amenable to analysis in addition to two measures of activity. A variety of genetic effects were revealed. All measures met the requirement that no strain effect should be present in the $W_r - V_r$ differences, but four of them, together with the activity-scores, gave regression coefficients for W_r on V_r which differed from unity, suggesting epistasis. No epistatic effects could be demonstrated for the following categories: sniffing at the wall, leaning, rearing, fur-grooming and walking. In these cases, the W_r, V_r graph revealed complete dominance for all categories, since the intercept did not differ significantly from zero. In the analysis of variance, additive genetic variance was established for all behaviour frequencies but for grooming. Dominance effects were found for almost all frequencies. If the dominance effects were not just irregular, they were ambidirectional or could not be specified by one of the subitems. Directional dominance was never found.

Intuitively, the following characters are expected to be related: grooming the fur and grooming the head; sniffing in the air and at the wall; leaning and rearing; walking, movements, and covered distances. In conformity with intuition the genetic analyses did not yield remarkable discrepancies within these classes. Rationally, however, there is no reason why a behaviour which is unitary within an ethological or psychological context should be genetically unitary, particularly if its genetic control is polygenic. That is, many independent genetic systems may underly the distinct behavioural units and no one-to-one correspondence need exist between these genetic systems and ethologically defined behaviours. Therefore, different ways of categorizing might produce quite different results and they cannot provide a very sound basis for generalizations. However, one general result was obtained by the analysis of the behavioural measures derived from our method of categorizing: a great similarity in the genetic determination of the frequencies was found with regard to the absence of directional dominance (b_1), suggesting that an intermediate level of phenotypic expression is adaptively superior to either extreme.

3.3.2. Passive avoidance

3.3.2.1. Apparatus and procedure. The apparatus used has been described by Jarvik and Kopp (1967). It consisted of a trough-shaped two-compartment box, one compartment being made of plexiglass sheets, the other of wood painted black. The plexiglass compartment measured 8×3 cm at the floor and widened to 8×12 cm at the top; the walls were 20 cm high. This compartment was illuminated by two bulbs of 150 W from a distance of 10 cm. The floor of the wooden compartment measured 18×3 cm and consisted of four metal sheets. At the top the compartment was 18×12 cm. There were four metal sheets (9×3 cm) attached to the two side walls. The compartments were connected by a hole ($\varnothing$ 2.5 cm), 1 cm above the floor, which could be closed by a guillotine door. At distances of 2 and 12 cm from the hole, two photoelectric cells were placed which controlled a timer. The latter transmitted a pulse to a shock generator which gave an electric shock of 35V/100KΩ for 5 sec to the metal sheets in the larger compartment. In the single learning trial, the animal was placed into the smaller compartment, facing the hole, and latency to entrance into the dark area was recorded. After the punishing shock the animal was left undisturbed for 24 hr. The next day the animal was put again into the apparatus and latency was measured.

3.3.2.2. Results and discussion. Three litters of each genotype were tested. The following variables were analyzed: latency in the training and in the test trial and the difference between them. The latency observed in the learning trial may be regarded as a measure for dark preference; the difference is an index of learning. The analyses of variance (Table 3.3) indicated additive genetic variance for dark preference. Dominance and reciprocal effects were found for all variables; the inbred strains differed with respect to the number of dominant alleles they carry. Directional dominance (b_1) was not demonstrated. Since neither variable showed a strain effect for the

TABLE 3.3

| | a | Tested against block interaction: | | | b | c | d | W_r+V_r | W_r-V_r | b_{Jr} |
		b_1	b_2	b_3						
Training	5.94'	6.29	9.24''	14.80''	11.70''	5.13'	3.23'	0.84	0.56	0.65
Test	1.31	9.08	15.79''	14.97''	13.75''	3.74'	5.94''	0.83	1.22	1.44
Difference	1.30	7.15	8.53''	5.92''	6.98''	3.86'	7.49''	0.43	1.39	1.15
	error	t_b	interc.	t_{Jr}	F_{hr}	dominance	parent means	h^2	DGD	k
Training	0.29	1.18	−3178.''	2.23'	0.92	H5CBZ	H5CBZ	0.43	0.90	51
Test	0.19	2.34'	−2422.''	7.65''	1.64	C5HZB	H5CZB	0.35	0.80	64
Difference	0.22	0.67	−974.	5.22''	1.08	CH5ZB	H5CZB	—	—	—

The genetic analyses of the latencies in the training and the test trial of a passive avoidance task, and the difference between these two latencies. ': $P \leq 0.05$; '': $P \leq 0.01$. k is the estimated number of genes involved.

differences $W_r - V_f$, the additivity-dominance model seems to be adequate. But for the latencies in the test trial, the slope of the regression of W_r on V_r exceeds 1.0; epistatic interaction is probably present. The significantly negative intercept of the joint regression line points to overdominance for dark preference. For this variable, there is a suggestion of directional dominance in that the rank-order of $W_r + V_r$ and parental means is the same.

The albino BALB mice show the greatest increase in latency between the learning trial and the test trial, a finding in accordance with the data of Winston et al. (1967). We used the difference between the latencies as a learning index. The animals learned to avoid the dark compartment; the mean latency in the test trial was about 3.5 times longer than that in the learning trial. Intuitively, an advantage should have existed in evolutionary history for fast learning to avoid dangerous places. This intuition is not confirmed by directional dominance for the learning index.

3.3.3. Runway

3.3.3.1. Apparatus and procedure. After having been tested for passive avoidance, the animals were subjected to twenty massed trials in an escape runway. The escape corridor consisted of a start-box ($10 \times 10 \times 30$ cm), a straight alley ($102 \times 10 \times 30$ cm), and a goal-box ($10 \times 10 \times 30$ cm). The grid floor of the start-box and the corridor consisted of 117 electrifiable rods (35V/100KΩ). The floor of the goal-box and all walls of the apparatus were made of wood. The start-box and the corridor were connected by a plexiglass guillotine door. At the beginning and the end of the corridor photo-cells were placed 100 cm apart, controlling a timer. In each trial the animal was placed in the start-box and the guillotine door was raised immediately. The intertrial time was approximately 30 sec. If an animal did not reach the goal-box within 100 sec, it was led over to it, and a missing score was recorded.

3.3.3.2. Results and discussion. The mean escape latencies are given graphically in Fig. 3.1. After a rapid decrease over the first few trials, the mean latencies of the F_1 hybrids approach an asymptote, but the parental means fluctuate. The latter result is mainly due to the performances of the C3H/HeJ and the CBA strains, the former even showed latencies increasing over trials. The decreasing rank-order of the individual parental curves was: C3H/He, CBA, C57BL, BALB and C3H/Z. Exactly the same rank-order

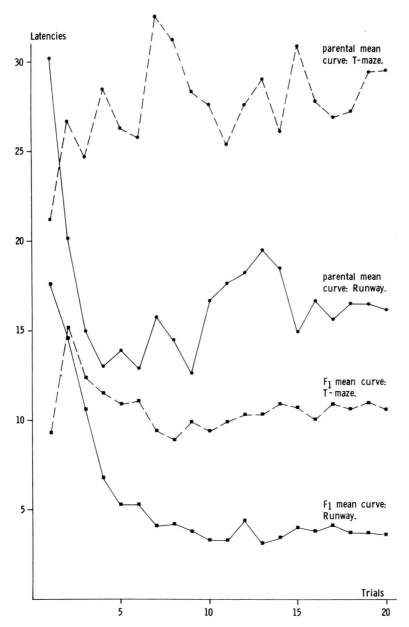

Fig. 3.1. The parental and hybrid mean curves for running times in a shock runway and an electrified T-maze.

TABLE 3.4

Tested against block interaction:

Trial	a	b_1	b_2	b_3	b	c	d	W_r+V_r	W_r-V_r	b_{jr}
1	87.22″	23.86′	9.65″	4.58′	10.23′	0.42	3.53′	6.56′	0.21	0.90
2	6.64′	17.91	2.84	4.47′	4.22″	6.71′	5.29″	2.04	1.15	0.90
3	9.55″	5.58	0.59	2.39	1.61	12.86″	8.69″	1.36	0.23	1.09
4	17.62″	12.44	12.68″	9.10″	11.90″	21.06″	8.28″	2.57	1.66	0.96
5	6.73′	68.91′	5.97	3.26	8.53″	6.01′	4.52′	1.32	1.61	0.62
6	7.86″	527.06″	10.89″	10.13″	14.83″	11.04″	19.56″	1.09	1.35	0.89
7	8.23″	80.05′	7.12′	4.22′	10.96″	5.43′	3.78′	1.82	1.45	0.71
8	16.73″	157.76″	15.24″	4.54′	16.47″	4.05′	6.24′	6.25′	1.33	0.63
9	20.05″	84.63′	22.32″	2.60	17.03″	4.21′	3.95′	3.03	2.00	0.89
10	18.33″	436.23″	17.36″	2.85	21.27″	2.87	3.09′	2.93	2.10	0.91
11	19.76″	41.94″	16.91″	2.17	19.08″	4.44′	2.12	3.74	2.80	0.80
12	19.88″	35.60″	35.76″	5.01″	25.96″	9.13″	19.04″	4.28″	0.39	1.04
13	12.15″	52.43″	20.36″	0.35	26.03″	1.62	8.15″	2.06	6.33″	0.87
14	15.13″	50.98′	22.96″	1.48	23.85″	5.48′	2.67	2.34	1.97	0.90
15	21.29″	111.53″	33.96″	2.70	24.72′	4.93′	3.15′	9.24″	1.25	0.87
16	26.09″	53.80′	39.60′	1.93	29.43″	3.96′	4.72′	8.44″	4.04′	0.80
17	28.78″	80.47′	65.90′	6.20′	38.48″	7.31″	8.43″	23.86″	2.31	0.85
18	21.09″	381.05″	25.67″	2.78	26.25″	4.13′	5.68″	14.97″	1.59	0.81
19	28.02″	147.42″	25.28″	3.06	24.63″	4.68′	5.31″	17.21″	1.08	0.85
20	31.34″	125.83″	52.60″	1.30	29.22″	2.61	4.03′	54.33″	1.29	0.90

Trial	Error	t_b	Interc.	t_{Jr}	F_{hr}	Domi-nance	Parent means	h^2	DGD	k
1	0.09	1.05	−2.6	9.86"	5.86'	BZCH5	ZBC5H	0.55	0.96	29
2	0.18	0.57	17.6	5.00"	0.18	BHZ5C	ZB5HC	—	—	—
3	0.08	1.11	−29.1	13.87"	4.00'	BZ5CH	ZB5CH	—	—	—
4	0.06	0.59	0.6	15.39"	0.14	BZ5CH	Z5BCH	0.32	0.75	16
5	0.07	5.43"	20.1	8.80"	1.54	B5ZCH	Z5BCH	0.22	0.65	14
6	0.11	1.06	2.4	8.24"	0.59	BZ5CH	Z5BCH	0.39	0.63	28
7	0.10	2.95"	11.8	7.09"	0.82	ZB5CH	Z5BCH	0.08	0.84	3
8	0.10	3.57"	15.6	6.15"	1.08	ZB5CH	ZB5CH	0.14	0.91	3
9	0.11	1.05	1.4	8.46"	0.77	Z5BCH	ZB5CH	0.16	0.90	3
10	0.08	1.07	8.8	11.38"	0.27	Z5BCH	ZB5CH	0.06	0.92	2
11	0.06	3.11"	2.6	12.43"	1.04	5ZBCH	ZB5CH	0.05	0.93	2
12	0.04	0.89	−9.2	25.17"	1.85	ZB5CH	ZB5CH	0.13	0.93	2
13	0.03	4.06"	2.7	27.42"	2.10	Z5BCH	ZB5CH	0.04	0.92	2
14	0.05	2.10'	16.9	19.19"	0.61	5ZBCH	ZB5CH	0.08	0.92	2
15	0.07	1.93'	13.3	12.65"	1.19	5ZBCH	ZB5CH	0.19	0.95	3
16	0.05	3.95"	17.8	15.79"	0.33	Z5BCH	ZB5CH	0.17	0.96	3
17	0.07	2.16'	9.5	12.09"	0.45	Z5BCH	ZB5CH	0.11	0.95	2
18	0.08	2.38'	4.8	9.84"	0.11	Z5BCH	ZB5CH	0.16	0.94	3
19	0.06	2.33'	9.6	13.65"	0.46	Z5BCH	ZB5CH	0.20	0.95	3
20	0.05	2.21'	10.3	18.97"	2.72	Z5BCH	ZB5CH	0.18	0.97	2

The genetic analyses of the running times in an electrified escape runway. ': $P \leq 0.05$; ": $P \leq 0.01$. k is the estimated number of genes involved.

was found for the curves of F_1 hybrids having one parent in common. In all cases the F_1 hybrids were superior to their common parents, suggesting hybrid vigour. The analyses regard repeated measurements and may not form independent tests. Since, however, the results of the independent analyses of the trials showed an overwhelming similarity, one may draw inferences for escape performance in general. Additive genetic variation was revealed for all trials. The inbred strains differed in breeding value for escape performance. For nineteen trials dominance effects (b) were present. On eighteen trials item b_2 was significant too, suggesting unequal distributions of dominant alleles over strains. The W_r, V_r graphs revealed that this result probably stems from the C3H/HeJ strain carrying by far the largest number of recessive alleles. The b_1 item was highly significant for all trials but for 2, 3 and 4; this directional dominance for low scores might be explained in terms of an adaptive superiority of fast escape performance. On eighteen trials maternal influences (c) were demonstrated, and random reciprocal differences (d) were found in all but two trials. The simple model is appropriate for the running times in the twenty trials in the runway as far as the tests on the differences $W_r - V_r$ are concerned; only in trials 13 and 16 there is a significant line effect (Table 3.4). For eleven trials the slope of the regression of W_r on V_r deviates significantly from 1.0; hence, in addition to directional dominance there is also epistatic interaction. In four additional cases the three regressions within blocks are heterogeneous. Four of the remaining five trials, however, yield much the same picture as in general is shown by the W_r, V_r graphs. The C3H/Z, BALB/c, and C57BL strains cluster at the lower part of the line, the CBA strain is intermediate, and the C3H/He strain carries most recessive alleles. The same distribution is found in the parental means, confirming the directional dominance found in the analysis of variance. The heritability estimates, h^2, show a decrease over trials, a finding which is in contradiction with the results of Broadhurst and Jinks (1961). These authors found heritability increasing over trials, reanalyzing Vicari's (1929) results on maze-learning. This change in heritability estimates does not support Tyler and McClearn's (1970) findings either.

The shortest escape times were determined for all animals and analyzed (Table 3.5). Apart from additive genetic variation among the inbred strains, dominance is demonstrated by the analysis of variance. The strains differed in the number of dominant alleles they carry (b_2). The F-ratio for directional dominance (b_1), however, reached only the 0.10 level of significance. The simple model is adequate since the test on $W_r - V_r$ differences does not reach significance, nor does the slope of the regression of W_r on V_r deviate from unity. The intercept indicates complete dominance for this characteristic.

The model is also adequate for the number of times the animals refuse to run. Apart from nonsignificance for the tests on the differences $W_r - V_r$ and the slope, the sum $W_r + V_r$ differs over strains, indicating that some strains carry more recessive alleles than others. The order of dominance almost entirely matches the magnitude of parental means, which suggests a directional dominance for low scores. The intercept does not differ from zero, which points to complete dominance. The direction in dominance is confirmed by b_1 in the analysis of variance and again most strains carry dominant alleles.

An alternative and perhaps more appropriate way of analyzing escape learning is the analysis of characteristics of the learning process itself. An attempt was made to evaluate this process by fitting an exponential equation $[y = a + b \times \exp(-cx)]$ to the learning curve of each genotype. The parameters of this equation may be identified in terms of psychological characteristics of the learning process. However, some of the genotype did not show decreasing escape latencies over trials; such a general trend, as exemplified by the decelerating decline of the curves of the F_1 hybrids (see Fig. 3.1), was observed for most learning curves. Moreover, the errors attached to the estimates of the parameters were not homogeneous across genotypes, thereby precluding an analysis of variance. Tyler and McClearn (1970) fitted a second degree polynomial regression to similar runway data. Performing a trend analysis – essentially the same procedure – on these data, a third degree polynomial function was found satisfactory. The results of the genetic analysis and the genetic correlations of the coefficients will be presented by Kerbusch (1974). The difficulty in finding an appropriate equation fitting the learning curves may arise in part from the inclusion of latencies shorter than 100 sec only. The latencies of animals which froze or escaped shock by sitting motionless on their hindlegs, touching one rod only, were left out. Since strain differences actually exist for these refusals to run, the data might be biased. Disregarding this shortcoming, strong directional dominance emerged for various aspects of successful learning, clearly indicating the adaptive value of this type of learning.

3.3.4. T-maze

3.3.4.1. Apparatus and procedure. The T-maze consisted of a start-box,

TABLE 3.5

	Tested against block interaction:									
	a	b_1	b_2	b_3	b	c	d	W_r+V_r	W_r-V_r	b_{Jr}
Refusals	21.06"	84.54'	36.77"	2.88	25.07"	4.35	4.56'	4.99'	1.90	0.97
Minimal time	8.20"	11.28	9.48"	3.13	9.28"	2.87	3.98'	1.31	1.11	0.97
	error	interc.	t_b	t_{Jr}	F_{hr}	dominance	parent means	h^2	DGD	k
Refusals	0.06	0.50	−0.04	15.03"	0.00	Z5BCH	ZB5CH	0.14	0.94	2
Minimal time	0.02	1.44	−0.42	51.29"	0.42	Z5BCH	ZBC5H	0.02	0.81	4

The genetic analyses of the number of refusals and the minimal escape time in a runway. ': $P \leqq 0.05$; ": $P \leqq 0.01$. k is the estimated number of genes involved.

a straight alley, two arms, and two end compartments, all of which had wooden walls, a removable plexiglass top, and a grid floor (257 rods). The start-box measured $15 \times 20 \times 20$ cm and gave access to the alley by a swing door. The alley was 102 cm long, 10 cm wide, and 20 cm high; at its entrance was a photo-cell which triggered a time-counter. A guillotine door was situated at the end of the alley. Both arms of the T-maze were perpendicular to the alley and measured $60 \times 10 \times 20$ cm. In front of both end-compartments, which were at right angles to the arms, photo-cells were present in order to stop the timer. Each end-compartment measured $15 \times 10 \times 20$ cm; a wooden box could be slid in so that the animal would be able to escape shock by jumping into it. A shock generator delivered a shock, $35V/100K\Omega$, to the rods. Before training, the animal would receive three trials with both wooden end-boxes in position; after these, 20 massed trials were applied with one end-box on the side the animal preferred least. If the animal made an error, it had to correct itself. Choices and latencies were recorded. If an animal did not reach the choice point within 100 sec it was led towards it and its latency remained unknown.

3.3.4.2. Results and discussion. In Table 3.6 the biometrical analyses are given for the latencies recorded in the 20 trials. For all trials the differences $W_r - V_r$ are constant for all strains and the slope of the regression of W_r on V_r deviates from unity only in trials 10, 12, and 19. The additivity-dominance model seems to be adequate and for most trials no epistatic interaction is found. Interpreting the W_r, V_r graph, the BALB strain is found to carry the largest number of dominant alleles, although it does not differ much from the C3H/Z strain, the C57BL/6J and the CBA strain are intermediate, and the C3H/HeJ strain carries the largest number of recessive alleles. A striking correspondence exists between the order of dominance and the order of means of the inbred strains, indicating directional dominance for low scores. The intercept of the regression line is negative for all trials, but nowhere it reaches significance; perhaps a slight dispersion of dominant alleles is present. For all trials, the analysis of variance provided evidence for additive genetic variation, thus, generally speaking, the inbred strains seem to differ in breeding value for running performance in a T-maze. Due to the large amount of non-additive variation present, the estimated heritabilities, however, are quite low, so that selection for running performance probably would not be very successful. The results of all trials indicate that dominance plays a role in the inheritance of running time. For 15 out of the 20 trials, directional dominance (item b_1) for fast running is established.

TABLE 3.6

Trial	Tested against block interaction:									
	a	b_1	b_2	b_3	b	c	d	W_r+V_r	W_r-V_r	b_{Jr}
1	13.64″	36.79′	5.60	7.48″	10.67″	0.53	8.08″	3.58′	3.10	0.92
2	4.06′	10.77	3.56	2.73	4.73″	5.39′	1.51	0.92	0.80	0.84
3	9.34″	12.15	3.32	5.29′	5.88″	1.36	2.44	1.48	0.52	0.85
4	9.70′	23.45′	5.48′	2.69	8.07″	1.03	2.46	1.41	1.48	0.99
5	12.00″	10.89	9.07″	4.95′	8.60″	4.19′	2.96	1.58	1.03	0.91
6	8.24″	91.99′	15.93″	4.96	18.55″	5.30′	1.90	5.24′	2.48	0.97
7	46.90″	47.13′	13.82″	4.27	19.50″	4.10′	5.47″	5.05′	0.60	0.90
8	16.87″	34.82′	4.78′	6.48″	9.96″	3.83	1.44	1.74	1.66	0.98
9	14.39′	19.49′	3.60	4.31′	7.08″	2.27	1.44	1.27	0.75	0.95
10	15.74″	17.04	4.41′	2.85	7.47″	1.70	2.71	1.30	2.61	0.80
11	19.66″	20.21′	5.83′	5.82′	8.96″	16.42″	3.16′	2.99	1.74	0.99
12	23.42″	12.93	6.42′	6.12″	8.62″	1.05	0.86	3.42	1.49	0.80
13	10.68″	27.74′	13.78″	1.56	10.98″	3.18	9.21″	2.74	0.48	1.01
14	4.89′	19.50′	3.08	2.78	5.49″	2.35	3.18′	1.33	1.73	1.05
15	4.04′	73.17′	4.19′	1.46	10.32″	2.47	3.20′	0.94	0.86	0.94
16	11.18″	77.41′	3.45	1.66	7.39″	2.31	3.83′	1.33	1.48	0.85
17	5.46′	19.25′	4.25′	1.40	6.30″	3.10	8.14″	1.04	1.09	0.72
18	8.50″	47.59″	7.15″	3.27	10.90″	4.69′	3.89″	1.52	1.56	0.95
19	12.40′	17.14	6.32′	3.47′	8.09″	2.02	6.74″	1.19	2.98	0.82
20	5.43′	73.06′	5.16′	2.32	8.46″	1.92	4.94″	1.53	2.66	0.95

Trial	Error	t_b	Interc.	t_{Jr}	F_{hr}	Domi-nance	Parent means	h^2	DGD	k
1	0.07	1.06	−18.1	12.70″	0.64	BZ5CH	ZB5CH	0.14	0.87	8
2	0.15	1.01	−32.4	5.48″	0.31	BZ5CH	ZB5CH	—	—	—
3	0.11	1.29	−46.7	7.37″	0.20	BZ5HC	BZ5HC	0.27	0.79	17
4	0.10	0.09	−87.6	10.37″	0.03	BZ5CH	ZB5CH	0.16	0.87	7
5	0.11	0.81	−60.1	8.59″	0.00	BZ5CH	ZB5CH	0.25	0.88	12
6	0.10	0.29	−21.0	9.84″	0.34	BZC5H	Z5CBH	0.21	0.90	7
7	0.06	1.68	−78.1	14.86″	0.76	B5ZCH	Z5BCH	0.05	0.94	3
8	0.06	0.37	−155.0	16.07″	0.67	BZ5CH	ZB5CH	0.04	0.89	4
9	0.08	0.60	−96.7	11.56″	0.50	BZ5CH	ZB5CH	0.07	0.86	5
10	0.08	2.31′	−68.1	9.50″	0.14	BZ5CH	ZB5CH	0.09	0.87	5
11	0.07	0.10	−57.5	14.75″	0.05	BZ5CH	ZB5CH	0.12	0.85	7
12	0.09	2.21′	−56.5	9.00″	0.20	BZ5CH	ZB5CH	0.12	0.91	5
13	0.09	0.14	−105.3	11.60″	2.66	BZ5CH	ZB5CH	0.12	0.93	5
14	0.17	0.31	−149.6	6.08″	0.08	BZ5CH	ZB5CH	0.16	0.83	8
15	0.08	0.78	−46.5	11.89″	0.03	BZC5H	ZBC5H	0.08	0.79	8
16	0.18	0.82	−97.2	4.68″	0.07	BZ5CH	ZB5CH	0.11	0.87	6
17	0.21	1.37	−61.9	3.44″	0.71	BZ5HC	ZB5HC	0.22	0.86	11
18	0.09	0.53	−60.1	11.19″	0.67	BZ5CH	ZB5CH	0.14	0.85	8
19	0.09	1.94′	−105.6	8.78″	1.25	BZ5CH	ZB5CH	0.11	0.88	6
20	0.15	0.32	−121.7	6.57″	0.08	BZC5H	ZB5CH	0.09	0.85	5

The genetic analyses of escape times in a T-maze. ′: $P \leq 0.05$; ″: $P \leq 0.01$. k is the estimated number of genes involved.

The analysis of the regression of W_r on V_r confirms that there is a directional dominance for fast running performance. Item b_2 is significant on most trials, suggesting unequal distribution of dominant alleles. In six cases irregular dominance factors (item b_3) show up. Since item c was significant for only six out of the twenty trials, maternal influences do not seem to play a very important role. The estimated minimum number of loci involved is consistently low.

For the number of errors made during twenty trails, the simple model seems appropriate since the differences $W_r - V_r$ do not differ significantly over strains. The slope of the regression of the covariances W_r on the variances V_r deviates from unity, hence epistatic interaction may be present. The significant strain effect for the sums $W_r + V_r$ indicates other effects beside the additive genetic variance shown by the analysis of variance. Random reciprocal differences and dominance were found. The dominance effects were specified as random dominance deviations (b_3) and directional dominance (b_1) for a small number of errors (see Table 3.7).

Although the F-ratio (3.31) for the differences $W_r - V_r$ is quite high for the number of refusals to run in the alley of the T-maze, it does not reach the 0.05 level of significance. The slope of regression of W_r on V_r is practically one and the intercept almost equals zero; therefore epistasis seems to be absent and the regression indicates complete dominance. The order of dominance as established by the regression corresponds quite well with the order of parental scores, since the C3H/Z, BALB, and C57BL strains cluster in both. Taken together with the directional dominance shown by the analysis of variance, this suggests directional dominance for low scores. The analysis of variance further reveals b_2 and b_3 effects and overall dominance (b). Additive genetic variance among the inbred strains is demonstrated.

For the T-maze performances, no clear learning phase existed as may also be seen from Fig. 3.1. Since all animals had had escape experience in the runway before they received training in the T-maze, it may be expected that no improvement occurred in the escape times. No distinction was made between times on correct and incorrect responses, since we failed to find a significant difference between times made in correct and incorrect trials.

3.4. General discussion and summary

In all situations except passive avoidance conditioning, additive variation

TABLE 3.7

	Tested against block interaction:									
	a	b_1	b_2	b_3	b	c	d	W_r+V_r	W_r-V_r	b_{Jr}
Refusals	13.70''	23.28'	6.34'	3.50'	10.13'	2.30	3.41	1.80	3.31	0.98
Errors	8.46''	38.38'	0.47	6.00''	4.61''	3.61	10.35''	5.00'	0.64	0.60
error		t_b	interc.	t_{Jr}	F_{hr}	domi-nance	parent means	h^2	DGD	k
Refusals	0.18	0.12	−2.3	5.50''	0.34	B5ZCH	Z5BCH	0.05	0.88	2
Errors	0.17	2.35'	0.5	3.57'''	0.26	5ZCBH	Z5BCH	—	—	—

The genetic analyses of the numbers of refusals and errors in a T-maze. ': $P \leqq 0.05$; '': $P \leqq 0.01$. k is the estimated number of genes involved.

was found for some of the behavioural parameters measured. About the presence or absence of dominance variation, especially directional dominance, firm conclusions can be drawn from a rather small diallel experiment like ours. In this respect, the various situations yielded very dissimilar results. For the behaviours measured in the open field no directional dominance could be demonstrated, but additive variation or ambidirectional dominance was found for most behavioural frequencies. These behaviours probably have a history of stabilizing selection. The findings for escape behaviour in the runway and the T-maze contrast markedly with this. In these situations, strong evidence was obtained for directional selection for fast performance in the past. Selection also favoured high escape tendency, as expressed by a small number of refusals to run in both situations, and making few errors in choosing the appropriate way out. Therefore, fast escaping and rapid escape learning may be regarded as more adaptive. Although no genetic correlations were calculated for the various traits observed in this study, the similarity of the results provided by the experiments with the runway and the T-maze suggests that common loci are involved in the control of these behaviours. Within the framework of the diallel method, Fulker et al. (1972) have designed a preliminary multivariate analysis of variance. By this analysis, a method is given to detect biological clusters in behaviour. This type of analysis should be further perfected in order to detect common genetical factors in behaviour.

References

Abeelen van, J. H. F. (1965): An ethological investigation of single-gene differences in mice. Thesis, (Brakkenstein, Nijmegen), Ch. IV.

Abeelen van, J. H. F. (1970): Genetics of rearing behavior in mice. *Behav. Genet.*, **1**, 71.

Breese, E. L. and K. Mather (1960): The organization of polygenic activity within a chromosome in *Drosophila*. II. Viability. *Heredity*, **14**, 375.

Broadhurst, P. L. (1960): Applications of biometrical genetics to the inheritance of behaviour. In: Experiments in Personality. Vol. 1. Eysenck, H. J., ed. (Routledge & Kegan Paul, London) pp. 1–102.

Broadhurst, P. L. and J. L. Jinks (1961): Biometrical genetics and behavior: reanalysis of published data. *Psychol. Bull.*, **58**, 337.

Fulker, D. W., J. Wilcock and P. L. Broadhurst (1972): Studies in genotype-environment interaction. I. Methodology and preliminary multivariate analysis of a diallel cross of eight strains of rat. *Behav. Genet.*, **2**, 261.

Hayman, B. I. (1954a): The analysis of variance of diallel crosses. *Biometrics*, **10**, 235.

Hayman, B. I. (1954b): The theory and analysis of diallel crosses. *Genetics*, **39**, 789.
Jarvik, M. E. and R. Kopp (1967): An improved one-trial passive avoidance learning situation. *Psychol. Rep.*, **21**, 221.
Kerbusch, J. M. L. (1974): Genetic analysis of exploratory behaviour, simple learning behaviour and cerebral AChE and ChE activities in mice by means of the diallel method. Thesis (Schippers, Nijmegen).
Mather, K. and J. L. Jinks (1971): Biometrical Genetics: The Study of Continuous Variation, 2nd ed. (Chapman and Hall, London), Ch. 9.
Messeri, P., A. Oliverio and D. Bovet (1972): Relations between avoidance and activity: a diallel study in mice. *Behav. Biol.*, **7**, 733.
Rick, J. T., G. Tunnicliff, G. A. Kerkut, D. W. Fulker, J. Wilcock and P. L. Broadhurst (1971): GABA production in brain cortex related to activity and avoidance in eight strains of rat. *Brain Res.*, **32**, 234.
Tyler, P. A. and G. E. McClearn (1970): A quantitative genetic analysis of runway learning in mice. *Behav. Genet.*, **1**, 57.
Vicari, E. M. (1929): Mode of inheritance of reaction time and degrees of learning in mice. *J. exp. Zool.*, **54**, 31.
Vossen, J. M. H. (1966): Exploratief gedrag en leergedrag bij de rat. Thesis, (Alberts, Sittard), Ch. III.
Winston, H. D., G. Lindzey and J. Connor (1967): Albinism and avoidance learning in mice. *J. comp. physiol. Psychol.*, **63**, 77.

CHAPTER 4

Applications of biometrical genetics to human behaviour

D. W. Fulker

The aim of applying biometrical genetics to human behaviour is to go beyond simple statements about the relative importance of genetic and environmental influences and provide information about the genetic architecture of the behaviour. That is, to determine whether the genes act additively, or non-additively as with dominance, to determine the level of dominance and whether it is uni- or ambi-directional, whether environmental influences are correlated or interact with the genotype, and how many loci control the behaviour.

While the investigation of the genetic architecture underlying continuous variation has reached a high degree of sophistication in animal studies, as may be seen from other chapters in Part I, the study of human variation is still at an initial stage. There are a number of reasons for this. Firstly, we are unable to design rigorous experiments with humans and are forced to accept the accidents of natural family groupings, twins, adoption, etc., where the unbiased nature of a laboratory experiment employing randomisation may be partially or even entirely lacking. Consequently, a careful and thorough sifting of human data is needed to establish its adequacy. Often, it must be admitted, the data are inadequate and there is little that can be done about it. Secondly, we do not have such adequate models that we can apply to human populations as we have for laboratory animals. The complexity of most human populations and the absence of experimental control suggest that models of more sophistication than have hitherto been developed by statistical geneticists are required. Thirdly, in order to apply the techniques and models that are available, we need very large, complex, and prohibitively expensive studies. It is the purpose of the present chapter to describe and illustrate the biometrical genetic approach and to indicate its scope and usefulness.

4.1. Statistical and genetical models

The basis of the biometrical genetic approach is a partitioning of the phenotypic variance (σ^2_{Total}) into genetic and environmental parts, G and E (Jinks and Fulker, 1970; Eaves and Jinks, 1972; Fulker, 1973). This partitioning is effected by an analysis of variance of various kinds of family groupings in which the mean squares are equated to the G's and E's of the model.

Consider the case of n pairs of siblings reared together in the same home, that is, the normal sibling family typical of our society. The analysis of variance partitions the total sum of squares into that between families and that within. These sums of squares, divided by their degrees of freedom, yield the mean squares (MS) of the analysis of variance. These, in turn, reflect two statistical features of the family situation, as shown below:

Source	df	Expected MS
Between families	n−1	$\sigma^2_W + 2\sigma^2_B$
Within families	n	σ^2_W

σ^2_W is the average within-family variance and reflects influences that tend to make siblings dissimilar. σ^2_B is the between-family variance or sibling covariance and reflects influences that make siblings alike, that is, shared influences. Together $\sigma^2_W + \sigma^2_B = \sigma^2_T$, the total phenotypic variance. It is these MS's to which the genetic and environmental models are fitted. If it is found that the σ^2_T's are the same for all the types of families in the investigation it is possible to fit the models to σ^2_B and σ^2_W in their more familiar standardised form, the intraclass correlation coefficient (r), since

$$r = \sigma^2_B/\sigma^2_T$$

and

$$1 - r = \sigma^2_W/\sigma^2_T.$$

The majority of investigators report their results in terms of correlations. However, when complex models are appropriate we cannot assume that the σ^2_T's are the same for all kinds of families and analysis in terms of correlations becomes difficult and clumsy, as will be seen in a later section of the chapter.

Contributing to σ^2_B and σ^2_W, which assess, respectively, the importance of influences that make siblings alike and different, are G_2 and E_2: the variance of genetic and environmental influences shared by siblings, and G_1 and E_1: the variance of genetic and environmental influences that are not shared.

The simplest model we could write is:

$$\sigma^2_B = G_2 + E_2$$
$$\sigma^2_W = G_1 + E_1$$

in which the effects of heredity and environment are additive. In order to disentangle the G's and E's, we need other kinds of families and to write their σ^2_B's and σ^2_W's in terms of the G's and E's appropriate to siblings. These expectations may be found in Table 4.1. It will be seen that each pair of σ^2_W and σ^2_B is expected to sum to the same quantity $\sigma^2_T = G_1 + G_2 + E_1 + E_2$. Therefore, provided all σ^2_T's are the same, each σ^2_B and σ^2_W may be replaced by r and $1-r$ and the analysis worked in terms of correlation coefficients instead of components of variance.

TABLE 4.1

Expectations of σ^2_W and σ^2_B for various kinds of families.

Type of family		Model			
		G_1	G_2	E_1	E_2
FS_T	σ^2_B	0	1	0	1
	σ^2_W	1	0	1	0
DZ_T	σ^2_B	0	1	0	1
	σ^2_W	1	0	1	0
MZ_T	σ^2_B	1	1	0	1
	σ^2_W	0	0	1	0
FS_A	σ^2_B	0	1	0	0
	σ^2_W	1	0	1	1
MZ_A	σ^2_B	1	1	0	0
	σ^2_W	0	0	1	1
U_T	σ^2_B	0	0	0	1
	σ^2_W	1	1	1	0

Key: Full siblings together (FS_T), Dizygotic twins together (DZ_T), Monozygotic twins together (MZ_T), Full siblings apart (FS_A), Monozygotic twins apart (MZ_A), and Unrelated individuals together (U_T).

4.2. Assumptions underlying the model

The above simple model makes a number of assumptions which are more or less plausible depending on the phenotype under investigation. In this section the assumptions will be stated and some statistical tests of their adequacy described.

An important assumption is that σ^2_T is the same for each kind of family. This may easily be tested by a heterogeneity of variance test on the σ^2_T's. The assumption may not be met for at least three different reasons. First, sampling may have been poor and the full range of the phenotype may not have been obtained for some of the types of families. Inadequate sampling would show up as an unsystematic heterogeneity of the σ^2_T's and perhaps in differences between means. Secondly, the elements of the model appropriate to some family types may not be appropriate to others. For example, E_1 may be much smaller for monozygotic (MZ_T) and dizygotic twins reared together (DZ_T) than for full siblings reared together (FS_T) due to a tendency for twins to share more experiences than siblings. A partial check on this would be to see if σ^2_W's differed for DZ_T and FS_T. A full check on the homogeneity of G's and E's between different kinds of families as distinct from poor sampling is probably not possible in all situations. It should be noted, however, that only E's should be affected since there are few plausible reasons why G's should differ between groups other than by poor sampling. If the E's are not comparable, then different kinds of E's can be fitted for different families. The third reason σ^2_T's may differ is perhaps the most interesting and deserves special attention. If there is any tendency during rearing for environmental influences to be differentially distributed to genotypes, a correlation between genetic and environmental influences may result and σ^2_T will no longer equal $G_1+G_2+E_1+E_2$. For example, parents of high IQ will show a tendency to have children who, in general, are brighter than average. At the same time they will probably provide a more intellectually stimulating environment for their children than parents of lower IQ. Consequently their children will score higher on tests of intellectual skills than they would have done if their environmental influences were random. Conversely, the children of parents of low IQ would score lower. This exaggeration of scores will appear as an increase in phenotypic variance. Formally the total variance is no longer

$$\sigma^2_T = G_1+G_2+E_1+E_2$$

but it is augmented by the covariance between g_2 and e_2 effects so that

$$\sigma^2_T = G_1 + G_2 + E_1 + E_2 + 2\text{cov } g_2 e_2.$$

Other more complex correlations are possible due to some kinds of selective placement and differential treatment of individuals within the family according to innate differences. Cattell (1960) discussed these correlations and developed models for dealing with them. More recently Jencks (1973) has suggested their importance in cognitive measures using the method of path analysis. An example of a powerful cov $g_2 e_2$ effect will be given later in the chapter.

When there are covariance effects of the $g_2 e_2$ kind, they can only occur for families reared together in the normal way. Foster families, such as monozygotic twins reared apart (MZ_A), full siblings reared apart (FS_A), and unrelated individuals reared together (U_T) will show little or no effect if fostering has been at random since e_2's are no longer systematically associated with g_2's. Consequently a test for cov $g_2 e_2$ is provided by a comparison of the σ^2_T's for families reared together and apart.

When we are fitting genetic models to variances and not to correlations, an overall test for the heterogeneity of σ^2_T's – whether due to inadequate sampling, heterogeneity of the model, or covariance effects – is supplied by a χ^2 test of goodness of fit of the model, provided we have sufficient different types of families to allow the model a reasonable chance of failing and a sufficient volume of data to afford a powerful test.

A second major assumption underlying the simple model is that the effects of the g's and e's are additive, that is, that there is no genotype-environment interaction ($G \times E$). Two approaches to the detection of $G \times E$ interaction have been employed. The first involves the use of MZ_A as pairs of isogenic individuals. The differences between members of each pair reflect a range of environmental effects for a reasonably sized sample. Consequently, if these differences are heterogeneous, or systematically related to genotype, as indicated by the average of the pair, we have evidence of $G \times E$. This approach is similar to that employing inbred lines of animals and their F_1's in which unequal within-genotype variances indicate a differential susceptibility to the effects of the environment. Jinks and Fulker (1970), Jensen (1970), and Eaves (1970) have used this approach. The general finding has been that there is little or no $G \times E$ for IQ but that there may be for some personality measures.

Another approach to detecting $G \times E$ is to see if the same models fit different populations or subdivisions of a single population. Scarr-Salapatek

(1971) used this approach to detect G × E in different socio-economic groups and among American blacks and whites for intellectual abilities. There was some suggestion that underprivileged groups were more susceptible to environmental influences than the more privileged groups, but the findings were of doubtful validity (Eaves and Jinks, 1972).

A remaining assumption is that fostering has been carried out at random, that is, that there is no selective placement. Some forms of selective placement will cause heterogeneous total variances. For example, if children are fostered predominantly to one section of the community, thus restricting the range of E_2, the σ^2_T for that group will be smaller than for non-fostered groups. However, other forms of selective placement will not affect the total variance, as when a child is reared by a relative. The extreme form of this type of placement would occur if the child were put back into its original home; then the phenotypic variance would clearly not be affected. To test for selective placement is often difficult because we need to know the trait-relevant variables in the parents. A correlation of zero between real and foster parents for these variables would indicate no selective placement. Often we are forced to guess the relevant variables in the absence of adequate psychological knowledge, and weak tests of placement effects result.

Provided the assumptions are met sufficiently to make estimation of G_1, G_2, E_1, and E_2 valid, we are in a position to interpret G_1 and G_2 in terms of gene effects.

4.2.1. Genetic expectations

Mather (1949) developed a genetical model of continuous variation based on that of Fisher (1918). The model places each locus on a scale according to whether it is homozygous and increasing (AA), homozygous and decreasing (aa), or heterozygous (Aa). The effect of AA is $+d$, that of aa is $-d$, and that of Aa is h, all measured from the average of AA and aa. In a randomly mating population where the frequency of A is u and the frequency of a is v, the total genetic variation is

$$G_1 + G_2 = \tfrac{1}{2}D_R + \tfrac{1}{4}H_R = 2\sum_i u_i v_i [d_i + (v_i - u_i)h_i]^2 + 4\sum_i u_i^2 v_i^2 h_i^2$$

where the summation signs operate over all loci influencing the trait. In terms of this model

$$G_1 = \tfrac{1}{4}D_R + \tfrac{3}{16} H_R$$

$$G_2 = \tfrac{1}{4}D_R + \tfrac{1}{16} H_R$$

Thus, we may use G_1 and G_2 to solve for D_R and H_R.

Where we have assortative mating, Fisher (1918) showed that the total variation is increased by $\tfrac{1}{2}(\tfrac{A}{1-A})D_R$, where A is the correlation between additive deviations of spouses. Then $G_1 = \tfrac{1}{4}D_R + \tfrac{3}{16}H_R$

$$G_2 = \tfrac{1}{4}D_R + \tfrac{1}{16}H_R + \tfrac{1}{2}\left(\frac{A}{1-A}\right)D_R$$

and we need a further statistic to solve for D_R, H_R, and A.

We may solve for these parameters if we know the marital correlation μ since

$$A = [\tfrac{1}{2}D_R + \tfrac{1}{2}(\frac{A}{1-A})D_R]\mu\sigma_T^2$$

which provides a further equation. Alternatively, if we have the parent-offspring correlation, $r_{p.o}$, we may use

$$\frac{r_{p.o}}{1+\mu} = \left[\tfrac{1}{4}D_R + \tfrac{1}{4}\left(\frac{A}{1-A}\right)D_R\right]\sigma_T^2$$

to give a solution.

Thus we see that when we have dominance but no assortative mating, $G_1 > G_2$, and when we have assortative mating but no dominance, $G_2 > G_1$. With both dominance and assortative mating, or neither, G_1 and G_2 will be similar in size.

If we are confident that the assumptions are met, and that therefore Fisher's model is appropriate, there are a number of other genetic expectations that may be used to gain information about gene effects. First, we may extend the family types to more remote relationships such as grandparents and grandchildren, cousins, uncle–nephew, etc., the expectations for which may be found in Fisher (1918), Burt and Howard (1956), and Eaves (1973). The genetical model is then fitted directly, there being no need to work first with G's.

Many other statistics are useful in exploring gene effects. For example, the result of inbreeding is to lower the mean expression of the trait if there is directional dominance, so that inbreeding depression may be used to detect the presence of dominance and its direction. If M_0 is the mean of the outbred population and M_f that of the inbred, then $M_0 - M_f = 2f\sum_i u_i v_i h_i$

after one generation of inbreeding, where f is the inbreeding coefficient of Wright. The partially inbred humans most readily available are the children of cousin marriages, where $f = \frac{1}{16}$.

Some third and fourth moments may also be useful in interpreting gene effects since all the moments have genetic expectations. However, large samples are required if third and fourth moments are to be at all accurately estimated, and moments above the third have been very little used. Two third moments are particularly useful. These are K_3, the sample skewness, and $\overline{K}_3$, the average within-family skewness. K_3 is a function of unequal allele frequencies, $u-v$, and a term indicating directional dominance, d^2h. Thus a significant K_3 is ambiguous in meaning. $\overline{K}_3$, however, is not sensitive to unequal allele frequencies, having the expectation

$$\overline{K}_3 = -3 \sum_i u_i^2 v_i^2 d_i^2 h_i.$$

Thus a significant and negative $\overline{K}_3$ indicates directional dominance for high expression of the trait.

When we have equal allele frequencies ($u = v$), then $K_3 = 4\overline{K}_3$. This relationship may be used, therefore, to test for equal allele frequencies. When $u = v$, the parameters D_R and H_R reduce to

$$D_R = \sum_i d^2_i$$
$$H_R = \sum_i h^2_i$$

so that $\sqrt{\frac{H_R}{D_R}}$ indicates the average level of dominance.

The range of the phenotype may also be used to estimate the number of loci influencing the trait. The extremes will approximate to the two genotypes: AABBCC....., the extreme high scorer, and aabbcc..., the low scorer. Thus $\frac{1}{2}$Range $= \sum_i d_i$, and n, the number of loci, may be estimated from $n = \frac{(\frac{1}{2}\text{Range})^2}{D_R}$, when $u = v$. This value will usually be a gross underestimate (see Mather and Jinks, 1971) and is still useful as a rough guide when $u \neq v$. The approach outlined above and the use of these various statistics will be illustrated in the next section.

4.3. Applications

4.3.1. General intelligence

One of the most intensively investigated behavioural measures from the biometrical point of view is IQ. Burt and Howard (1956) applied Fisher's approach to extensive correlation data, concluding that there was additive and dominance variation as well as variation for assortative mating. Broad heritability was in the region of 85 % and narrow heritability about 70 %. When the correlations were based on what they considered their best estimates of IQ, heritabilities were highest and there was no evidence of correlated genetic and environmental influences. While Burt and Howard's approach was valid it did not make an efficient use of the available information, nor did it allow any assessment of the errors involved in the estimation of the genetic and environmental parameters. Jinks and Fulker (1970) reanalysed Burt's correlations, using a more efficient procedure, and augmented his data with other findings on IQ, pursuing the biometrical analysis further.

The analysis of correlations is simpler to illustrate than that based on components of variance. However, it is only justified if the total variances for each type of family in the study are homogeneous, as shown in the previous section. Fortunately, it would appear that this is a reasonable assumption concerning Burt's data, although a thorough test of the assumption is not possible with the information available. Firstly, Burt and Howard's finding of no correlated genetic and environmental influences implies homogeneity of variances. Their method, based on path analysis, is not particularly sensitive and only approximate since it assumes that the correlation for which we are looking does not exist – in order that the observed phenotypic correlations may be considered to be based on the same total variances and be comparable. A direct test is the comparison of the total variance for FS_T and MZ_A, these being 225 and 234 respectively. The $F_{105,527} = 1.04$ and has a chance probability of 0.6. Since the most plausible source of heterogeneity, that is a positive correlation between genotype and environment, would cause the variance of the reared-apart group to be smaller, not larger, than the reared-together group, this small difference in variance is almost certainly due to sampling. Jencks (1973) also tested for a genotype-environment correlation in Burt's MZ_A data. He noted that the MZ_A were only "half" reared apart. That is, in general, one twin member was fostered while the other remained with its natural mother. This method of fostering, in the presence of correlated genetic and environmental influences, would result in a difference in variance between those reared at home and those fostered. Separating the data into these two groups he found no such difference. We may conclude, then, that there is no evidence of correlated genetic and environmental effects in Burt's data, although other sources of heterogeneity remain a possibility. Certainly its most likely source is absent.

The assumption of no selective placement is less certain. Its presence is a possibility for the U_T groups as we will see when we come to fit a G and E model. However, for the MZ_A group there seems little reason to suspect its presence. Burt (1966) convincingly shows that as far as socio-economic status (SES) is concerned there is absolutely no correlation between real and foster parents. While SES is only one environmental influence likely to modify IQ, the absence of selective placement with respect to this variable does suggest that no obvious attempt at selective placement took place.

The remaining assumption concerning the adequacy of a simple G and E model is that of no genotype-environment interaction ($G \times E$). Correlating twin pair sums with twin pair differences, for Burt's 53 MZ_A, gave a small nonsignificant correlation and Jensen (1970), collecting together all available IQ data on MZ_A, also found no correlation. There is no evidence, then, of heterogeneity of MZ_A differences, suggesting that the whole range of genotypes are equally susceptible to environmental influences. While this correlation test will only reveal a linear relationship, scatter diagrams of twin sums against differences also fail to reveal more complex trends. It remains possible, of course, that more complex interactions take place in which, for example, some high-scoring genotypes are more modifiable than low ones, but other high genotypes are less modifiable, cancelling out any obvious effect along the genotypic scale. However, it seems likely that no apparent form of $G \times E$ exists for IQ.

Having satisfied ourselves that the assumptions of a simple G and E model are reasonably met in Burt's data, this model may be fitted. The correlations for IQ (final assessment) are given, together with the model, in Table 4.2.

In order to estimate the three parameters G_1, G_2, and E_2 from the six correlations, a weighted multiple regression estimation procedure is used. The values of the six correlations are regressed onto the coefficients of the model, each correlation being weighted according to the precision with which it is determined. The precision of the correlation is taken as the inverse of its sampling variance, where $V(r) = \frac{(1-r^2)^2}{N}$, which for large samples is the amount of information, that is $I(r) = \frac{1}{V(r)}$. The regression technique uses weighted least squares where, instead of estimating parameters that minimise

TABLE 4.2

Observed and expected values of correlations for six kinds of families and the weights I(r) used in fitting the simple genetical model (IQ).

Type of family	Correlation (r)	Model G_1	G_2	E_2	V(r)	I(r)	Expected (r)	Observed minus Expected
MZ_T	0.92	1	1	1	0.000249	4016.1	0.9274	−0.0074
MZ_A	0.87	1	1	0	0.001113	898.5	0.8376	+0.0324
DZ_T	0.54	0	1	1	0.003945	253.5	0.5326	+0.0074
FS_T	0.53	0	1	1	0.001958	510.7	0.5326	−0.0026
FS_A	0.44	0	1	0	0.004305	232.2	0.4428	−0.0028
U_T	0.27	0	0	1	0.006321	158.2	0.0897	+0.1803

G_1 reflects genetic differences between siblings
G_2 reflects genetic similarities between siblings
E_2 reflects environmental similarities between siblings
E_1 reflects environmental differences between siblings;
 E_1 obtained as $1 - r(MZ_T)$

the sum of squares for the discrepancy between the observed correlations (0) and the expected values (E), as in least squares, $\sum I(0-E)^2 = \chi^2$ is minimised. The estimates are obtained by solving the equations $(\mathbf{C\, I\, C})^{-1}(\mathbf{C\, I\, o}) = \mathbf{g}$, where $\mathbf{C}$ is the matrix of coefficients of the model, $\mathbf{I}$ a diagonal matrix of the weights, $\mathbf{o}$ the vector of observations, and $\mathbf{g}$ the vector of estimates we wish to obtain. The standard errors of the estimates are obtained from the square roots of the leading diagonal of $(\mathbf{C\, I\, C})^{-1}$. This method has the advantages of giving approximate maximum likelihood estimates of the parameters, allows their standard errors to be calculated, and provides a χ^2 test of goodness of fit of the model. Applying this method, Jinks and Fulker obtained

$$G_1 = 0.39 \pm 0.03$$
$$G_2 = 0.44 \pm 0.04$$
$$E_1 = 0.08 \pm 0.01$$
$$E_2 = 0.09 \pm 0.03.$$

E_1 being obtained directly as $1 - MZ_T$ correlation, the standard error being that of the correlation.

Calculating the expected correlations on the basis of these estimates gave the values in column 8 of Table 4.2. All discrepancies between observed and expected values (column 9) are small except for U_T, and this discrepancy is responsible for the borderline failure of the model reflected in $\chi_3^2 = \sum I(0-E)^2 = 6.3$ with a probability between 0.05 and 0.10. Clearly the model only roughly fits the data, and one suspects that a placement effect, not allowed for in the model, is inflating the correlation for U_T, which should be a direct estimate of E_2. At this point there is a choice. We can either accept the model and bear in mind that it may be somewhat approximate, or drop the U_T correlation and refit the model. The latter course results in similar values of the parameters

$$G_1 = 0.40 \pm 0.03$$
$$G_2 = 0.47 \pm 0.04$$
$$E_1 = 0.07 \pm 0.01$$
$$E_2 = 0.06 \pm 0.03.$$

and in a nonsignificant $\chi_2^2 = 0.3$. The results are so close to those obtained when retaining U_T that it does not seem worth reducing the data; the first set of estimates will be used for further analysis. It is worth drawing attention to the reason for this similarity between the two sets of estimates. By using a weighted estimation procedure, and the correlation for U_T being small, it carries the smallest weight of any of the six correlations. Thus, although its lack of fit to the model is detected, the consistency of the remaining correlations is still given due emphasis.

Using these estimates of G_1 and G_2, and having some faith in their validity, we may attempt to fit Fisher's more directly genetical model, augmenting the G's with the marital correlation, μ, which is 0.39 in Burt's data. The following three equations may be used to solve for A, D_R, and H_R.

$$\mu = 0.39$$
$$G_1 = \tfrac{1}{4}D_R + \tfrac{3}{16}H_R = 0.39$$
$$G_2 = \tfrac{1}{4}D_R + \tfrac{1}{16}H_R + \tfrac{1}{2}(A/1-A)D_R = 0.44$$

These equations may be solved numerically by a simple iterative method using starting values obtained from the relationship $A = h_n^2 \mu$, where h_n^2 is the realisable heritability and is analogous to the narrow heritability of a randomly mating population. The expression for h_n^2 is $[\tfrac{1}{2}D_R + \tfrac{1}{2}(A/1-A)D_R]/$ Total variance. Thus the maximum value of $h_n^2 = G_1 + G_2$, which assumes $H_R = 0$ and lets D_R take its own value. Using these starting values and carrying out only six iterations, the estimates converge to

$$D_R = 0.96$$
$$H_R = 0.80$$
$$A = 0.25$$

so that $\sqrt{H_R/D_R} = 0.91$.

The value of $\sqrt{H_R/D_R}$ indicates, approximately, the average level of dominance, being h/d for a single locus with equal allele frequencies. Clearly there is substantial dominance variation, at least consistent with complete dominance. The broad heritability, $[\frac{1}{2}D_R + \frac{1}{4}H_R + \frac{1}{2}(A/1-A)D_R]/$Total variance, is 0.83 and the realisable or narrow heritability, $[\frac{1}{2}D_R + \frac{1}{2}(A/1-A)D_R]/$Total variance, is 0.63.

Jinks and Fulker (1970) used the parent-offspring correlation to augment G_1 and G_2 and estimate D_R and H_R, finding very similar values, and Fulker (1973) used a more direct method including the parent-offspring correlation and the six correlations in Table 4.2 with practically the same result. However, once we wish to include correlations for other kinds of relatives, such as cousins and uncle-nephew, in order to extend the validity of the model,

TABLE 4.3

Correlations for IQ from Burt (1966) and Jencks (1973).

Relationship	Burt	Jencks*
Parent-child together	0.49	0.55
Parent-child apart	—	0.45
Grandparent-grandchild	0.33	—
MZ_T	0.92	0.97
MZ_A	0.87	0.75
Like-sex DZ_T	0.55	0.70
Unlike-sex DZ_T	0.52	—
FS_T	0.53	0.59
FS_A	0.44	—
Uncle-niece etc.	0.34	—
1st cousins	0.28	—
2nd cousins	0.16	—
Fosterparent-fosterchild together	0.19	0.28
U_T	0.27	0.38
Marital	0.39	0.57

* Corrected for unreliability and differences between sample means.

the estimation procedure becomes unsatisfactory since extra G's, beyond G_1 and G_2, have to be written in the expectations in order to retain linear models. Fisher's model, however, is non-linear so that a non-linear weighted regression technique would be more efficient. Eaves (see Jinks and Eaves, 1974) has utilised a procedure based on the Newton-Raphson numerical method of solving non-linear equations. This method was applied to all the available familial correlations of Burt shown in Table 4.3, fitting Fisher's genetical model directly.

Eaves fitted two models, making different assumptions about the importance of E_2 in the expectations of parents and offspring. In the first, he (Eaves personal communication) assumed that no common environment operated on the parent-offspring relationship, as did Jinks and Fulker. With this assumption he obtained the following estimates:

$$D_R = 0.88 \pm 0.14$$
$$H_R = 0.74 \pm 0.20$$
$$A = 0.34 \pm 0.07$$
$$\mu = 0.42 \pm 0.06$$
$$E_2 = 0.07 \pm 0.02$$
$$E_1 = 0.09 \pm 0.02$$

which gives $\sqrt{H_R/D_R} = 0.92$

and $\chi_9^2 = 12.93$, $P = 0.10-0.20$ for goodness of fit of the model. Thus the model is seen to fit well and to be in excellent agreement with the estimates of Jinks and Fulker given previously. In the second model, an E_2 was included in the expectation for parents and offspring on the assumption that some shared environment seems likely and that it would, in many instances, be similar to the shared environments of ordinary siblings. Perhaps the E_2 for parents and offspring would be smaller than that for siblings, but in that case the discrepancy would show up during the model-fitting procedure. Using this model, the estimates were:

$$D_R = 0.57 \pm 0.17$$
$$H_R = 1.15 \pm 0.25$$
$$A = 0.47 \pm 0.10$$
$$\mu = 0.41 \pm 0.08$$
$$E_2 = 0.10 \pm 0.03$$
$$E_1 = 0.08 \pm 0.02$$

which gives $\sqrt{H_R/D_R} = 1.42$

and $\chi_9^2 = 8.96$, $P = 0.50-0.30$. These estimates still indicate substantial dominance variation but are only in reasonable agreement with the previous ones. However, the assumption of E_2 applying to Burt's parent-offspring correlation, which was adopted for comparison with Jenck's data, also shown in Table 4.3, is unreasonable since Burt (1966) has shown that the correlation is very similar whether the parent is measured as a child, before experiences are shared, or later as an adult.

In order to further test the validity of the findings from Burt's correlations to other IQ data, Jinks and Eaves (1974) fitted the second model to the comprehensive set of correlations, mainly from American studies, brought together by Jencks (1973). Here the results were similar to Burt's, assuming the second model, being

$$D_R = 0.48 \pm 0.10$$
$$H_R = 1.37 \pm 0.11$$
$$A = 0.30 \pm 0.11$$
$$\mu = 0.57 \pm 0.02$$
$$E_2 = 0.29 \pm 0.02$$
$$E_1 = 0.03 \pm 0.02$$

which gives $\sqrt{H_R/D_R} = 1.69$ and $\chi_4^2 = 6.63$, $P = 0.20-0.10$.

They concluded that ignored placement effects may be increasing the apparent importance of dominance over that in Burt's data. However, their findings from the data of both Burt and Jencks confirm the findings of Jinks and Fulker that there is substantial dominance variation, probably complete, and that assortative mating is important. The nonsignificant χ^2 for Jencks' data confirms the absence of correlated genetic and environmental influences for IQ, contrary to Jencks' suggestion.

Eaves (1973) has further validated this model using the very extensive pedigree data of Reed and Reed (1965). These are based on 80,000 individuals for whom detailed family trees are available. Eaves took a sample of 3558 individuals from a single generation and classified them hierarchically by common ancestry, first by parents, then by grandparents, up to great-great-great-grandparents. He then subjected these five levels of hierarchy to an analysis of variance and equated the observed mean squares at each level to their genetical expectation according to Fisher's model.

Various models were fitted, involving common environments (E_2), dominance, and several levels of assortative mating, as well as additive variation and random environment (E_1). The most striking finding was that the influence of assortative mating was marked and required by all models that fitted. Moreover, the level of assortative mating, as reflected in the parameter A = 0.27, the correlation between additive genetic values for spouses, was very close to that found by Jinks and Fulker in Burt's data where A was 0.26. Unfortunately, this design does not provide a very powerful method of detecting dominance and a nonsignificant H_R was indicated. However, Eaves was able to show that a complete level of dominance was as consistent with the data as $H_R = 0$, its inclusion in the model hardly changing the goodness of fit except to improve it slightly. Consequently we may conclude that the findings from Reed and Reed's data are strikingly consistent with those from Burt's, given the limitations of the design. The consistency of the findings with those from Jenck's data is good in broad outline but less so regarding the absolute values of the parameters. Having rather good estimates of the genetical parameters controlling IQ, which we now know are reasonably realistic for much of the available IQ data, Jinks and Fulker went on to estimate the direction of dominance, the minimum number of loci, and the average allele frequency.

A significant value of H_R indicates dominance variation, but whether the dominance is directional or not requires other methods. The most powerful method for detecting directional dominance is a comparison of an inbred and outbred group of subjects. Schull and Neel (1965) found, in a large sample of Japanese children, that those resulting from marriage between cousins had significantly lower IQ's than those whose parents were not related. Such a result strongly suggests directional dominance for high IQ.

In Reed and Reed's data, the analysis of skewness also indicates dominance for high IQ. Using 689 sibling families of size three or more, Jinks and Fulker found that the average within-family skewness, $\overline{K}_3$, was -257 ± 121; $P < 0.05$. The overall skewness, $K_3 = -1077$ is almost exactly $4\overline{K}_3$, which suggests equal allele frequencies on average for the genes controlling IQ. If the allele frequencies are equal, then the parameter D_R takes a simple form in terms of gene effects

$$D_R = \sum_i d_i^2$$

and the number of loci may be estimated as

$$n = (\tfrac{1}{2} \text{ Range})^2 / D_R.$$

Taking the range for IQ as at least 150 gave a value of n = 22, which is a minimal estimate and may possibly indicate that the units of segregation resolved by these biometrical analyses are whole chromosomes, there being

23 pairs in man.

The genetical model obtained for IQ may be further validated by two predictions: the regression of children's IQ's towards the population mean and the frequency of subnormality in the children of subnormal parents.

TABLE 4.4

Mean IQ's of father and children according to fathers' SES
(Data from a study by Burt; Eysenck, 1971).

Father's SES	Higher Professional	Lower Professional	Clerical	Skilled	Semi-Skilled	Unskilled
N	120	1240	4880	10320	13000	10440
Father's IQ	139.7	130.6	115.9	108.2	97.8	84.9
Children's IQ	120.8	114.7	107.8	104.6	98.9	92.6
Children's estimated IQ $\mu = 0.39$, $h^2_n = 0.71$	119.6	115.0	107.9	104.1	98.9	92.8

Table 4.4 shows Burt's IQ data for 40,000 children and their fathers, grouped by SES. A narrow heritability of 0.71 was used to predict the regression, a value that differs slightly from that of 0.66 obtained by Jinks and Eaves (1974) but which is the best estimate obtained by Jinks and Fulker (1970) and Fulker (1973). Together with the marital correlation $\mu = 0.39$ the estimated offspring score was obtained from that of the fathers by the application of two regression equations. First the mother's score was predicted from that of the father, using the equation

$$\text{mother's score} = \mu \, (\text{father's score} - 100) + 100$$

after which the midparent score, that is the average of mothers and fathers, was calculated for each SES group. The predicted offspring score was then calculated from

$$\text{offspring score} = h^2_n \, (\text{midparent score} - 100) + 100$$

which is shown in the final row of Table 4.4. These estimated offspring scores, predicted on the basis of an appropriate genetical model, compare closely with the observed values in the row above.

To predict the frequency of subnormality in the offspring of subnormals, Reed and Reed's (1965) extensive IQ data were used. The authors give the frequency of subnormality (defined as IQ $\leq$ 69) among the children of two subnormal parents as 39%, and of one subnormal and one unspecified parent as 16%. Taking the observed population mean of 104 and that of the subnormal parents as 60, the predicted offspring mean was calculated using appropriate regression equations, exactly as before. Then normal probability tables were used to convert this mean to a percentage. Using values of $h_n^2 = 0.71$ and $\mu = 0.39$, the predicted frequency was 17% against an observed frequency of 16% when only one parent was subnormal, and 38% against 39% observed when both parents were subnormal. Such good agreement suggests that a substantial part of mental subnormality, particularly the less severe kind, is due to segregation of the kind of polygenic system previously described. This finding is in agreement with that of Penrose (1963) who found that heritabilities derived from a polygenic model accurately predicted the IQ's of siblings, half-siblings, and nephews and nieces for patients with IQ's above 50, but that for the relatives of the severely subnormal, with IQ's below 50, the predicted results were much worse than actually observed. His findings indicate that the severely subnormal were the result of environmental traumas and interacting gene effects, such as occur when severely deleterious recessives influence a trait.

One further point emerges from the analysis of IQ. The finding of dominance for high IQ suggests that it has been an important component of fitness during man's evolutionary history. The arguments relating genetical architecture to natural selection are given in chapter 2 in this section. Briefly, Mather (1953) has argued that traits showing strong directional dominance will have been subject to strongly directional natural selection, while those showing mainly additive variation or ambidirectional dominance will have been selected for an intermediate optimum. IQ, then, would appear to reflect a measure of some biological importance.

4.3.2. Educational attainments

The biometrical analysis of IQ according to a simple G and E model yields fairly consistent results across several bodies of data. However, when we apply a similar approach to Burt's data on educational attainments, it is clear that a simple model is not valid. Table 4.5 shows the correlations for a composite of simple attainments involving reading and arithmetic tests. Applying the weighted least squares estimation procedure gave the following values:

TABLE 4.5

Correlations for six kinds of families and a single genetical model (educational attainments).

Type of family	Correlation	Model G_1	G_2	E_2
MZ_T	0.98	1	1	1
MZ_A	0.62	1	1	0
DZ_T	0.83	0	1	1
FS_T	0.80	0	1	1
FS_A	0.53	0	1	0
U_T	0.54	0	0	1

$$G_1 = 0.17 \pm 0.02$$
$$G_2 = 0.53 \pm 0.05$$
$$E_1 = 0.20 \pm 0.01$$
$$E_2 = 0.28 \pm 0.05$$

The difference between G_1 and G_2 is highly significant (normal deviate = 6.12; $P < 0.001$), G_2 being more than three times the magnitude of G_1. Such a large difference is almost impossible on any simple genetical model, an improbably high degree of assortative mating being required to bring it about. A more reasonable explanation is that correlated genetic and environmental effects, which the simple model ignores, are responsible for this large discrepancy. Such effects, if not allowed for in the model, would inflate G_2 and E_2 in Burt's study where individuals reared apart were, in fact, only "half" reared apart (one individual was fostered and the other remained with its biological parents). The model for this situation is difficult to put in tabular form since the expectations are no longer linear. The reared-together and reared-apart correlations, which have different denominators, are given below.

$r(MZ_T) = (G_1 + G_2 + E_2 + 2 \text{ Cov } g_2 e_2)/(G_1 + G_2 + E_1 + E_2 + 2 \text{ Cov } g_2 e_2)$
$r(FS_T) = r(DZ_T)$
$r(DZ_T) = (G_2 + E_2 + 2 \text{ Cov } g_2 e_2)/(G_1 + G_2 + E_1 + E_2 + 2 \text{ Cov } g_2 e_2)$
$r(MZ_A) = (G_1 + G_2 + \text{Cov } g_2 e_2)/(G_1 + G_2 + E_1 + E_2 + \text{Cov } g_2 e_2)$
$r(FS_A) = (G_2 + \text{Cov } g_2 e_2)/(G_1 + G_2 + E_1 + E_2 + \text{Cov } g_2 e_2)$
$r(U_T) = (E_2 + \text{Cov } g_2 e_2)/(G_1 + G_2 + E_1 + E_2 + \text{Cov } g_2 e_2).$

The correlations given in Table 4.5 were substituted into these equations and an approximate non-linear weighted regression procedure was used to estimate the G's and E's. This procedure involved iterations from close initial values and then several possible combinations of systematically varied estimates. Those estimates to which the residual χ^2 was most sensitive were fixed first, the less sensitive ones later. Although this procedure does not guarantee a minimum χ^2, the nonsignificant value of $\chi_2^2 = 1.53$, $P < 0.20$, was deemed small enough for the estimates to be close to those obtained by more sophisticated methods. The following estimates were obtained:

$$G_1 = 0.17$$
$$G_2 = 0.17$$
$$E_1 = 0.02$$
$$E_2 = 0.22$$
$$2\,\text{Cov}\,g_2 e_2 = 0.42.$$

Now G_2 and G_1 take more realistic values but clearly the picture for educational attainments is very different from that for IQ. Only 34% of the variation is directly attributable to genetic causes, 64% being due to differences in family environment and its correlated genetic effects. We do not know what influences contribute to the large E_2, but the value suggests that considerable scope exists for the manipulation of attainments should these influences be identified. If the detrimental influences contributing to E_2 could be prevented, or their effects ameliorated, much of the covariance which accentuates the performance of the disfavoured genotypes, would disappear. As a result a great deal of poor attainment would disappear.

The genetic picture for attainments is not clear from these estimates. We can still equate G_1 and G_2 to their appropriate expectations, but the value of A in the equation

$$A = (\text{Realisable heritability})\mu$$

is uncertain since we do not know what the form of this heritability should be. For example, does it include covariance effects, either in its numerator or denominator, or not? Certain inferences, however, are possible. Since $\mu = 0.68$ for general attainments there should be some influence due to assortative mating reflected in G_2. But $G_2 = G_1$, therefore there must be some dominance variation contributing to G_1. That is, $G_1 = G_2$ in the presence of assortative mating indicates both dominance and assortative-mating variance, a picture closely resembling that of IQ. The interesting possibility exists that the genetic variation for general attainments is simply that for IQ overlain by more environmental influences. That is, as far as the genetic picture is concerned, general attainments simply provide an unreliable measure of IQ. If it is appropriate to write

$$A = \mu[\tfrac{1}{2}D_R + \tfrac{1}{2}(A/1-A)D_R]/\text{Total variance}$$

then it is possible to solve for D_R, H_R, and $\tfrac{1}{2}(A/1-A)D_R$, combining the above equation with those for G_1 and G_2 as for IQ. The following estimates were obtained:

$$D_R = 0.42$$
$$H_R = 0.35$$
$$\tfrac{1}{2}(A/1-A)D_R = 0.04$$
$$\sqrt{H_R/D_R} = 0.91$$

The similarity between the genetic control of general attainments and IQ is now striking, the dominance ratios being exactly the same, and the assortative-mating variance, as a fraction of total G, being very similar for the two measures (13 % for attainments and 18 % for IQ). However, the validity of this analysis depends on Fisher's model being correctly applied, and it must be admitted that the theoretical justification is not clear.

The analysis in this section illustrates the superiority of working with raw variances rather than correlations. The non-linear aspect of the model follows simply from having to allow for the two different variances in the denominator of the correlations for groups reared together and apart. By using the between-group variances in the place of correlations, the model could not only be fitted as before but its validity tested by showing that the total variances for the groups differ. It is to be regretted that the raw data of Burt's studies no longer exist (Jensen, 1974) so that a more rigorous analysis of the genetic relationship between IQ and attainments cannot be carried out.

4.3.3. Schizophrenia

The biometrical analysis of discontinuous measures may be carried out by using tetrachoric correlations derived from contingency tables. For example, if those with IQ greater than 100 are labelled bright, and those with IQ less than 100, dull, we can set up for fathers and sons, say, a 2×2 contingency table giving the proportion of dull and bright sons born to both dull fathers and bright fathers, respectively. If it can be assumed that the underlying continuous variable (in this case IQ) is normally distributed, the parent-offspring correlation for IQ is estimated by the tetrachoric correlation appropriate to this table. With correlations for sufficient groups of relatives we could then carry out the kind of analysis illustrated in the previous sections. For IQ we would expect very similar results using either the familiar intraclass correlation or its tetrachoric counterpart, the only difference being that the latter would probably be subject to greater sampling errors.

While it would be absurd to use the tetrachoric correlation for the analysis of IQ, since we already have the continuous measure, it can be useful where the underlying variable has not been, or cannot be, measured. Many abnormal behaviours, such as schizophrenia, involve variables of this kind. In the case of schizophrenia, the status of the underlying variable is not clear. If the schizophrenic is viewed as someone who was once quite indistinguishable from the rest of the population, but who subsequently developed very extreme and bizarre behaviour, the underlying variable is probably best seen as one involving liability to schizophrenia. Only with sufficient units of liability does an individual pass the threshold and become qualitatively different. On the other hand, if the schizophrenic is viewed as someone whose behaviour was always unusual and who has now become in need of medical attention, then we are employing the notion of a non-latent variable. Presumably schizophrenia involves both kinds of variables. In either case the threshold analysis may be used: in the former situation the threshold being real, in the latter a semantic convenience.

The threshold model of liability to disease was developed by Falconer (1965) using regression analysis, and applied to a number of disorders (Falconer, 1967). Gottesman and Shields (1967) applied this model to twin data on schizophrenia, concluding that there was a substantial genetic component in liability. By considering data from other relatives, they attempted a rudimentary gene-effect analysis, finding no evidence of dominance (Gottesman and Shields, 1968). Fulker (1973) illustrated the biometrical genetic approach by fitting genetical models to tetrachoric correlations calculated from concordance rates for MZ and DZ twins and separated parents and offspring (PO_A). Only recent studies, where sampling bias was minimal, and diagnosis reasonably objective and uniform, were used. The studies and their concordance rates are shown in Table 4.6.

Gottesman and Shields' (1966) twin data were collected during 16 years of continuous admissions to the Maudsley Hospital, London. Harvald and Hauge's (1965) twin data were collected through State records of twin

TABLE 4.6

Concordance rates for recent schizophrenia studies.

Source	Concordance		
	MZ_T	DZ_T	PO_A
Gottesman and Shields (1966)	10/24	3/33	...
Harvald and Hauge (1965)	2/7	3/59	...
Tienari (1963)*	1/16	...	...
Kringlen (1966)	17/55	14/178	...
Heston (1966)	...	...	5/47
Rosenthal et al. (1968)	...	...	3/54
Heterogeneity χ^2	3.64 (ns)	0.57 (ns)	0.77 (ns)
df	3	2	1
% Concordance	29.41	7.41	7.92
Tetrachoric correl. based on p = 1.14 %	0.76±0.04	0.35±0.01	0.37±0.02

* Given in Slater and Cowie (1971). p denotes population incidence. (From Fulker, 1973).

births in Denmark, Tienari's (1963) study using similar records in Finland. Kringlen's (1966) twins were obtained by coinciding records of twin births and psychoticism in Norway. The two studies involving PO_A were included because they allowed the effects of common environment (E_2) to be assessed. Heston's (1966) study was carried out through Oregon State hospitals in the U.S.A. and Rosenthal's (1968) in Denmark, using State records of adoption.

The concordances in Table 4.6 were tested for homogeneity of samples within the three groups to see if they could be legitimately pooled. In no case did the χ^2's give reason to suspect heterogeneity. The pooled concordance rates are given in the lower part of the Table.

These pooled concordance rates were used, together with a population incidence of 1.14%, to calculate tetrachoric correlations using a suitable approximate formula from Penrose (see Edwards, 1969).

$$r = (0.57 \log_{10} k)/(-\log_{10} p - 0.44 \log_{10} k - 0.26)$$

where p is the population incidence and k the relative incidence in relatives, k = concordance/p, concordance stated as a decimal fraction. The model fitted to these three correlations is given in Table 4.7.

This model includes additive variation (D_R), dominance variation (H_R), and common environment (E_2). The effect of assortative mating has been

TABLE 4.7

A simple model for schizophrenia.

Type of family	Correlation	S.E.	Model D_R	H_R	E_2
MZ_T	0.76	0.04	$\frac{1}{2}$	$\frac{1}{4}$	1
DZ_T	0.35	0.01	$\frac{1}{4}$	1/16	1
PO_A	0.37	0.02	$\frac{1}{4}$	0	0

(Adapted from Fulker, 1973).

excluded from the model, there being only a 2% concordance among spouses. This figure implies a marital correlation of about 0.1 which, on Fisher's model, is low enough to warrant excluding its effect. The estimates obtained were:

$$D_R = 1.48 \pm 0.14$$
$$H_R = 0.20 \pm 0.48$$
$$E_1 = 0.24 \pm 0.06$$
$$E_2 = -0.03 \pm 0.03$$

indicating that neither dominance variation nor home environment play much part in liability to schizophrenia. By refitting a simple model, excluding these nonsignificant effects, the following estimates were obtained:

$$D_R = 1.46 \pm 0.12$$
$$E_1 = 0.27 \pm 0.06$$

with a nonsignificant residual $\chi_2^2 = 1.64$, indicating a good fit of the data to the model. Both the broad and narrow heritability for liability to schizophrenia are given by

$$\tfrac{1}{2}D_R/(\tfrac{1}{2}D_R + E_1) = 73\%.$$

An approximate estimate of the number of loci controlling liability may be obtained from the expression

$$\text{Number of loci} = (\tfrac{1}{2} \text{Range})^2/D_R.$$

While the range of liability is not known, it must be at least twice the normal deviate corresponding to the mean value for schizophrenics taken as about 1% of the population. Falconer (1965) gives tables for calculating means of individuals falling under parts of the normal curve. Using these tables, the range corresponding to p = 1.14% is greater than 5.2 standard deviations which gives a number of loci in excess of 4. Since this is a gross underestimate, polygenic rather than single-gene control is clearly indicated.

While the adequacy of these estimates depends on the adequacy of the concordance rates, and these to some extent must reflect the uncertainties of diagnosis, the findings do seem consistent with other findings on schizophrenia. For example, the unimportance of home or shared environment has been noted by both Heston (1966) and Rosenthal (see Kety et al., 1971) in their studies where concordance rates for separated parents and offspring (shown in Table 4.6) were no different from rates for parents and offspring reared together, a group not included in the present analysis. Fischer (1971) used an ingenious design to show the unimportance of a schizophrenic parent in the home environment, a design involving the children of MZ twins discordant for schizophrenia. The children would have an equal probability of becoming schizophrenic for genetic reasons, only differing in whether the parent manifested schizophrenia or not. The age-corrected risk for the children was found to be 9.5% where the parent was schizophrenic and 12.3% where the parent was not. These figures strongly suggest that schizophrenia in the parent is not an important environmental determinant of risk in their children.

The 73% heritability of liability gives an accurate prediction of risk in the children of two schizophrenic parents. Using a population incidence of 1.14%. the mean of this group, in standard deviation scores, is obtained as 2.62 from Falconer's (1965) tables. Thus the midparent value is also 2.62 when both parents are schizophrenic. The regression equation predicting offspring mean is

$$\text{offspring mean} = h_n^2 \times \text{midparent value}$$
$$= 0.73 \times 2.62$$
$$= 1.91 \text{ standard deviations.}$$

The threshold value corresponding to an incidence of 1.14% is 2.28. Thus the offspring are 0.37 (= 2.28 − 1.91) standard deviations from the threshold. This value corresponds to 35.3% area under the normal curve and is the predicted risk for the children of two schizophrenic parents. Slater and Cowie (1971), collecting together all available data, found a risk of 49/134, or 36.6%, a figure in striking agreement with that predicted by the calculations above.

4.4. Conclusion

The analyses described in the three sections above have served to illustrate something of the scope of the biometrical genetic approach. It is unfortunate, however, that so little raw data are available to enable a demonstration of its full power. As shown, without raw data the analysis of correlations does not allow a full test of the assumptions underlying the model, means and variances, as well as correlations, being required. However, given these limitations, the approach reveals numerous insights into the genetic foundations of behaviour and provides a powerful basic methodology capable of considerable elaboration.

References

Burt, C. (1966): The genetic determination of differences in intelligence: A study of monozygotic twins reared together and apart. *Brit. J. Psychol.*, **57**, 137.

Burt, C. and Howard, M. (1956): The multifactorial theory of inheritance and its application to intelligence. *Brit. J. statist. Psychol.*, **8**, 95.

Cattell, R. B. (1960): The multiple abstract variance analysis equations and solutions for nature-nurture research on continuous variables. *Psychol. Rev.*, **67**, 353.

Eaves, L. J. (1970): Aspects of human psychogenetics. Ph.D. thesis, University of Birmingham, England.

Eaves, L. J. (1973): Assortative mating and intelligence: an analysis of pedigree data. *Heredity*, **30**, 199.

Eaves, L. J. and Jinks, J. L. (1972): Insignificance of evidence for differences in heritability of IQ between races and social classes. *Nature*, **240**, 84.

Edwards, J. H. (1969): Familial predisposition in man. *Brit. Med. Bull.*, **25**, 58.

Eysenck, H. J. (1971): *Race, Intelligence and Education*. Temple-Smith, London.

Falconer, D. S. (1965): The inheritance of liability to certain diseases estimated from the incidence in relatives. *Ann. Hum. Genet.*, **29**, 51.

Falconer, D. S. (1967): The inheritance of liability to diseases with variable age of onset, with particular reference to diabetes mellitus. *Ann. Hum. Genet.*, **31**, 1.

Fischer, M. (1971): Psychoses in the offspring of schizophrenic monozygotic twins and their normal co-twins. *Brit. J. Psychiat.*, **118**, 43.

Fisher, R. A. (1918): The correlation between relatives on the supposition of mendelian inheritance. *Transact. Roy. Soc. (Edinb.)*, **52**, 399.

Fulker, D. W. (1973): A biometrical genetic approach to intelligence and schizophrenia. *Soc. Biol.*, **20**, 266.

Gottesman, I. I. and Shields, J. (1966): Contributions of twin studies to perspectives in schizophrenia. In: Progress in Experimental Personality Research, pp. 1–84. B. A. Maher, ed. Academic Press, New York.

Gottesman, I. I. and Shields, J. (1967): A polygenic theory of schizophrenia. *Proc. Nat. Acad. Sci.*, **58**, 199.
Gottesman, I. I. and Shields, J. (1968): In pursuit of the schizophrenic genotype. In: Progress in Human Behaviour Genetics, pp. 67–105. S. G. Vandenberg, ed. John Hopkins Press, Baltimore.
Harvald, B. and Hauge, M. (1965): Hereditary factors elucidated by twin studies. In: Genetics and Epidemiology of Chronic Diseases, pp. 61–76. J. V. Neel, M. W. Shaw and W. J. Shull, eds. Public Health Service publication number 1163, Washington, D.C.
Heston, L. L. (1966): Psychiatric disorders in foster home reared children of schizophrenic mothers. *Brit. J. Psychiat.*, **112**, 819.
Jencks, C. (1973): *Inequality: A Reassessment of the Effect of Famil y and Schooling in America.* Allen Lane, London.
Jensen, A. R. (1970): IQ's of identical twins reared apart. *Behav. Genet.*, 1, 133.
Jensen, A. R. (1974): Kinship correlations reported by Sir Cyril Burt. *Behav. Genet.* 4, 1.
Jinks, J. L. and Eaves, L. J. (1974): IQ and inequality. *Nature*, **248**, 287.
Jinks, J. L. and Fulker, D. W. (1970): Comparison of the biometrical genetical, MAVA, and classical approaches to the analysis of human behavior. *Psychol. Bull.*, **73**, 311.
Kety, S. S., Rosenthal, D., Wender, P. H. and Schulsinger, F. (1971): Mental illness in the biological and adoptive families of adopted schizophrenics. *Amer. J. Psychiat.*, **128**, 302.
Kringlen, E. (1966): Schizophrenia in twins: An epidemiological-clinical study. *Psychiatry*, **29**, 172.
Mather, K. (1949): *Biometrical Genetics: The Study of Continuous Variation.* Methuen. London.
Mather, K. (1953): The genetical structure of populations. *Symp. Soc. Exp. Biol.*, **7**, 66,
Mather, K. and Jinks, J. L. (1971): *Biometrical Genetics: The Study of Continuous Variation*, Second edition. Chapman and Hall, London.
Penrose, L. S. (1963): *The Biology of Mental Defect*, 3rd edition. Sidgwick and Jackson London.
Reed, E. W. and Reed, S. C. (1965): *Mental Retardation: A Family Study.* Saunders, London.
Rosenthal, D., Wender, P. H., Kety, S. S., Schulsinger, F., Welner, J. and Østergard, L. (1968): Schizophrenics' offspring reared in adoptive homes. *J. Psychiat. Res.*, **6**, 377.
Scarr-Salapatek, S. (1971): Race, social class and IQ. *Science*, **174**, 1285.
Schull, W. J. and Neel, J. V. (1965): *The Effects of Inbreeding on Japanese Children.* Harper and Row, New York.
Slater, E. and Cowie, V. (1971): *The Genetics of Mental Disorders.* Oxford Univ. Press, London.
Tienari, P. (1943): Psychiatric illnesses in identical twins. *Acta psychiat. scand.*, Suppl. 171.

CHAPTER 5

The genetic basis of evolutionary changes in behaviour patterns

D. Franck

5.1. Introduction

For evolution to operate, there must exist genetic differences within a single species. Selective breeding of extreme phenotypes has shown, in a great number of studies, that phenotypic behavioural differences within one species can, to a considerable extent, be ascribed to genetic differences. Statistically significant differences have been obtained for a great variety of behavioural traits after only a few generations of selection. The majority of such experiments, as might be expected, have used species with a very rapid succession of generations like *Drosophila*, the mouse, or the rat. For example, genetic influence has been demonstrated for the following quantitative behavioural traits: intensity and sign of the phototactic (Hirsch and Boudreau, 1958; Hadler, 1964) and geotactic behaviour (Hirsch and Erlenmeyer-Kimling, 1961) of *Drosophila melanogaster*, intensity of the optomotor reaction of *Drosophila melanogaster* (Siegel, 1967), mating speed (time until copulation, following presentation of male and female to one another) of *Drosophila melanogaster* and *D. simulans* (Manning, 1961, 1968), intraspecific aggressiveness in mice (Lagerspetz and Lagerspetz, 1971), and performance of rats in certain learning experiments (Tryon, 1940, 1942; Heron, 1935, 1941; Thompson, 1954; Bignami and Bovet, 1965). Fuller and Thompson (1960) have presented an extensive review of the results of selection experiments.

Artificial selection in general results in a reduction of variation, as compared with the ancestral strain. A certain proportion of the individual differences nevertheless remains, in spite of continued selection, because it is determined by nongenetic influences. By comparison of the original variation with that which remains, an estimate can be made of the magnitude of the genetic contribution to the variation originally observed. The "heritability value" so obtained varies between zero and one. Hirsch (1970) has emphasized that a heritability estimate signifies far less than is sometimes assumed. Since the gene frequencies differ in different populations and, moreover, change from one generation to the next, the heritability value is related only to a single generation of a given population. By alteration of the environmental factors, the nongenetically determined component of individual behavioural differences, and thus the heritability value, can be considerably changed. High or low heritability implies nothing about how a behavioural trait might have developed under changed environmental conditions.

Despite the limited information given by heritability estimates, selection experiments confirm impressively the assumption that there exists considerable genetic variation within a species with respect to behaviour. If the environmental influences upon a population change, a change in the genetic situation will appear by way of natural selection. Thus the genetic variation, as Manning (1963) has put it, provides the raw material for the process of natural selection. The selection experiment may be considered as a phylogenetic model experiment, in which natural selection is replaced by artificial selection. Even when environmental conditions remain the same, evolutionary processes may operate if new kinds of recombinants arise or if the gene pool of the population is altered by mutations. As yet, only very few data on the mutability of behavioural traits are available. Newcombe and McGregor (1964) found a reduction of maze-learning ability after ancestral X-ray irradiation. Holzberg and Schröder (1972) observed that the male descendants of convict cichlid fishes (*Cichlasoma nigrofasciatum*) which had been irradiated with X-rays showed a diminution of aggressiveness, though their general locomotor activity was unchanged.

Evolutionary changes of behaviour can act as isolating mechanisms. It has been demonstrated by selective breeding that ethological isolating mechanisms can in principle be produced or at least strengthened by selection, acting upon the existing genetic variation. Partial sexual isolating between two stocks of *Drosophila melanogaster* could be achieved by removing all heterozygotes (Knight et al., 1956). The ethological isolating mechanisms of the twin species *Drosophila pseudoobscura* and *D. persimilis*, initially only weakly developed, could be considerably strengthened by removing the F_1 hybrids (Koopman, 1950; Kessler, 1966).

In contrast to the many selective breeding experiments, there are surprisingly few cases in which an intensive genetic analysis using crossing experiments has been attempted. But only when the segregation and recombination of behavioural units in the F_2 and backcross generations can be followed, the term "behavioural genetics", with its strict implications, can properly be applied. The ethologist interested in the evolution of behaviour will turn his attention primarily to the earliest steps in the process of evolution of new traits. Therefore he is primarily interested in the genetic analysis of behavioural differences in animals which are as closely related as possible. One may expect to find a relatively simple genetic situation in ecologic races or geographical subspecies of one species, or in very closely related species. With the evolution of rigid isolating mechanisms, of course, the crossing experiment cannot be applied any longer.

The genetic analysis of evolutionary changes implies that one of the two forms investigated is the more primitive phylogenetically, and the other the more highly evolved. But even the more primitive form has undergone a process of evolution following the divergence of the more evolved one. Even if one assumes that the more primitive behaviour has remained phenotypically constant, the underlying genetic situation may have changed. Therefore a genetic analysis of evolutionary changes is, in the strict sense, impossible.

5.2. Hybrid behaviour

5.2.1. Methodological limitations

Only a very few detailed genetic analyses of behavioural differences have been made. The main reason for this lies in the fact that the appearance of behavioural traits, in contrast to morphological characteristics, is not continuous in time; rather, they can be observed only under favourable environmental conditions. It may take weeks of observation to establish with considerable certainty whether a hybrid has a particular behavioural characteristic within its repertoire or not. A further difficulty is presented by the fact that qualitative behavioural differences within a species are rare. In general, intraspecific behavioural differences are only quantitative, e.g. differences in frequency of occurrence, or intensity. Furthermore, the mean values measured are not only encumbered with statistical uncertainty, but they are also dependent upon variable environmental conditions, which cannot always be kept sufficiently constant even under standardized laboratory conditions. Thus the genetic analysis of quantitative behavioural differences encounters severe methodological difficulties.

At first glance, the simplest approach appears to be the genetic analysis of qualitative behavioural differences between various species, e.g. with regard to fixed action patterns. Fixed action patterns are easily recognizable behavioural differences of constant form, which are just as characteristic for a species as are morphological characteristics. But it is possible only in a few favourable cases to produce species hybrids, and these, moreover, usually prove to be sterile. If in addition one takes into account the usual prerequisites for crossing experiments, such as easy breeding under laboratory conditions, rapid succession of generations, and adequately large numbers of descendants, only relatively few favourable experimental objects for investigations into behavioural genetics remain.

5.2.2. Demonstration of the genetic basis of behavioural differences

Many investigations of the behaviour of species hybrids show clearly that species-specific behavioural differences can, to a considerable extent, be ascribed to genetic differences, though further evidence about the genetic basis of the behavioural differences is impossible.

In some cases the behaviour of natural hybrids in the wild has been investigated. These results, however, permit only limited inference, since the genetic contributions of the two ancestral forms are unknown. The mating-calls of hybrids of various North American anurans proved to be more or less intermediate in comparison to those of the parent species. In an overlap zone of *Microhyla olivacea* and *M. carolinensis*, the calls of the presumed hybrids were intermediate with respect to duration and mid-point frequency (Blair, 1955). Similarly, the calls of a hybrid between *Bufo americanus* and *B. woodhousei* were intermediate in various respects (Blair, 1956). The mating calls of hybrids between *Bufo a. americanus* and *B. woodhousei fowleri* had intermediate pulse rates (Zweifel, 1968). In this case the hybrid character of the animals was not always morphologically recognizable, so that the pulse rates might be a more sensitive criterion for hybridization than morphological characteristics. The hybrid nature of a male *Anolis* lizard (*A. trinitatis* × *A. aeneus*), observed by Gorman (1969) in Trinidad, could be reliably confirmed on the basis of morphological, cytological, and biochemical criteria. The territorial threat behaviour directed toward intruders was almost exactly intermediate with respect to the mean interval between successive flicks of the tail, while two characteristics of the hybrid "bobbing sequence" corresponded to those of the two parent species, respectively. Even differences in the social structure of two species of baboons proved to be largely genetically determined. Nagel (1971, 1972) investigated the social behaviour of baboons in a hybrid zone of *Papio hamadryas* and *P. anubis*. In contrast to *anubis*, *hamadryas* forms harems, and the males keep their females together by means of a "guarding" behaviour. In the hybrid group, most of the harems were more or less unstable, evidently because the hybrids lacked the efficient and complete guarding technique of *hamadryas* males. The behaviour of the hybrids showed far greater individual differences than did that of the parental species, and was correlated not with the ecological situation but with the genetic composition as inferred from the external appearance.

The number of reports on the behaviour of hybrids raised in captivity, since the early publications of Lorenz (1941), Leyhausen (1950), Osche (1952), Hinde (1956), and Hörmann-Heck (1957), has increased considerably in recent years. The genetic basis of behavioural differences is most evident when clearly recognizable behavioural elements of the two parental forms appear combined in the hybrid. The F_1 hybrids of the poeciliid fishes *Xiphophorus helleri* and *X. montezumae cortezi*, for example, displayed mating behaviour which was identical to that of pure *helleri* males when they were strongly sexually excited, whereas with less sexual excitation they showed courtship patterns of *montezumae* (Franck, 1970). The F_1 hybrids of substrate- and mouth-breeding cichlid fishes of the genus *Tilapia* displayed elements of the brood-care behaviour of both parent species in a quite disorderly manner (Heinrich, 1967). A corresponding observation on nest material carrying behaviour was made by Dilger (1959, 1962a, b) for F_1 hybrids of the parrots *Agapornis fischeri* and *A. roseicollis*. The first species transports the nest material in its bill, while the second carries it tucked between the feathers of the lower back or rump. The hybrids attempted to transport the nest material in their feathers, but also carried it to the nest in their bills. In various other bird hybrids it was found that both parental species contributed elements to the courtship behaviour of the hybrids (*Tadorna tadorna* × *Mergus merganser* – F_1; Lind and Poulsen, 1963; *Anas platyrhynchos* × *A. acuta* – F_2; Sharpe and Johnsgard, 1966; domestic chicken × *Phasianus colchicus* – F_1; Stadie, 1967; F_1 and F_2 hybrids between various *Streptopelia* species; Davies, 1970). A corresponding result was found by Stadie (1969) for the aggressive and courtship behaviour of F_1 hybrids of *Gallus sonnerati* × domestic chicken.

If the two parent species differ qualitatively (i.e., certain behaviour patterns appear in only one of the species and are entirely lacking in the other), such patterns may appear in the hybrids either unchanged or more or less incomplete or weakened, or they may even be lacking altogether. Negative results of observations should, however, be interpreted with caution, since after a very long period of observation it may turn out that the behaviour pattern actually does occasionally appear. Examples of all three of the above modes of inheritance are quite numerous in the literature. Particular components of behaviour patterns, described as a unity from a comparative ethological standpoint, may be associated with entirely different modes of inheritance. Thus the sword bending in the courtship of *Xiphophorus montezumae* males does appear in the above mentioned F_1 hybrids, but the taxis component of the same pattern – the adoption of a frontal position at right angles in front of the female – does not (Franck, 1970).

If a given behavioural pattern appears in both parental species, but in somewhat different form, again various components or parameters of the behaviour pattern can be inherited in different modes. An F_1 hybrid of the crickets *Acheta rubens* and *A. veletis* had a song of intermediate pulse-rate but with chirps which were characteristic for *veletis* (Bigelow, 1960). The crowing of F_1 hybrids of the domestic chicken × *Phasianus colchicus* consisted, as in the chicken, of four syllables, but it was similar to that of the pheasant in that the separation between the syllables was more marked, and no syllable was emphasized (Stadie, 1967). In the bowing display of *Streptopelia* doves it was apparent that different elements of a single fixed action pattern were inherited in different ways. At the top of the bow the position was often reminiscent of that of one parental species and at the lowest point of the bow, of the other (Davies, 1970).

Species frequently differ from one another in ways which can only be described quantitatively – for example, with respect to the frequency of occurrence of certain elements of behaviour. The genetic basis of such quantitative differences becomes probable if in the F_1 generation intermediate values can be demonstrated. For several parameters of the songs of some different insect species the values in the F_1 generation lay between those for the parental species (chirping rate in *Nemobius allardi* × *N. tinnulus*; Fulton, 1933; average duration of the chirps in *Chorthippus brunneus* × *C.*

biguttulus; Perdeck, 1957; duration of single calls of the male in *Euscelis lineolatus* × *E. plebejus*; Strübing, 1963). Crossing experiments with cichlid fishes of the genus *Tilapia* gave similar results for the so-called contact behaviour of the young toward the parents (cf. p. 129). The frequency of the contacts and, even more strikingly, their durations were more or less intermediate (Peters, 1965; Bauer, 1968). Moreover, the frequency of spawning movements in F_1 hybrids of *Tilapia tholloni* and *T. nilotica* was intermediate between those of the parent species (Heinrich, 1967). Results for intraspecific aggressive behaviour are also consistent with these findings. In the poeciliid fish *Poecilia sphenops* the frequency of aggressive behaviour differs between the cave-dwelling form and the form living above ground; the cave form shows an almost complete loss of aggressive behaviour, whereas the values for F_1 hybrids proved to be intermediate (Parzefall, in preparation).

Occasionally, behavioural patterns in the F_1 generation are considerably more (or less) marked than in either parent species. For example, Stadie (1969) observed that F_1 cocks of *Gallus sonnerati* × domestic chicken crowed more frequently during the pauses in threat behaviour than did those of either parent species. Similarly, in the bowing display, some *Streptopelia* hybrids bowed either more or less deeply than did the doves in either parent species (Davies, 1970). Furthermore, behavioural characteristics which are common to both parent species may be completely disorganized in the F_1 hybrids. While the songs of *Streptopelia senegalensis* and *S. roseogrisea* var. *risoria* are markedly rhythmic, the hybrids have no consistent rhythm at all (Lade and Thorpe, 1964). In ducks, Ramsay (1961) found that courtship sequences which were common to both parent species could be considerably changed in the hybrids.

Sometimes the behaviours studied appear in the hybrids in a presumably ancestral form. On the basis of their observations of anatid hybrids, Heinroth (1911) and Lorenz (1959) applied the concept of atavism to animal behaviour. Lind and Poulsen (1963) interpreted numerous behavioural traits in the courtship of a shelduck (*Tadorna tadorna*) × goosander (*Mergus merganser*) hybrid as atavisms. The fact that a few highly ritualized behavioural patterns of the parent species appeared in the hybrid in an incomplete or simplified form was explained by reversion to a more primitive, less ritualized condition. The shelduck feeds by dabbling, which is regarded as primitive, the goosander, on the other hand, by diving. Since the hybrid showed feeding by dabbling only, this was interpreted as a reversion to the ancestral behaviour. In contrast to this intergeneric hybrid, hybrids of the mallard (*Anas platyrhynchos*) and pintail (*A. acuta*) were observed by Sharpe and Johnsgard (1966) to have no behavioural patterns which were not typical for one of the two parental species. Dilger (1962b) and Buckley (1969) observed that their hybrids of *Agapornis* often tried to tuck nest material at many locations of the body which were inappropriate for successful rump-carrying, and which had never been used by either parent. Since tucking elsewhere than on the rump is the normal behaviour of some more primitive *Agapornis* species and since it also appears in *Loriculus galgulus*, the behaviour of the hybrid was interpreted as atavistic. Steiner (1966) has pointed out, however, that use of the term atavism is justified only if the recapitulation of the ancestral condition can be explained embryologically or genetically. It is certainly doubtful whether such behaviour patterns of the hybrids, however primitive they may appear phenotypically, can actually be explained by a recurrence of the ancestral gene pattern. A more plausible explanation is that disruption of the parental behavioural pattern in the hybrid gives rise to forms of behaviour which only accidentally resemble the supposed primitive form.

The various ways in which behaviour, which in the parents has reached a high degree of phylogenetic adaptation, is transmitted to the hybrid often results in a complete loss of the biological significance of the behaviour. In these cases, the hybrid behaviour presents impressive demonstrations that genetically determined defects of behaviour cannot always readily be compensated by adaptive learning processes. In hybrids of substrate-breeding and mouth-breeding species of *Tilapia*, elements of the two parental types of behaviour alternated in a completely disorderly manner, so that successful breeding was quite impossible (Heinrich, 1967). One reason for the inability of Dilger's *Agapornis* hybrids to carry nest material tucked between the feathers was that a behavioural component called "tremble-shoving" was lacking (Buckley, 1969). Although the birds carried the material in their bills with increasing frequency, even after three years they still tried occasionally – and still without success – to transport it between their feathers (Dilger, 1959, 1962a).

Occasionally, incomplete behaviour patterns in hybrids corresponded to an increase in the frequency of unritualized displacement activities, as compared with the parent species; examples include bill-dipping and bill-shaking, head-shaking, body-shaking, and drinking in the courtship behaviour of the shelduck × goosander hybrid observed by Lind and Poulsen (1963), or rump-preening and head-scratching in the incomplete tucking sequences of *Agapornis* hybrids (Buckley, 1969).

Insofar as the reciprocal F_1 hybrids have been investigated, the results have usually been consistent (Osche, 1952; Perdeck, 1957; Strübing, 1963; Davies, 1970; Franck, 1970), but discrepancies have also been found. The duration of contact behaviour of young F_1 hybrids of *Tilapia nilotica* ♀ × *T. heudeloti macrocephala* ♂ lay approximately midway between the values for the two parent species, but in the reciprocal cross a strong expression of the *heudeloti* component was observed (Bauer, 1968). F_1 hybrids of *Drosophila pseudoobscura* ♀ × *D. persimilis* ♂ produced songs which were clearly classifiable as *pseudoobscura*-type, whereas the reciprocal cross produced *persimilis*-type songs. These results suggest that the most important genes responsible for the differences in song of the two species are located on the X chromosome (Ewing, 1969).

Nonuniform behaviour of species hybrids in the F_1 generation has been mentioned only by Stadie (1969). The ritualized aggressive and courtship behaviour of F_1 cocks of *Gallus sonnerati* × domestic chicken was in some animals reminiscent of *G. sonnerati*, and in others of the domestic chicken. This is possibly a consequence of the use of one domesticated form, which might be assumed to be more heterozygous for the participating genes.

Even without a detailed genetic analysis it is often possible to find further evidence for the genetic basis of species-specific behavioural differences by observing F_2 animals and the offspring of backcrosses. Thus, Sharpe and Johnsgard (1966) found in eleven F_2 hybrids of mallard × pintail a significant positive correlation between the inheritance of behavioural and plumage characteristics. Animals similar to mallards in plumage appearance also showed a corresponding courtship behaviour, and *vice versa*. On the other hand, Franck (1970) failed to demonstrate a corresponding correlation in *Xiphophorus* hybrids. A further indication of the genetic basis of species-specific behavioural differences is the larger variability in the F_2 generation as compared with the F_1, which is to be expected under the assumption of segregation and independent assortment of the relevant genes (e.g., Fulton, 1933; Schemmel, 1967; Franck, 1970; Parzefall, in preparation; Sharpe and Johnsgard, 1966; Davies, 1970). Finally, the greater similarity of backcross hybrids to the species to which the backcross was made is a further argument in favour of the genetic determination of the behavioural differences investigated (Franck, 1970; Lade and Thorpe, 1964).

5.2.3. The genetic analysis of behavioural differences

Only rarely the methodological prerequisites have been available for an attempt at genetic analyses of evolutionary changes. It is not by chance that until now all studies which satisfy this demand have been carried out with invertebrates or with fish. With these animals it is relatively easy to breed an adequate number of individuals in a short time.

Osche (1952) attempted a genetic analysis of the "waving" behaviour of larval nematodes, which is common in the genus *Rhabditis* and presumably aids the spreading of the worms by insects. The two ecological races *Rhabditis inermis inermis* and *R. i. inermoides* differ most strikingly in the waving behaviour of the larvae. While waving larvae are lacking in the subspecies *inermis*, they are found in *inermoides*. The F_1 generation, which showed waving behaviour, was backcrossed twice with *inermis*. The F_3B descendants of individual F_2B females were bred for several further generations. Then each culture was investigated for waving behaviour. Twenty-nine cultures showed waving, whereas 26 cultures did not. The ratio of approximately 1:1 suggests monofactorial inheritance. But since neither the F_2 generation was investigated nor the backcrosses to *inermoides*, the results must be interpreted with caution. Furthermore, the interpretation becomes more precarious because not the behaviour of individuals, but that of whole cultures was studied. A culture was classified as "waving" if at least a few animals showed waving behaviour.

A very careful analysis was carried out by Hörmann-Heck (1957) with species hybrids of the crickets *Gryllus campestris* and *G. bimaculatus*. The larvae differed distinctly, without overlapping, in intensity of their fighting behaviour. The results for the F_2 generation and two backcrosses indicated monofactorial inheritance. A corresponding result was found for the frequency of the trembling movements of the antennae in postcopulation display, which again is very different in the two species, without overlap among the variants. Pre-courtship sounds are characteristic of the courtship of *bimaculatus* and are entirely lacking in *campestris*; here, too, the results are in favour of the assumption of monofactorial inheritance. Oscillating movements of the head and prothorax during copulation are characteristic of *campestris* and lacking in *bimaculatus*. This is the only behavioural difference investigated in these two species for which the inheritance appears to be more complicated and which is evidently determined by several genes. Unfortunately, the final link in the chain of evidence for monofactorial determination of the three behavioural differences first cited is missing: No test was made of whether the supposedly homozygous individuals, which according to theory are to be expected in the F_2 and backcross generations, did in fact breed true. Because of the great theoretical significance of these findings, a further study is urgently required.

A genetic analysis of the juvenile contact behaviour in *Tilapia* species was attempted by Bauer (1968). *Tilapia nilotica* is a species in which the female incubates the young in her mouth. After swimming freely, the young continue to be led about by the mother; if there is danger they are taken into the mouth. Their contact behaviour is evidently connected with this relationship. Now and then, an individual young will swim from the swarm to the mother (or to an appropriate dummy) and carry out a brief contact. In *Tilapia heudeloti macrocephala* the mouth-breeding is taken over by the male. The young leave the mouth of the father at a very advanced stage of development and are independent immediately. Accordingly, *heudeloti* shows in dummy experiments only a very poorly developed contact behaviour. The variants within the two parental generations overlap only slightly. For the genetic analysis, the two reciprocal F_1 generations, the F_2 generation, and all four backcrosses were used. The interpretation of the results is difficult because of the large variation in the contact behaviour of *nilotica* and the differing results in the reciprocal F_1 generations (see p. 125). Bauer (1968) does not exclude the possibility of monogenic inheritance; in any case, he claims that relatively few genes are involved since the range of variability in the F_2 is almost as wide as that of the parent species.

The Mexican characiid fish *Astyanax mexicanus* has also been used in investigations of behavioural genetics. This species includes a series of cave populations consisting of animals which have reduced eyes and are more or less depigmented. Originally, the cave-dwellers were considered to be a separate genus and were given the generic name *Anoptichthys*. Since, however, no isolating mechanisms are known, the epigean river fish and his cave derivates must be one species. The epigean form shows a well-developed fright reaction, elicited by an alarm substance of a wounded conspecific, while in the cave-dwelling form (for which I will use the name *Anoptichthys*) the ability to react to alarm substance is lacking. All 43 F_1 hybrids tested reacted to alarm substance. Among the 210 F_2 hybrids, on the other hand, 13 fish were without a detectable fright reaction, as were 8 of the 55 hybrids resulting from the backcross to *Anoptichthys* (Pfeiffer, 1966). Thus it has been demonstrated that the degenerative evolution of the cave form has already given rise to a polygenically determined behavioural difference. Pfeiffer (1966) assumes that two dominant genetic factors segregating independently are involved. However, he emphasizes that even after extensive experiments one can state only with reservation that animals lack the fright reaction as programmed at the genetic level. Further, it proved difficult to judge the behaviour of the F_2 generation objectively, since the reaction is influenced by visual stimuli and since the eye development in the F_2 generation is widely variable. While the presence of polygenic inheritance in the present case seems to be granted, further information about the number of participating genes is not available with certainty.

In an investigation of the feeding behaviour of *Astyanax mexicanus*, Schemmel (1967) avoided the influence of visual stimuli by blinding the experimental animals. While a blinded *Astyanax* in search of food on the substrate stands vertically, *Anoptichthys* searches the bottom at an angle of 45°. All the F_1 hybrids took the position of *Astyanax*. In the F_2 generation however much individual variation was observed. Most of the F_2 individuals behaved more or less intermediately, while the behaviour of others was like that of one or the other parental form. These behavioural differences were not associated with the distribution of taste buds. Even though a detailed analysis is lacking, the results are favouring the assumption of a cumulative interaction of several genes.

Poeciliid fishes of the genus *Xiphophorus* proved to be a favourable object for the genetic analysis of behavioural differences between closely related species (Franck, 1969, 1970). The three swordtail species (= subgenus *Xiphophorus*) and five platy species (= subgenus *Platypoecilus*) can, with slight limitations, easily be crossed with one another in the laboratory, including crosses between swordtails and platys. The F_1 hybrids proved in many cases to be fertile. Despite the evidently very close relationship of all these species, the courtship displays of the males, in particular, have diverged rapidly in evolution, so that the various species differ not only quantitatively in this respect, but also qualitatively.

After preliminary investigations of numerous other hybrids, the two species *X. helleri* and *X. montezumae cortezi* were chosen for a genetic analysis, since on the one hand they are particularly closely related and, on the other, show clearly distinguishable behavioural patterns.

In the "backing courtship" of *X. helleri*, the male swims very quickly in front of the female. The dorsal and caudal fin are spread, as is the ventral fin, which is turned toward the female. The male then swims more slowly backward toward the female with fins folded, and subsequently makes another very quick forward movement. If the female does not flee, the male alternates several times in succession between forward and backward phases. The orientation of the courting male to the female is extremely variable in *helleri*. A starting position in front of the female is most frequent, but it is not uncommon for the male to swim behind the female, so that it is oriented antiparallel to the female, and then proceeds to engage in "backing from behind". The alternating *helleri*-type courtship is almost entirely lacking in the strain of *X. montezumae cortezi* investigated. In the first phase of the *montezumae* courtship the male approaches the female, simultaneously bending the sword-like caudal appendage away from the female. This "sword bending" is characteristic of all *montezumae* forms and does not appear in any other species of *Xiphophorus*. In the next phase of courtship, the male swims in a "frontal position" at right angles to the female and continues to bend the sword away; the ventral fin turned toward the female is raised. Raising of the ventral fin is quite strictly associated with taking up the frontal position. In the subsequent phases of courtship the male swims backwards toward the female, but does not alternate between forward and backward phases as occurs in the *helleri* courtship.

The 27 F_1 hybrids of *helleri* ♀ × *montezumae* ♂ studied clearly combined behavioural characteristics of maternal and paternal origin. The entire *helleri* courtship, in all its details, was passed on by females of the parent generation to their F_1 sons. The *montezumae* courtship, on the other hand, was only incompletely developed in the F_1 males. All 27 F_1 males displayed sword bending but never simultaneously with the frontal position. In contrast to the parent species, raising of the ventral fin was not connected to the frontal position at right angles to the female. The genetic analysis was aided by the fact that the *helleri* and *montezumae* elements were not superposed in the behaviour pattern of the F_1 hybrids but, rather, were associated with different levels of sexual motivation. The *montezumae* elements appeared at a relatively low level of sexual motivation; as reactivity increased they were replaced by a typical alternating *helleri* courtship. The F_1 hybrids thus switched between *montezumae*- and *helleri*-type courtship patterns. Only in a brief transition phase were *montezumae* and *helleri* elements superimposed. The total behavioural pattern thus produced was very similar to that of *X. montezumae montezumae* (cf. Franck, 1968). The main difference was the absence of the frontal position.

For the genetic analysis, 37 F_2 hybrids, 8 male hybrids of a backcross with *helleri*, and 26 male hybrids of backcrosses with *montezumae* (12 F_2B, 9 F_3B, 5 F_4B) were used. Three behavioural characteristics seemed particularly appropriate for analysis, since they could be observed very frequently in one parental species and were completely lacking in the other: "backing from behind" (a characteristic of *helleri*), "sword bending" (a characteristic of *montezumae*), and the adoption of a "frontal position" at right angles in front of the female, which is associated with sword bending in *montezumae*. All three of these characteristics segregated and recombined independently.

The results in each case may be summarized as follows (cf. Franck, 1970):

1. Backing from behind was observed in 35 F_2 males and was absent in only two males. With monofactorial inheritance, this ratio should have been 3:1. The difference to the theoretical 3:1 ratio is statistically significant at the level $P < 0.02$. Since it is not possible to decide with absolute certainty whether the two males which were never seen to perform backing from behind actually lacked this behaviour entirely, further information about the number of genes involved cannot be obtained. With certainty, the behavioural difference is determined polygenically.

2. Sword bending was also present in 35 F_2 males and lacking in two. This behavioural difference, too, is determined polygenically. Here again, no further conclusions about the number of genes can be drawn. For the two males without sword bending, the observations were continued over a long period. One finally did show signs which could be interpreted as sword bending at a very low intensity. Even without considering males displaying no sword bending at all, the intensity measured as percent of sword bending at full intensity, varied much more in the F_2 than in the F_1 generation. Further, a quantitative analysis of sword bending showed that in the backcross with *helleri* a marked decrease in intensity and frequency appeared in males actually displaying sword bending, whereas in the backcrosses with *montezumae* there was an increase. Therefore, since the behavioural pattern did not appear alternately at full strength or not at all but, rather, its intensity and frequency seemed to depend upon predominance of the *montezumae* or *helleri* genome, it may be supposed that there is – at least in part – a cumulative effect of the genes involved.

3. The frontal position, at right angles in front of the female, was completely suppressed in the F_1 generation and appeared in only one male of the F_2 generation, evidently with lowered intensity. Therefore this behavioural difference is also determined polygenically. At first sight, one might explain the result by cooperation of two recessive alleles, but in that case,

in the generation resulting from a backcross with *montezumae*, only every fourth male ought to show the frontal position. In the case of a larger number of recessive alleles, the proportion of F_2B males exhibiting the frontal position should be even smaller. In fact, however, 10 out of 12 backcross males showed this behavioural pattern. The departure from expectation is highly significant ($P < 0.001$) in spite of the small numbers. The surprisingly frequent appearance of the frontal position in the generation produced by a backcross with *montezumae* rules out the possibility to explain the lack of this behaviour pattern in the F_1 generation by its dependence upon recessive alleles. A possible explanation is that the manifestation of the frontal position in the F_1 is prevented by the effects of restriction genes (modifiers) from the *helleri* genome.

The investigations of *Xiphophorus* hybrids, apart from demonstrating the genetic basis of numerous interspecific behavioural differences, show further that the presence or absence of clearly distinguishable behavioural units cannot be ascribed to single genetic units, but rather that the qualitative behavioural differences investigated are determined polygenically.

5.3. Discussion

5.3.1. What does hybrid behaviour say about the problem of genetic and environmental determination of behaviour traits?

Numerous behavioural studies using species hybrids or forms which have been brought about by a process of evolution have demonstrated that the observed behavioural differences could at least in part be ascribed to genetic differences. Thus the investigations confirm that the observed behavioural differences can be considered as a result of evolution. But a particular behavioural difference cannot be described as determined purely genetically unless its dependence upon a specific number of Mendelian factors can be demonstrated. Moreover, such a conclusion is valid only for the environmental conditions of the experiment. Any alteration of these may affect the behavioural difference investigated. In order to exclude environmental influences, the geneticist attempts to keep the environmental conditions as invariant as possible. As a consequence, his findings cannot clarify any dependence of the behavioural differences upon the environment.

It has often been emphasized that crossing experiments are concerned only with behaviour differences. It would be quite incorrect to dismiss this argument as sophistry. Even the demonstration of monofactorial inheritance of behaviour differences does not mean that environmental conditions or learning processes do not participate in the ontogenesis of behaviour in an individual. For example, the genetic finding that differences among species-specific behavioural patterns are entirely ascribable to particular genes does not rule out the possibility that social experience may be required for the expression of the genes in a given individual, unless the acquisition of social experience has been excluded. This problem can be solved only by social-deprivation experiments. Genetic and ontogenetic studies are not alternatives, but have to complement one another.

Quite commonly in the literature, the meaningless statement is made that a behaviour pattern is determined genetically or by the environment. The realization by geneticists that genes control only the reaction norm with respect to the environment has still not thoroughly pervaded the field of behavioural sciences, nor has the realization by ethologists, obvious to a biologist, that even learned forms of behaviour are not without a genetic basis. Lorenz (1961) stressed that the disposition to learn must be considered as a result of evolution in just the same way as are fixed action patterns. Thus both the genome and the environment are always involved in the ontogenesis of a behavioural trait. The task of behavioural genetics is to study the genetic alterability of behaviour under given environmental conditions. This by no means makes superfluous those investigations into the degree to which behaviour can be environmentally influenced that keep the genetic situation as constant as possible.

5.3.2. *The polygenic basis of evolutionary changes*

The intraspecific genetic component of behavioural variation which can be demonstrated in selection experiments generally concerns relatively inconspicuous quantitative differences. Hirsch and coworkers (most recently Hirsch and Ksander, 1969) were able, by means of a careful crossing analysis, to estimate the relative influences of single chromosomes upon individual differences in the development of geotactic behaviour in *Drosophila melanogaster*. The three chromosomes studied, II, III, and X (the small chromosome IV was not considered), all changed their influence on the geotactic behaviour in the course of the selection experiments. Thus, the intraspecific individual differences with respect to intensity and sign of the geotactic reaction depend on several genes which are distributed over most of the *Drosophila* genome. If evolutionary alterations would proceed from such distributed patterns of genes, it would not be surprising that behavioural differences are polygenic in nature even in the earliest stages of evolution. Efforts toward a genetic analysis of behavioural differences produced by evolutionary processes, sparse though they still are, in general tend to speak in favour of polygenic inheritance, regardless of whether the behavioural differences concerned are those between ecologic races of the same species or between closely related species. Contrasting with this view are the results of Hörmann-Heck (1957) in crickets, where three of the four characteristic differences investigated showed monofactorial inheritance. Here, though, one could in fact be dealing with gene-blocks consisting of a complex of closely linked single genes. In the waving behaviour of *Rhabditis* (Osche, 1952) the genetic situation could be similar. Demonstration of monofactorial inheritance, of course, does not mean that the individual ontogenesis of the behaviour is dependent only on one pair of alleles. The genes controlling ontogenesis of either parental form cannot be specified in crossing experiments. In the case of monofactorial inheritance, therefore, only one of the many genes participating in ontogenesis is known, because the two ancestral forms differ in this one gene only.

A great many discoveries from the comparative study of behaviour are in favour of the idea that single mutations generally bring about only a slight, inconspicuous change in behaviour and therefore are usually not recognizable. Qualitative behavioural differences within a species and, moreover, in many cases even between closely related species are rare. On the other hand, quantitative behavioural differences (e.g., differences in frequency or intensity of single behavioural patterns) are quite common. The same phenomenon can be observed in domestication, which may be considered as a large-scale model experiment in evolutionary biology. Compared with the vast number of quantitative behavioural differences, the qualitative differences are rather rare (e.g., Immelmann, 1962, 1963; Kruijt, 1964; Herre and Röhrs, 1970; Zimen, 1971). Scott and Fuller (1965), in their extensive investigations of genetically determined behavioural differences between various races and breeds of dogs, also came to the conclusion that it is impossible to find even one behavioural difference by which two forms could be distinguished clearly and without overlap. Through artificial selective breeding, certain behavioural traits may be enhanced or weakened, but nothing new has arisen.

Manning (1965), on the basis of similar considerations, developed the hypothesis that single genes always bring about only small threshold alterations which accumulate in the course of evolution. If the frequency of a behaviour decreases, it is because of a diminished reactivity, and thus an increase in threshold, which only in extreme cases may result in the complete disappearance of the behaviour under normal environmental conditions. Manning argues that even a great many other evolutionary changes can be ascribed to threshold changes. For example, the change of emphasis within a pattern could result from threshold changes on the motor side of a mechanism, such that particular muscle groups come into action earlier, or later, for a shorter or longer time, and so on. However, we are still a long way from a physiological understanding of those changes which can be described as threshold changes. It must be admitted that we have so far been unable in a single case to make even moderately precise inferences about the action of the genes. The best hope for advances in this area lies in a closer collaboration between behavioural genetics and neurophysiology.

Although single genes generally bring about only slight quantitative changes in behaviour, a few genes with comparatively large effects are known. In mouse strains many, usually recessive, behavioural mutations are known, which through defects in the vestibular and cochlear apparatus or localizable malformations in the nervous system produce characteristic (loco-)motor anomalies (Grüneberg, 1952; Deol, this volume). In the honeybee there are presumably two independently segregating and recombining genes which cause a breakdown of the resistance to American foulbrood disease. One gene prevents opening of the infected cells by the worker bees and the other removal of the dead larvae (Rothenbuhler, 1964). The selective disadvantage of these genetic changes is obvious.

In summary, there is no doubt that species-specific behaviour patterns are extraordinarily stable genetically. The appearance of genetic changes which impair to a catastrophic degree the probability of survival of their carrier is a rare event. An illuminating explanation of this fact may be the existence of complex polyfactorial modes of inheritance. One must assume that the genetic system underlying species-specific behaviour is structured in such a way that single mutations can drastically alter it only rarely, or not at all. Where several genes interact cumulatively, single changes in the gene-system would result only in quantitative changes in behaviour. A single mutational event could never entirely extinguish the behavioural difference. Thus far, the discoveries of behavioural genetics indicate that evolutionary changes in behaviour patterns arise, in general, step by step through complex alterations in a system of genic interactions which is most effectively stabilized.

5.4. Summary

Evolutionary processes operate on a basis of genetic variation. Direct information on the genetic background of behavioural evolution can be obtained by studying the behaviour of hybrids, particularly species hybrids. Since under natural conditions such hybrids occur rarely and, moreover, are of unknown descent, a detailed genetic analysis becomes virtually impossible in them. Hybridization experiments under laboratory conditions have also their methodological limitations, and evidence that genetic factors control the behavioural differences as observed in a given environment does not exclude the possibility that environmental factors and learning processes, too, are involved in the ontogenesis of these behaviour patterns. Despite these difficulties, it is demonstrated that in some cases, using a suitable object, a genetic analysis is feasible. A short overview of our results with different *Xiphophorus* species is given and the genetic steps in evolutionary change are discussed. The polygenic basis of small behavioural differences will prevent rapid evolutionary changes. Interspecific behavioural differences are stabilized to the extent that they rarely can be eliminated by single-gene mutations.

References

Bauer, J. (1968): Vergleichende Untersuchungen zum Kontaktverhalten verschiedener Arten der Gattung *Tilapia* (Cichlidae, Pisces) und ihrer Bastarde. *Z. Tierpsychol.*, **25**, 22.

Bigelow, R. S. (1960): Interspecific hybrids and speciation in the genus *Acheta*. *Can. J. Zool.*, **38**, 509.

Bignami, G. and Bovet, D. (1965): Expérience de sélection par rapport à une réaction conditionnée d'evitement chez le rat. *C.R. Acad. Sci. (Paris)*, **260**, 1239.

Blair, W. F. (1955): Mating call and stage of speciation in the *Microhyla olivacea-M. carolinensis* complex. *Evolution*, **9**, 469.

Blair, W. F. (1956): The mating calls of hybrid toads. *Tex. J. Sci.*, **8**, 350.

Buckley, P. A. (1969): Disruption of species-typical behavior patterns in F_1 hybrid *Agapornis* parrots. *Z. Tierpsychol.*, **26**, 737.

Davies, S. J. J. F. (1970): Patterns of inheritance in the bowing display and associated behaviour of some hybrid *Streptopelia* doves. *Behaviour*, **36**, 187.

Dilger, W. C. (1959): Nest material carrying behavior of F_1 hybrids between *Agapornis fischeri* and *A. roseicollis*. *Anat. Rec.*, **134**, 554 (abstract).

Dilger, W. C. (1962a): The behavior of lovebirds. *Sci. Amer.*, **206**, 88.

Dilger, W. C. (1962b): Behavior and genetics. In: Roots of Behavior. E. L. Bliss, ed. (Harper and Brothers, New York) pp. 35–47.

Ewing, A. W. (1969): The genetic basis of sound production in *Drosophila pseudoobscura* and *D. persimilis*. *Anim. Behav.*, **17**, 555.

Franck, D. (1968): Weitere Untersuchungen zur vergleichenden Ethologie der Gattung *Xiphophorus* (Pisces). *Behaviour*, **30**, 76.

Franck, D. (1969): Genetische Grundlagen der Evolution tierischer Verhaltensweisen. *Zool. Anz.*, **183**, 31.

Franck, D. (1970): Verhaltensgenetische Untersuchungen an Artbastarden der Gattung *Xiphophorus* (Pisces). *Z. Tierpsychol.*, **27**, 1.

Fuller, J. L. and Thompson, W. R. (1960): Behavior Genetics. (Wiley, New York) pp. 1–396.

Fulton, B. B. (1933): Inheritance of song in hybrids of two subspecies of *Nemobius fasciatus* (Orthoptera). *Ann. Entom. Soc. Amer.*, **26**, 368.

Gorman, G. C. (1969): Intermediate territorial display of a hybrid *Anolis* lizard (Sauria: Iguanidae). *Z. Tierpsychol.*, **26**, 390.

Grüneberg, H. (1952): The Genetics of the Mouse, 2nd ed. (Nijhoff, The Hague) pp. 1–650.

Hadler, N. M. (1964): Genetic influence on phototaxis in *Drosophila melanogaster*. *Biol. Bull.*, **126**, 264.

Heinrich, W. (1967): Untersuchungen zum Sexualverhalten in der Gattung *Tilapia* (Cichlidae, Teleostei) und bei Artbastarden. *Z. Tierpsychol.*, **24**, 684.

Heinroth, O. (1911): Beiträge zur Biologie, namentlich Ethologie und Psychologie der Anatiden. *Verh. V. Int. Ornithol. Kongr.*, Berlin. pp. 589–702.

Heron, W. T. (1935): The inheritance of maze learning ability in rats. *J. comp. Psychol.*, **19**, 77.

Heron, W. T. (1941): The inheritance of brightness and dullness in maze learning ability in the rat. *J. genet. Psychol.*, **59**, 41.

Herre, W. and Röhrs, M. (1970): Experimentelle Beiträge zur Stammesgeschichte der Vögel. Ergebnisse zoologischer Domestikationsforschung. *J. Orn.*, **111**, 1.

Hinde, R. A. (1956): The behaviour of certain cardueline F_1 interspecies hybrids. *Behaviour*, **9**, 202.

Hirsch, J. (1970): Behavior-genetic analysis and its biosocial consequences. *Seminars in Psychiatry*, **2**, 89.

Hirsch, J. and Boudreau, J. C. (1958): The heritability of phototaxis in a population of *Drosophila melanogaster*. *J. comp. physiol. Psychol.*, **51**, 647.

Hirsch, J. and Erlenmeyer-Kimling, L. (1961): Sign of taxis as a property of the genotype. *Science*, **134**, 835.

Hirsch, J. and Ksander, G. (1969): Studies in experimental behavior genetics: V. Negative geotaxis and further chromosome analyses in *Drosophila melanogaster*. *J. comp. physiol. Psychol.*, **67**, 118.

Holzberg, S. and Schröder, J. H. (1972): Behavioural mutagenesis in the convict cichlid fish, *Cichlasoma nigrofasciatum*. I. The reduction of male aggressiveness in the first postirradiation generation. *Mutation Res.*, **16**, 289.

Hörmann-Heck, S. v. (1957): Untersuchungen über den Erbgang einiger Verhaltensweisen bei Grillenbastarden (*Gryllus campestris* ∼ *Gryllus bimaculatus*). *Z. Tierpsychol.*, **14**, 137.

Immelmann, K. (1962): Vergleichende Beobachtungen über das Verhalten domestizierter Zebrafinken in Europa und ihrer wilden Stammform in Australien. *Z. Tierzücht.*, **77**, 198.

Immelmann, K. (1963): Wie verändert die Domestikation das Verhalten der Tiere? Allgemeine Erkenntnisse aus der Beobachtung domestizierterPrachtfinken. *Umschau*, **63**, 590.

Kessler, S. (1966): Selection for and against ethological isolation between *Drosophila pseudoobscura* and *Drosophila persimilis*. *Evolution*, **20**, 634.

Knight, G. R., Robertson, A. and Waddington, C. H. (1956): Selection for sexual isolation within a species. *Evolution*, **10**, 14.

Koopman, K. F. (1950): Natural selection for reproductive isolation between *Drosophila pseudoobscura* and *Drosophila persimilis*. *Evolution*, **4**, 135.

Kruijt, J. P. (1964): Ontogeny of social behaviour in Burmese Red Jungle fowl. *Behaviour, Suppl. XII*, 1.

Lade, B. I. and Thorpe, W. H. (1964): Dove songs as innately coded patterns of specific behaviour. *Nature*, **202**, 366.

Lagerspetz, K. M. J. and Lagerspetz, K. Y. H. (1971): Changes in the aggressiveness of mice resulting from selective breeding, learning and social isolation. *Scand. J. Psychol.*, **12**, 241.

Leyhausen, P. (1950): Beobachtungen an Löwen-Tiger-Bastarden mit einigen Bemerkungen zur Systematik der Großkatzen. *Z. Tierpsychol.*, **7**, 46.

Lind, H. and Poulsen, H. (1963): On the morphology and behaviour of a hybrid between goosander and shelduck (*Mergus merganser* × *Tadorna tadorna*). *Z. Tierpsychol.*, **20**, 558.

Lorenz, K. (1941): Vergleichende Bewegungsstudien an Anatinen. *J. Ornith.*, **89**, 194.

Lorenz, K. (1959): Psychologie und Stammesgeschichte. In: Die Evolution der Organismen, Vol. 1. G. Heberer, ed. (Fischer, Stuttgart) pp. 131–172.

Lorenz, K. (1961): Phylogenetische Anpassung und adaptive Modifikation des Verhaltens. *Z. Tierpsychol.*, **18**, 139.

Manning, A. (1961): The effects of artificial selection for mating speed in *Drosophila melanogaster*. *Anim. Behav.*, **9**, 82.

Manning, A. (1963): Evolutionary changes and behaviour genetics. Proc. XIth Internat. Congr. Genetics, vol. 3. (Oxford, Pergamon) pp. 807–815.

Manning, A. (1968): The effects of artificial selection for slow mating in *Drosophila simulans*. I. The behavioural changes. *Anim. Behav.*, **16**, 108.

Nagel, U. (1971): Social organization in a baboon hybrid zone. Proc. 3rd. Internat. Congr. Prim., vol. 3 (Karger, Basel) pp. 48–57.

Nagel, U. (1972): Verhaltensbeobachtungen an freilebenden Pavian-Bastarden. *Umschau*, **72**, 560.

Newcombe, H. B. and J. F. McGregor (1964): Learning ability and physical well-being in offspring from rat populations irradiated over many generations. *Genetics*, **50**, 1065.

Osche, G. (1952): Die Bedeutung der Osmoregulation und des Winkverhaltens für freilebende Nematoden. *Z. Morph. Ökol. Tiere*, **41**, 54.

Parzefall, J. (in preparation). Zur Genetik des Aggressionsverhaltens. Untersuchungen an Bastarden ober- und unterirdischer Populationen von *Poecilia sphenops* (Pisces).

Perdeck, A. C. (1957): The isolating value of specific song patterns in two sibling species of grasshoppers (*Chorthippus brunneus* and *C. biguttulus*). *Behaviour*, **12**, 1.

Peters, H. M. (1965): Angeborenes Verhalten bei Buntbarschen. *Umschau*, **65**, 665 und 711.

Pfeiffer, W. (1966): Über die Vererbung der Schreckreaktion bei *Astyanax* (Characidae, Pisces). *Z. Vererbungsl.*, **98**, 97.

Ramsay, A. O. (1961): Behaviour of some hybrids in the mallard group. *Anim. Behav.*, **9**, 104.

Rothenbuhler, W. C. (1964): Behavior genetics of nest cleaning in honey bees. IV. Responses of F_1 and backcross generations to disease-killed brood. *Amer. Zool.*, **4**, 111.

Schemmel, C. (1967): Vergleichende Untersuchungen an den Hautsinnesorganen ober- und unterirdisch lebender *Astyanax*-Formen, ein Beitrag zur Evolution der Cavernicolen. *Z. Morph. Tiere*, **61**, 255.

Scott, J. P. and Fuller, J. L. (1965): Genetics and the Social Behavior of the Dog. (Univ. Chicago Press, Chicago and London) pp. 1–468.

Sharpe, R. S. and Johnsgard, P. A. (1966): Inheritance of behavioral characters in F_2 mallard × pintail (*Anas platyrhynchos* × *Anas acuta*) hybrids. *Behaviour*, **27**, 259.

Siegel, I. H. (1967): Heritability and threshold determinations of the optomotor response in *Drosophila melanogaster*. *Anim. Behav.*, **15**, 299.

Stadie, C. (1967): Verhaltensweisen von Gattungsbastarden *Phasianus colchicus* × *Gallus gallus f. domestica* im Vergleich mit denen der Ausgangsarten. *Zool. Anz.*, *31st Suppl.*, 493.

Stadie, C. (1969): Vergleichende Beobachtungen an Verhaltensweisen verschiedener Wildhuhnarten der Gattung *Gallus*. *Zool. Anz.*, **183**, 13.

Steiner, H. (1966): Atavismen bei Artbastarden und ihre Bedeutung zur Feststellung von Verwandtschaftsbeziehungen. Kreuzungsergebnisse innerhalb der Singvogelfamilie der Spermestidae. *Rev. Suisse Zool.*, **73**, 321.

Strübing, H. (1963): Lautäußerungen von *Euscelis*-Bastarden (Homoptera-Auchenorhyncha). *Verh. Dtsch. Zool. Ges.*, *Zool. Anz.* 27, Suppl. 268.

Thompson, W. R. (1954): The inheritance and development of intelligence. *Res. nerv. ment. Dis.*, **33**, 209.

Tryon, R. C. (1940): Genetic differences in maze learning in rats. In: National Society for the Study of Education, 39th yearbook. (Public School Publishing Co., Bloomington, Ill.) pp. 111–119.

Tryon, R. C. (1942): Individual differences. In: Comparative Psychology. F. A. Moss, ed. (Prentice-Hall, New York) pp. 330–365.

Zimen, E. (1971): Wölfe und Königspudel, vergleichende Verhaltensbeobachtungen. Studies in Ethology. Piper, München, pp. 1–257.

Zweifel, R. G. (1968): Effects of temperature, body size, and hybridization on mating calls of toads, *Bufo a. americanus* and *Bufo woodhousii fowleri*. *Copeia* **1968**, pp. 269–285.

CHAPTER 6

A behavioural variant in the three-spined stickleback

P. Sevenster and M. 't Hart

6.1. Introduction

In two laboratory strains of sticklebacks (*Gasterosteus aculeatus* L.) we found a deviation in behaviour which seemed to offer an interesting subject for genetical analysis. Although in sticklebacks the span of a generation is much longer than that in *Drosophila*, it can be made shorter than in Mendel's peas under suitable laboratory conditions (5–7 months). Thus we found it quite feasible to carry out genetical work with this species. Before reporting on our study, we intend to provide some general information on the stickleback's biology, especially on its breeding cycle. Since this cycle has been described frequently, we will restrict ourselves to the briefest outline and refer the interested reader to ter Pelkwijk and Tinbergen (1937) and van Iersel (1953).

6.1.2. Breeding cycle

In spring, the male sticklebacks become brightly coloured and establish territories in shallow waters. They dig a nest-pit by removing mouthfuls of sand and deposit plant fragments and similar material on the bottom of this pit. The material is glued together with mucous threads which are secreted from the kidney and drawn over the nest by the male in a sliding sort of motion. Since he also repeatedly bores the snout into the nest at one side, and in a constant direction, the nest acquires an entrance. After building has gone on for some time (a matter of hours), the male on one occasion will not withdraw from the entrance after boring but, by beating his tail vigorously, wriggle through the nest and leave it at the other side. It is with this movement of creeping through that we will be concerned in this chapter; a more detailed description follows below. The nest has now become a tunnel with an exit and an entrance and it is ready to receive a female for spawning.

Females wander about individually or in shoals; when ripe they are induced by the males of the territories they come across to follow them to the nest and enter the tunnel to spawn. After spawning, the female leaves by the exit and the male goes through as well, releasing his sperm in the tunnel. The male then chases the female from his territory, and he is now considered as having entered the "parental phase" in which he guards the eggs and ventilates them until hatching, which occurs in a week or so. With this period we are not concerned; an extensive treatment can be found in van Iersel (1953). The entire breeding cycle – from the establishment of a territory till hatching – can easily be observed under laboratory conditions in a comparatively small aquarium. In this paper we will deal with the "sexual phase" only, i.e. the period between the first occurrence of creeping through (when the nest can be said to be completed) and the presence of fertilized eggs. Accordingly, by preventing spawning, the sexual phase can be prolonged (usually for some weeks) until the male builds a new nest. Throughout this period the male remains ready to court a ripe female in his territory.

6.1.3. Courtship

In this species, too, courtship is a complicated sequence of interactions between male and female which we shall not analyze in detail. Here the term is used exclusively for the behaviour of a male towards a ripe female which is presented in a glass tube at some distance (30 cm) from the nest. This is a standard method to prevent the female from following the male and to keep the initial stages of the male's courtship going on for some time. Even in the glass tube, the ripe female will adopt her courtship posture when the male approaches: head-up and turning towards the male, she presents a fairly constant stimulation. Typically, the male reacts to this situation by shuttling back and forth between the nest and the female; he usually approaches the female with a "zig-zag dance", but returns to the nest on a smooth course. When close to the female, he may show some aggression, bumping or snapping at her, and at the nest-site various activities may be displayed. Of these, only creeping through will be dealt with at some length, since it is the centre of our investigation.

6.1.4. Creeping through

In the sexual phase creeping through (Fig. 6.1), which marks its beginning, is repeated from time to time. It commonly follows boring, and it might be considered a continuation of boring. There is, however, a rather marked, abrupt transition, suggesting itself as the proper start of creeping through. In boring, the tail is moving to and fro in a high frequency and with a low amplitude, almost giving an impression of quivering. When creeping through follows, this abruptly changes into rather slow, clearly distinct beats of large amplitude, creating an impression of a considerable effort. These tail-beats come in bouts alternating with pauses in which the fish lies perfectly still, usually with the head already protruding from the exit. Finally the male leaves the nest in a gliding motion, propelling himself with fluttering pectoral fins, the tail trailing behind. The moment the tail has left the exit is taken as the termination of creeping through, since the fish then resumes normal locomotion. On the basis of these criteria the duration of the activity can be accurately measured.

Fig. 6.1. Creeping through.

An important point is the way in which the male's entire behaviour is influenced by creeping through. There are pronounced fluctuations in a number of activities, contingent on the occurrence of creeping through (van Iersel, 1953; Sevenster, 1961; Nelson, 1965). The interval between two such occurrences may be one to several hours, depending on the individual and on external circumstances. Successive intervals in one animal are fairly constant, so that the fluctuations mentioned are cyclic. The interval is considerably shortened, however (or the cycle accelerated), during intensive courtship, when it becomes about 10–60 minutes, the shortest interval ever found being 7 minutes. Often creeping through occurs immediately after a female is presented. The duration of creeping through is also shortened in courtship: about 10 sec instead of the usual 20 sec or more.

6.1.5. "Double" creeping through

So far we can establish that the act of creeping through is always followed by a "refractory" period of at least 7 min in which no creeping through occurs. However, in one laboratory strain a male was discovered which regularly crept through two, or even three or four times in quick succession, usually during the same visit to the nest. In that case, after leaving the exit, he would turn around and immediately pass through the tunnel again so that no more than a few seconds elapsed between the successive passages. In other cases the fish would leave the nest after the first passage and creep through again on the next visit after an interval of about 15–20 seconds. However, after such a bout of creeping through the male entered a refractory period of normal duration. In this respect a bout thus seems to correspond to the single passage in a normal occurrence of creeping through and it is therefore considered as one occurrence but in an aberrant form called "double creeping through", in which "double" means "two or more passages in quick succession". It should be stressed that in the same fish single passages were also frequently observed.

The aberrant fish was used for raising the next generation of our strain and it soon became apparent that practically all males of this strain showed "double creeping through" on occasion. This suggested a genetical basis for the aberration. Meanwhile, however, it was found that also in another laboratory strain (8th generation) the males often showed "double creeping through", which seemed to suggest that laboratory conditions were somehow responsible for the phenomenon. It is the aim of this study to investigate some factors underlying the double creeping through.

6.1.6. Morphological features and their geographical distribution

The taxonomy of *Gasterosteus aculeatus* L. is still rather incomplete. A number of types are distinguished according to the number of lateral bony plates (Bertin, 1925). Three of these types are found in N.W. Europe: forma trachura with plates from the pectoral fin to the tail (29–42 on the two sides

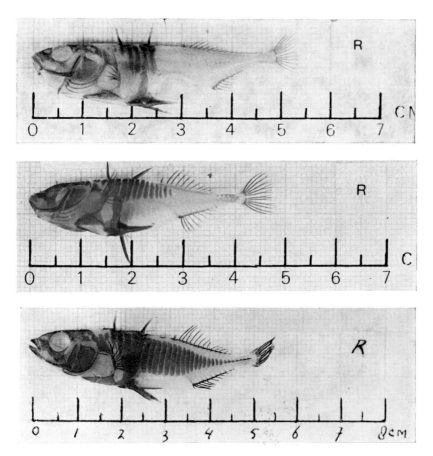

Fig. 6.2. Stained preparations; forma leiura (top); forma semi-armata (middle); forma trachura (bottom).

together; Heuts, 1947), forma leiura with only a few plates near the pectoral fin (1–9), and forma semi-armata with plates extending from the pectoral fin about halfway to the tail (9–29); see Fig. 6.2. It has been shown that forma semi-armata is the hybrid of f. trachura and f. leiura (Münzing, 1959). The coastal populations of N.W. Europe are polymorphic with respect to plate number: they comprise all three forms in variable proportions. These populations are anadromic: most of the fish winter in the sea and the estuaries, and in the spring they return to fresh water for breeding. The populations living further inland do not migrate and comprise only the forma leiura (Heuts, 1947; Münzing, 1959).

The polymorphic coastal populations differ from the monomorphic inland populations also in other respects. We found a considerable difference in the colour pattern of the skin, which is especially striking in the young (Fig. 6.3). Young of coastal origin show at most a faint pattern of bars;

Fig. 6.3. Fish of coastal populations: T-strain (top) and fish of inland populations: L-strain (bottom).

young of inland populations show a more contrasting, but also more chequered pattern of blotches rather than bars.

6.2. Material and methods

We have made use of four strains of sticklebacks: (1) The first derives from fish originally caught in a brook near Vaassen, central part of The Netherlands. This strain is called V; it consists entirely of fish of the leiura form, and they show the blotched pattern characteristic of inland populations. It is normal with respect to creeping through. (2) The second strain is called L; it derives from the same population in Vaassen, but differs from V by a high incidence of creeping through. (3) The third strain derives from fish of the coastal populations, caught near Den Helder, N.W. part of The Netherlands. The strain is called T and it consists of trachura forms only and shows the inconspicuous barred pattern; it is characterized by frequent double creeping through. (4) A number of individuals caught in the wild from a coastal population in the island of Tholen, S.W. part of The Netherlands. They were normal with respect to creeping through, and will be called W.

In the following the numbers of the generations are indicated by a subscript, e.g. T_8 is the eighth laboratory-raised generation of strain T. The fish were reared in planted tanks of various sizes (often $70 \times 40 \times 50$ cm), and fed on live *Artemia*, *Tubifex*, and *Chironomus* larvae. They were kept at a temperature of about 20 °C, with artificial light for 16 hr a day (ca. 7:30–23:30). Under these summer-like conditions it takes most animals about 6–8 months to become reproductive, the record being $3\frac{1}{2}$ months for the first nest being built; usually in each generation a small number of fish lags behind. The onset of reproduction in the males finds its expression in their coloration (blue eyes, red underside) and in their behaviour (territorial aggression, nestbuilding). Males showing such signs were transferred from the holding tanks each to a separate tank of $60 \times 35 \times 40$ cm. These tanks were provided with a sand bottom, vegetation and nesting material, and the summer-like conditions were maintained. As soon as a male had finished a nest he was exposed to a courtship test twice a day. The test consisted of a two minutes' presentation of a ripe female in a glass tube, as described in the introduction. Every day the first test was made between 9:00 and 10:30, the second between 15:30 and 17:00. During each test the male's behaviour was registered, especially the occurrences of creeping through. This was repeated until the male had shown creeping through in ten of these tests. If among these 10 occurrences there was at least one "double" occurrence, that male was considered to be an aberrant individual and classed as a "maniac"; if not, he was classed as "normal". Then usually the male was killed and preserved for examination of the lateral plates. The females were usually kept until they died and then also preserved.

Initially, plate numbers were assessed without any special technique, but later a staining method with alizarin was applied. The eyes and entrails were removed; the animals were immersed in 4% formalin for 24 hr and then rinsed in running water for another 24 hr. The next two days they were macerated and bleached in a 3–4% KOH solution in distilled water with a few drops of 30% H_2O_2 added. Alizarin powder was dissolved in distilled water till saturation and 1–2% KOH and again alizarin added until a deep wine red colour was obtained. In this solution the fish were left for 3–5 hr and they were subsequently transferred to glycerine and left there until the bony plates had become a bright red and no pigment was left in the muscular tissues. After this treatment it is possible to count the plates accurately and conveniently, even the tiny ones above the opercula. Photographs were routinely made of the stained specimens to facilitate storing for future reference.

In each generation a small number of males (2–6) and a somewhat larger number of females (4–12) were used for raising the next generation. These were not carefully selected, but there has certainly been a tendency to pick large, vital individuals which became reproductive at an early age. In all strains breeding became increasingly difficult over the successive generations. For this reason the T_{10} generation had to be crossed with some wild-caught individuals, which resulted in an improvement of breeding capacity for the next eight generations. Lack of proper food and deficient rearing conditions caused the regrettable loss of certain crosses and made our programme less complete than we had planned.

6.3. Results

Before proceeding to investigate the heredity of "double creeping through", we had already collected relevant data on this phenomenon in the three laboratory strains. These data will be summarized first.

As described in the introduction, the aberration was first noticed in L_1. The male in question was used to raise the next generation and from then

Behavioural variant in three-spined stickleback 149

TABLE 6.1

Creeping through in successive generations of the L-strain, occurring in 2-minutes' courtship tests.

Generation	No. of fish tested	No. of tests	No. of maniacs	No. of tests	% maniacs	Occurrences of creeping through: total	maniacs	'double'	Frequency per test: maniacs	normals	Ratio of 'doubles' in maniacs
L_2	10	50	7	35	70	42	30	15	0.86	0.80	0.50
L_3	10	110	8	91	80	94	80	44	0.88	0.74	0.55
L_4	15	444	13	354	87	393	323	93	0.91	0.78	0.29
L_5	5	141	5	141	100	111	111	20	0.79	—	0.18
L_6	15	224	10	134	67	141	91	33	0.68	0.56	0.36
L_7	5	68	4	48	80	58	46	26	0.96	0.60	0.57
L_8	12	201	8	130	67	177	120	40	0.92	0.80	0.33
L_9	9	81	8	70	89	75	65	21	0.93	0.91	0.32
L_{10}	9	106	6	72	67	89	60	14	0.83	0.85	0.23
L_{11}	15	165	14	155	93	150	140	64	0.90	1.00	0.46
L_{12}	21	217	18	182	86	198	168	62	0.92	0.86	0.37
L_{13}	25	303	22	253	88	244	222	57	0.88	0.44	0.26
L_{14}	20	256	18	220	90	200	180	68	0.82	0.56	0.38
L_{15}	12	143	11	132	92	120	110	41	0.83	0.91	0.37
L_{16}	4	56	3	42	75	40	30	14	0.71	0.71	0.47
Total	187	2565	155	2059	83	2132	1776	612	0.86	0.70	0.34
										(0.83)*	

* The frequency per test is the number of occurrences divided by the number of tests. Ratio of 'doubles' is the number of double occurrences divided by total number of occurrences (in maniacs).

a number of animals was tested in each generation. However, the testing method had not yet been standardized to the same extent as in our later genetical experiments. Courtship tests were made when an opportunity presented itself, so that some generations were more thoroughly tested than others. Some animals were tested only a few times and others more often than required. Nevertheless, the outcome of the earlier tests appeared to be quite comparable to our later findings, so that they are all included in Table 6.1.

Strain L. In all, 187 males were tested (Table 6.1). In 2132 tests, out of a total of 2565, creeping through occurred, that is 0.83 per test. (The maximum frequency is 1.0 per test of 2 min, since an occurrence of creeping through – whether normal or involving multiple passages through the nest – is followed by an interval of at least 7 min; see above). Of the 2132 occurrences registered, 612 were "double" (including threefold, fourfold, etc.); a ratio of 0.29. This means that, on an average, we could observe about 3 double occurrences in 10 tests with creeping through. The adopted criterion therefore seems acceptable. Using that criterion, 155 males (83%) had to

TABLE 6.1a

Numbers of second, third, etc. passages through the nest following the first in maniacs of strain L.

Generation	No. of maniacs	Numbers of creeping through comprising:					
		1st	2nd	3rd	4th	5th	6th passage
L_2	7	30	15	1			
L_3	8	80	44	11	4		
L_4	13	323	93				
L_5	5	111	20	2			
L_6	10	91	33	3			
L_7	4	46	26	3	2		
L_8	8	120	40	6			
L_9	8	65	21	4			
L_{10}	6	60	14				
L_{11}	14	140	64	8	1		
L_{12}	18	168	62	10	6	1	1
L_{13}	22	222	57	5	1	1	
L_{14}	18	180	68	12	1		
L_{15}	11	110	41	7			
L_{16}	3	30	14				
Total	155	1776	612	72	15	2	1

be classed as "maniacs". In L_4, L_5, and L_8, testing was continued far beyond the criterion, but the number of maniacs does not change if we take only the first ten occurrences into account, namely 13, 5, and 8, respectively (compare the Table). This again strongly supports the validity of our criterion. Probably, therefore, the pronounced fluctuations in the percentages of maniacs over successive generations are a real phenomenon and not merely due to insufficient testing. Apart from these fluctuations there seems to be a slight increase in the percentage of maniacs. The frequency of occurrences per test, finally, also varies from one generation to another; this may reflect the variable rearing conditions, but it may also be an effect of inbreeding. Further evidence is needed here. This frequency is lower, though not quite consistently, in normals than in maniacs.

On many occasions the "double" occurrences were in fact multiple ones, comprising up to six passages, as Table 6.1a shows. (The record so far has been eight successive passages). Individual maniacs differed considerably in this respect, e.g. some were apt to creep through four times, others never more than twice.

Strain T. In Table 6.2 the results for strain T are summarized. Tests were carried out in three generations and in the standard fashion, so that of each male 10 occurrences of creeping through are available. None of the double occurrences consisted of more than two passages. The percentage of maniacs is as high as in strain L, but the frequency of creeping through per test is lower (0.69 as compared to 0.83), and did not differ between maniacs and normals. The ratio of double occurrences is similar to that of L: 0.29 as compared to 0.34.

Strain V. In Table 6.3 the data for the V-strain are presented. This strain was derived from fish of the Vaassen population, and the male parents (V_0) had been thoroughly tested for occurrence of double creeping through: all were found normal. The Table shows that in four subsequent generations no double creeping through occurred. Yet the frequency of creeping through per test is as high as in L (0.83) and in fact higher than in the normals of L (0.70). It is therefore not just a strong tendency to creep through, in itself, which leads to double occurrences. This is also apparent from the data of strain T, where the frequency of creeping through is comparatively low, but the percentage of maniacs high.

Since the three strains were raised under practically identical conditions, and nevertheless differ with respect to the phenomenon of creeping through, a genetical basis of this phenomenon is indicated. To investigate this in more

TABLE 6.2
Creeping through in the T-strain.

Generation	No. of fish tested	No. of tests	No. of maniacs	No. of tests	% maniacs	Occurrences of creeping through:			Frequency per test:		Ratio of 'doubles' in maniacs
						total	maniacs	'double'	maniacs	normals	
T_7	16	228	13	189	81	160	130	37	0.69	0.77	0.28
T_8	6	90	5	72	83	60	50	13	0.70	0.55	0.26
T_9	3	42	3	42	100	30	30	10	0.71	—	0.33
Total	25	360	21	303	84	250	210	60	0.69	0.70	0.29
									(0.69)*		

* Average for maniacs+normals.

TABLE 6.3

Creeping through in the V-strain.

Generation	No. of fish tested	No. of tests	No. of maniacs	Occurrences of creeping through: total	'double'	Frequency per test
V_1	8	93	0	78	0	0.84
V_2	5	54	0	46	0	0.85
V_3	10	123	0	100	0	0.81
V_4	17	195	0	160	0	0.82
Total	40	465		384		0.83

detail, we made crosses between the strains. For each cross three males and three to six females were used as a rule.

Crosses $T \times V$ and $T \times W$. In crossing T with V the reciprocals were bred, namely $T_8 \times V_1$ and $V_1 \times T_8$, in which notation the male parent is written first. The data are presented in Table 6.4. In 8 males no double creeping through was observed and there was no difference between the reciprocals. The frequency of creeping through per test is 0.68, which is typical for the T-strain (0.69) and unlike the V-strain (0.83).

In the F_2 generation one of the reciprocals was lost; only $(V_1 \times T_8) \times (V_1 \times T_8)$ is available. Of these, 43 males were tested. There were 10 maniacs (23 %), suggesting a Mendelian ratio of 25 %. The maniacs showed a creeping through frequency of 0.78, which is somewhat below the frequency in the V-strain (0.83; Table 6.3), but obviously higher than that in maniacs of the T-strain. The F_2-normals had a frequency slightly higher than that of the T-strain. On the whole, therefore, this frequency of the F_2 seems to be about intermediate between T and V. There were 22 double occurrences, a ratio of 0.22, which is somewhat lower than that in the T-strain maniacs (0.29).

One backcross, $V_1 \times (V_1 \times T_8)$, was available, of which ten males were tested. No maniac was found, which is in agreement with a Mendelian inheritance. The frequency of creeping through per test was 0.73, identical to that of the F_2 normals. Unfortunately the other backcross was lost. We therefore decided to test an additional cross of T_{10} to females from Tholen (W_0).

Of the resulting F_1, 16 males were tested and no maniacs were found. The frequency of occurrence was high: 0.80. In the F_2 generation 36 males

TABLE 6.4

Creeping through in crosses and back-crosses of strains T and V, and T and W.

Generation	No. of fish tested	No. of tests	No. of maniacs	No. of tests	% maniacs	Occurrences of creeping through: total	maniacs	'double'	Frequency per test: maniacs	normals	Ratio of 'doubles' in maniacs
F_1											
$V_1 \times T_8$	5	58	0			40				0.69	
$T_8 \times V_1$	3	46	0			31				0.67	
Total	8	104	0			71				0.68	
F_2											
$(V_1 \times T_8) \times (V_1 \times T_8)$	43	581	10	128	23	429	100	22	0.78	0.73	0.22
										(0.74)*	
Backcross											
$V_1 \times (V_1 \times T_8)$	10	124	0			91				0.73	
F_1											
$T_{10} \times W_0$	16	199	0			160				0.80	
F_2											
$(T_{10} \times W_0) \times (T_{10} \times W_0)$	36	492	9	98	25	360	90	25	0.92	0.68	0.28
										(0.73)*	

* Between parentheses: averages for maniacs+normals.

Behavioural variant in three-spined stickleback   155

TABLE 6.5

Creeping through in crosses of strains L and V.

Generation	No. of fish tested	No. of tests	No. of maniacs	No. of tests	% maniacs	Occurrences of creeping through: total	maniacs	'double'	Frequency per test: maniacs	normals	Ratio of 'doubles' in maniacs
F_1											
$L_9 \times V_2$	10	124	2	21	20	101	20	2	0.95	0.79	0.10
$V_2 \times L_9$	10	118	2	20	20	102	20	4	1.00	0.84	0.20
Total	20	242	4	41	20	203	40	6	0.98	0.81	0.15
									(0.84)*		
F_2											
$(L_9 \times V_2) \times (L_9 \times V_2)$	27	314	10	117	37	270	111	33	0.95	0.81	0.30
$(V_2 \times L_9) \times (V_2 \times L_9)$	48	537	18	175	37.5	475	162	56	0.93	0.86	0.35
Total	75	851	28	292		745	273	89	0.94	0.84	0.32
									(0.87)*		

* Between parentheses: average for maniacs+normals.

could be tested and 9 turned out be maniacs: exactly 25 %. The maniacs showed a very high frequency of creeping through: 0.92 per test, and the normals scored much lower: 0.68. The ratio of double occurrences – of which two were in fact triples – is much the same as that in the T-strain (0.28 and 0.29, resp.). Hence, in the $T \times V$ and $T \times W$ crosses, the F_1 generations (and the backcross of F_1 to V) did not yield any maniacs, whereas in the F_2 generations virtually 25 % were maniacs. This is compatible with unifactorial inheritance of the maniac aberration, which thus would be recessive.

Crosses $L \times V$. Reciprocal crosses between L_9 and V_2 were made. In both F_1 generations maniacs were found; the data are given in Table 6.5. There is no difference between the reciprocals. The frequency of occurrence per test is the same or even higher than those of the parental strains. The ratio of double occurrences (0.15) is conspicuously low.

The F_2 crosses did not show differences between reciprocals. The percentage of maniacs is 37. The situation is apparently more complicated than the clear-cut Mendelian ratios in the $T \times V$ crosses. However, we assume that the unifactorial determination does apply also in this case. This implies that there is a manifestation of the recessive allele, in the heterozygous condition, of about 20 %. The very low ratio of double occurrences in the supposedly heterozygous maniacs could be taken as evidence for this view. On the basis of this assumption we may expect that the same applies for the heterozygotes of the F_2 generation, that is, for half the number of animals tested. This would produce, apart from the 25 % of the homozygous maniacs, an additional 10 % (20 % of half the sample) of heterozygous maniacs, bringing the total percentage of maniacs to 35 %. This is reasonably close to the 37 % actually found. We can also start the calculation from the other end, taking the 37 % as the more reliable value, since it is based on more individuals. This would indicate an extra percentage of heterozygous maniacs of $37\% - 25\% = 12\%$, and therefore a manifestation of $2 \times 12\% = 24\%$ in the F_1 instead of the 20 % actually found. Both ways the data seem to be in reasonable agreement with our assumption.

Ontogenetic factor. From our crosses it is evident that a genetical difference exists between normal sticklebacks and maniacs. However, occasionally we have come across wild-caught individuals that showed double creeping through. Whenever we could trace their origin, it appeared that they had been exposed to the summer-like laboratory conditions, and had thus become reproductive, at an early age. In batches of young fish caught in the autumn or in early winter and immediately transferred to summer-like conditions in the laboratory, often a few maniacs turned up. We decided to test the possible influence of this factor more systematically. Animals of T_7 were kept under winter-like conditions (8 hr of light per day; 10 °C) for four months in large storage tanks. Consequently they became reproductive at a much later age than the controls which were kept under the usual summer-like conditions. Of the 12 "retarded" males tested, none showed double creeping through, whereas of 16 controls 13 were found to be maniacs.

Another experiment of this type, but more standardized, was carried out with L_{14} and L_{15}. A group of young animals was randomly selected two months after hatching and exposed to an artificial winter (as above) for four months. Another group was kept under summer-like conditions all the time. The size of tanks ($60 \times 40 \times 35$), feeding, and density (20 fish per tank) were identical for the two groups. The retarded group was exposed to summer-like conditions after their four months' artificial winter and as soon as reproductive males appeared they were transferred to experimental tanks. Only at that time males of the control group were transferred from their storage tanks into separate experimental tanks. The two groups were simultaneously allowed to build nests and then tested. Of 14 L_{14} controls, 13 appeared to be maniacs, whereas of the 11 retarded males only one was a maniac; he double-crept through only once. Of 11 L_{15} controls, 10 were maniacs, and of the 11 retarded L_{15} males none. Finally, it was established that young from retarded L_{14} and L_{15} animals became maniacs after maturation under summer-like conditions. Evidently, a four months' period of short day-length and low temperature during ontogeny tends to suppress the expression of a genetical disposition for double creeping through.

Inheritance of plate number. Since T and V differ with respect to plate number, the crosses between these strains enabled us to study the inheritance of this feature as a sideline. In collecting and presenting the pertinent data we followed Münzing (1959). The crosses were made between T_8 and V_1; therefore, only the data for these generations and their immediate offspring are given in Figs. 6.4 and 6.5 as frequency distributions of plate number. There is a clear-cut difference: all T_8 and T_9 individuals are of the forma trachura, all V_1 and V_2 individuals of the forma Leiura. The two reciprocal F_1 hybrids are intermediate, all fish being of the forma semi-armata (Fig. 6.6). In the F_2 generation $(V_1 \times T_8) \times (V_1 \times T_8)$, there seems to exist a deviation: there is a surplus of fish with high plate number and semi-armata

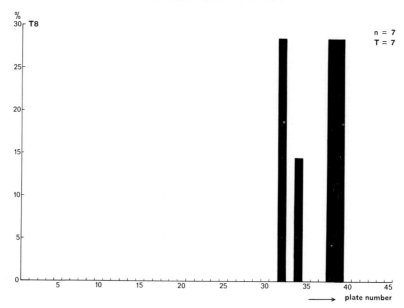

Fig. 6.4. Frequency distribution of plate numbers in T_8 and T_9.

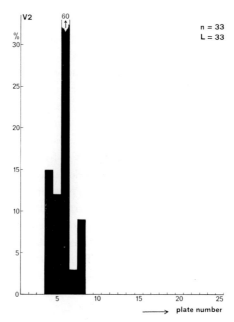

Fig. 6.5. Frequency distribution of plate numbers in V_1 and V_2.

Behavioural variant in three-spined stickleback 159

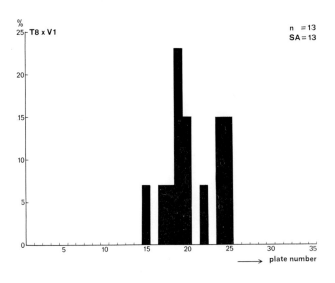

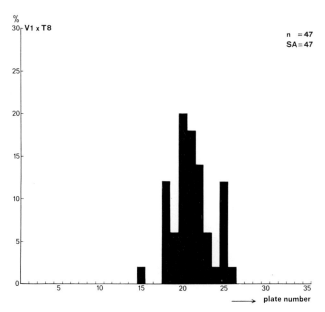

Fig. 6.6. Frequency distribution of plate numbers in $V_1 \times T_8$ and $T_8 \times V_1$.

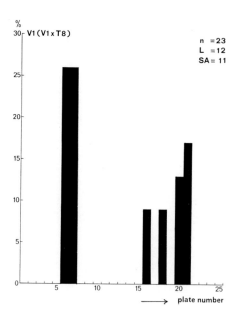
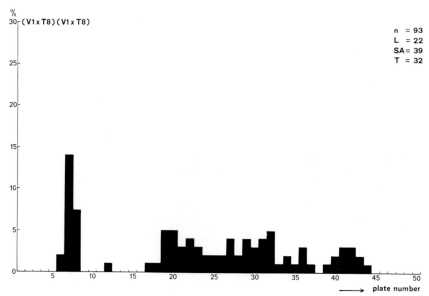

Fig. 6.7. Frequency distribution of plate numbers in $V_1 \times (V_1 \times T_8)$ and $(V_1 \times T_8) \times (V_1 \times T_8)$.

seems to merge with trachura. The distribution in Fig. 6.7 (22 leiura, 39 semi-armata, and 32 trachura) does not differ from the expected ratio $1:2:1$ (χ^2; $0.10 < P < 0.20$). Smaller samples from this generation, however, showed a much better agreement with the expectation: 74 fish were divided into 16 trachura, 35 semi-armata, and 23 leiura; of a group of 43 males which were tested for creeping through, 9 were trachura, 22 semi-armata, and 12 leiura. The backcross $V_1 \times (V_1 \times T_8)$ consisted of semi-armata and leiura in equal numbers (Fig. 6.7), and hence confirms the expectation.

All in all, the data agree with Münzing's view that the lateral armouring in this species is controlled by a single pair of alleles, one allele for high plate number, and one for low, the heterozygote being intermediate. But it seems likely that other factors may interfere with this simple system; otherwise, a clearer segregation would be expected. This is also suggested by Münzing.

Inheritance of colour pattern. T and V differ also with respect to the colour pattern (compare Fig. 6.3). Unfortunately it is not always possible to establish this difference. It is most pronounced in young or halfgrown, live fish. Even then reservations have to be made because the condition of the fish and the illumination in the tank may cause variations. In dead animals, fixation in formalin may often enhance the pattern. Thus it has been possible to study the inheritance of the pattern in $T \times V$ crosses. The blotched pattern of V appeared to be dominant over the vaguely barred pattern of T: all the F_1 young of $T_8 \times V_1$, $V_1 \times T_8$, and of the backcross $V_1 \times (V_1 \times T_8)$ presented the blotched appearance at some stage of their development in the rearing tanks. In the F_2 segregation occurred and the fish from one sample were examined individually: 58 were found to be blotched, 21 barred. In 74 fish from this sample, plate numbers were also counted (see above). It appeared that the 16 trachura fish comprised 12 blotched and 4 barred individuals, the 35 semi-armata comprised 22 blotched and 13 barred, and the 23 leiura comprised 19 blotched and 4 barred, which seems an approximation of the expected dihybrid ratio for independent segregation: $3:1:6:2:3:1$ (the difference from this expected ratio is nonsignificant: χ^2; $0.30 < P < 0.50$) We may conclude that colour pattern and plate number are independently transmitted.

Independent segregation of traits. In the F_2 generation $(V_1 \times T_8) \times (V_1 \times T_8)$ 43 males were tested for creeping through. In these males the lateral plates were counted, and 9 were found to belong to forma trachura, 22 to semi-armata, and 12 to leiura, as we have seen above. For the 33 normal males the distribution was 7 trachura, 18 semi-armata, and 8 leiura; for the 10 maniacs it was 2 trachura, 4 semi-armata, and 4 leiura. This distribution is compatible with the dihybrid ratio with independent segregation, $3:6:3:1:2:1$ (χ^2; $P > 0.05$). We conclude that double creeping through and plate number are independently transmitted.

Unfortunately an independent segregation of double creeping through and colour pattern could not be established quantitatively. From our qualitative observations such an independence seems likely.

6.4. Discussion

In normal stickleback males, the act of creeping through involves only a single-passage through the nest, followed by a refractory period of at least 7 min, but usually 15 min or more. To account for this refractory period, Nelson (1965) postulated a threshold which is raised by actual creeping through. In our aberrant animals, called "maniacs", creeping through frequently consists of two or more passages in quick succession well within the two minutes of a courtship test, and it is only on this succession of passages that a refractory period ensues. In terms of Nelson's model, there seems to be a defect in the threshold-raising mechanism in these maniacs so that it often takes several passages to produce a refractory period. We shall not enter any further into the nature of this aberration. One of us ('t Hart) will publish his work on the underlying mechanism elsewhere. However, one of his results is relevant to our conclusions: a sudden shortening of the nest tunnel of a wild-caught male shortens his refractory period but does not result in double creeping through. This also holds for males of the V strain (all 17 V_4 males have been tested for this). But the same procedure does induce double creeping through in animals of the L strain classed "normal" on the basis of our criterion. This is important for two reasons. First, it shows that it is not the length of the tunnel or the size of the nest which turns a male into a maniac; moreover, a wild-caught male whose nest has been replaced by a maniac's nest does not creep through double. Maniacs, therefore, do not creep through double because they build aberrant nests. Second, the tunnel-shortening experiments indicate that "normals" of our L strain also have a disposition to double creeping through, unlike fish of the V strain. This leads us to assume that all animals of the

L strain are genetically maniacs. This is supported by the observation that no obvious difference in the percentage of maniacs exists between fish descending from a typical maniac and those from a "normal" male of either the T or the L strain, although we have no definite proof on this point. A third argument is that it would be unlikely to find the Mendelian ratio in our crosses if the parental strains had not been homogeneous for the factor concerned. On these considerations we accept this homogeneity in our L, V, and T strain.

Why, then, did we consistently find an appreciable number of "normals" in nearly all our samples? It has already been argued that this cannot be due to insufficient testing: prolonged testing of L_4, L_5, and L_8 did not change the percentage of maniacs. Nor is it due to the age at which the animals were tested: a group of L_2 males tested at 9 to 11 months of age still included 80 % maniacs. "Normals" are therefore not males which were tested too late. On the whole we got the impression that "normals" are somewhat less vigorous than maniacs. They are often slow in nest building, court less intensively, are shy, and are less brightly coloured. This explains probably also why the incidence of creeping through per test is lower for normals than for maniacs in nearly all samples. It has already been argued that a high incidence in itself does not determine creeping through. Yet, within the strains there is an obvious correlation between frequent creeping through and being a maniac. Evidently there is still much to be investigated on this point, but it seems that in "normals" the fenotypic expression of the maniac genotype is reduced to zero by factors connected with general "vigour". Probably the same factors can be held responsible for the variation in expression in the maniacs (varying ratio of "doubles"; varying number of passages). That these factors may be genetic and dependent on inbreeding seems to emerge from the following.

In all our crosses a striking heterosis became evident: the hybrid fish grew up much better and were more uniform both in size and in vigour. Yet the rearing conditions were the same as those for the inbred strains. The heterozygosity must have had a favourable effect on general vigour; a lack of vigour in the "normals" of the T- and L-strains may therefore be due to inbreeding and originate from homozygosity. This might be connected with the pronounced fluctuations in the percentage of maniacs over the successive generations of the L-strain, as already pointed out. In spite of these fluctuations there seems to be a slight increase in the percentage of maniacs in the breeding history of the L-strain, possibly due to selection. In choosing males for breeding we tended to prefer early breeders, fast nest-builders, and to some extent established maniacs. This might have increased the percentage of maniacs, i.e. the penetrance of double creeping through, by selection for a favourable genetic background.

It is interesting to note that our selection of early breeders did not bring about a shortening of generation span, or, at best, very little: In the T_{1-5} generations it took an average of 187 days for the first nest to be built, in the T_{6-10} generations 171 days, in L_{1-5} this average was 156 days, in L_{9-14} 143 days, and finally, in V_1 155 days. The values are in agreement with data presented by Baggerman (1958) for wild-caught young under constant summer-like conditions.

The genetic disposition to creep through double does not express itself in fish which have been exposed to an artificial winter. Further, wild-caught fish may show the aberration, but only when exposed to summer-like conditions at an early age. It would appear therefore that in the wild the phenomenon of double creeping through never occurs. (Recently, in January and March, we found young sticklebacks in the field and these might perhaps become prematurely reproductive in the summer). It remains an open question how the genetic factor affects the animals in the field and why it is maintained in at least three populations (Vaassen, den Helder, and Tholen). So far we cannot distinguish between retarded maniacs and retarded V-animals, but this has not yet been analyzed in detail. It will be especially interesting to study the heterozygotes more closely, since the situation in the wild suggests a stable polymorphism.

Finally we should like to stress that some very rare exceptions have been observed. Two males of the V-strain have been seen to creep through double, though outside the range of conditions chosen for our tests: one after his nest had been replaced several times, one in the parental phase.

We conclude that a major genetic factor is involved in the aberration of double creeping through. This factor shows simple Mendelian inheritance, the normal condition being dominant, and there is no evidence for sex-linkage. Its transmission is independent of that of the factor for plate number. Its expression depends on daylength and/or temperature during ontogeny. Other environmental factors, too, may affect its expression.

We may point out that the phenomenon discussed presents a clear illustration of the interaction of genetical and environmental influences. In the comparison of L- and V-animals under summer-like conditions double creeping through appears as a genetically determined behavioural variant.

On the other hand, in the retardation experiments within the L-strain the same variant appears to be environmentally determined.

References

Baggerman, B. (1958): An experimental study on the timing of breeding and migration in the three-spined stickleback (*Gasterosteus aculeatus* L.). *Arch. Néerl. Zool.*, **12**, 150.

Bertin, L. (1925): Recherches bionomiques, biométriques et systématiques sur les épinoches (*Gasterosteidae*). *Ann. Inst. Oceanogr. Monaco*, **2**, 1.

Hart, M. 't and Sevenster, P. (1970): Gedragsgenetisch onderzoek bij de driedoornstekelbaars *Gasterosteus aculeatus* L. *Genen Phaenen*, **14**, 19.

Heuts, M. J. (1944): Laterale bepantsering en groei bij *Gasterosteus aculeatus* L. *Natuurwetensch. Tijdschrift* **26**, (1), 40–52.

Heuts, M. J. (1947a): The phenotypical variability of *Gasterosteus aculeatus* L. populations in Belgium. *Med. Kon. Akad. Wetensch.*, **9**, (25), 1.

Heuts, M. J. (1947b): Experimental studies on adaptive evolution in *Gasterosteus aculeatus* L. *Evolution*, **1**, 89.

Iersel, J. A. A. van (1953): An analysis of the parental behaviour of the male three-spined stickleback (*Gasterosteus aculeatus* L.). *Behaviour, Suppl.* **3**, 1.

Münzing, J. (1959): Biologie, Variabilität und Genetik von *Gasterosteus aculeatus* L. (Pisces). Untersuchungen im Elbegebiet. *Internat. Rev. gesammt. Hydrobiol.*, **44**/3, 317.

Münzing, J. (1962): Die Populationen der marinen Wanderform von *Gasterosteus aculeatus* L. (Pisces) an den holländischen und deutschen Nordseeküsten. *Neth. J. Sea Res.*, **1**, 508.

Nelson, K. (1965): After-effects of courtship in the male three-spined stickleback. *Z. Vergl. Physiol.*, **50**, 569.

Pelkwijk, J. J. ter and Tinbergen, N. (1937): Eine reizbiologische Analyse einiger Verhaltensweisen von *Gasterosteus aculeatus* L. *Z. Tierpsychol.*, **1**, 193.

Sevenster, P. (1961): A causal analysis of a displacement activity (fanning in *Gasterosteus aculeatus*). *Behaviour, Suppl.* **9**, 1.

CHAPTER 7

The role of sexual selection in the maintenance of the genetical heterogeneity of *Drosophila* populations and its genetic basis

E. Boesiger

7.1. Introduction

The concept of sexual selection has been defined by Darwin in his "On the Origin of Species by Means of Natural Selection". According to his definition, sexual selection differs from natural selection in that it "depends, not on a struggle for existence, but on a struggle between the males for possession of the females; the result is not death to the unsuccessful competitor, but few or no offspring... in many cases, victory will depend not on general vigour, but on having special weapons, confined to the male sex". He considers the weapons of the males, but also their specific sexual traits, like the mane of the lion, the shoulder-pad of the boar, the hooked jaw of the male salmon, to be as important for victory as the sword or spear. When males and females differ in structure, colour, or ornament, he postulates that such differences have been mainly caused by sexual selection, but he adds that not all sex differences can be attributed to sexual selection. In his famous "The Descent of Man", containing an extensive discussion of sexual selection, Darwin states very clearly: "But in most cases it is scarcely possible to distinguish between the effects of natural and sexual selection", thus anticipating the critique of the concept of sexual selection. As he himself admitted, sexual selection does not exclusively depend on a struggle between males for the possession of females; secondary sexual traits do not evolve through the pressure of sexual selection alone, and there is no clear-cut boundary between sexual selection and natural selection. It is now quite clear that many characteristics of the sexual dimorphism which are specific for males are mainly used in the defence of a territory, which is very important in natural selection. This is so, for example, in the case of the conspicuous red breast of the robin or the male nightingale's song, used in the first place for territory marking, but also for attracting and stimulating females. In other cases, for example *Philomachus pugnax*, well studied by Selous (1929) in Holland, the male plumage and the male courtship dances might well be an effect of Darwinian sexual selection in the original sense of the concept.

The founder of the formal genetics of *Drosophila melanogaster*, Morgan (1909, 1932), severely opposed the Darwinian idea of sexual selection. Since one finds in most species equal proportions of males and females, he postulated that nearly all males have the same opportunity to copulate and, hence, that there is no sexual selection. This argument is fallacious. In polygamous species of birds and mammals, only a minority of males, which may be as low as 10 %, participate in the reproduction of the species. This means that conditions for sexual selection are present. In monogamous species of birds and mammals, the fraction of reproducing males is higher, but by no means 100 %. Even in populations of invertebrates, for example, insects, where complete panmixia is assumed by many authors, there exists sexual selection, which is the main subject of this chapter. Darwin wondered whether sexual selection exists in the lower species. Sturtevant, a collaborator of Morgan's, showed that copulations do not happen at random in Drosophilids. Morgan (1932) has underestimated the behavioural capacities of insects, since he states: "In insects, where some remarkable cases of sexual dimorphism are known, it is doubtful whether the psychic development is sufficient to ascribe to them the necessary powers of discrimination and choice". Discussing then the problem of intersexes and the factors responsible for sex determination, and taking into account the experimental transformation of genetical males into functional females by hormones, Morgan says that ". . . if it could be shown that the same hormones that affect the secondary sexual characters are important in regulating the sexual behaviour of the male, which is essential to reproduction, the secondary sexual characters would then be regarded as by-products of the hormones and would not call for explanation of a different kind". Morgan is well-known as a pioneer geneticist but, after all, he was primarily an embryologist and for this reason he was well-aware of the great importance of organizers and mediators in development and physiology. It is not the purpose of this chapter to discuss that problem at length, but it seemed interesting to present the somewhat ambivalent views of a great geneticist. His postulate concerning the secondary sexual characters ("... nevertheless, in so far as the experimental evidence with birds has shown that it is in the gonads rather than in the brain that the actual change originates",...) is characteristic for his typological thinking. It is not a question of choosing between the hormones or the brain as the determinants of the evolution of the secondary sexual characters. Both are interacting through the hypothalamus in higher species and both have of course a genetical basis. It is clear that the coding genes are not the sole determinants of the structure and the behaviour of an organism; the latter may vary according to the diverse influences of the environment. In higher species, the plasticity in behaviour is considerable but not unlimited, even in man, but in lower organisms the genetic determinants of behaviour are predominant. It is evident, however, that for evolutionary changes within a species over successive generations only shifts in the gene pool are relevant. The main forces of evolution are natural and sexual selection in addition to some secondary effects of random events, such as genetic drift.

Even if it is not possible to distinguish clearly between the roles of natural and sexual selection in the evolution of a particular secondary sexual character, it seems useful to maintain both terms. The two concepts have of course been further elaborated since Darwin; sexual selection and natural selection are complex phenomena. One aspect of sexual selection, in the original Darwinian sense, is the evolution of secondary sexual characters. Another even more important aspect is the reproductive isolation between species by sexual selection. From an evolutionary point of view, sexual isolation is the chief function of sexual selection, that is to say, of sexual discrimination. There is a third important function of sexual selection, *viz.* the maintenance of genetical heterogeneity in populations as a consequence of the advantage of heterozygotes in sexual selection. This has been experimentally demonstrated in Drosophilids (Boesiger, 1958, 1962, 1967). This chapter deals with these effects of sexual selection on the heterogeneity of *Drosophila* populations and reports on the genetic basis of sexual vigour of males.

7.2. *Methodology*

Different techniques were used for studying sexual selection in Drosophilids.
 1. Population cages. Large populations of virgin females and of males carrying appropriate recessive or dominant markers were placed together

for a given time in a population cage. After the experiment, each female was put in a separate vial so that her offspring revealed by which type of male or males she had been fertilized. A deviation from the frequencies of fertilization expected according to panmixia indicates the degree and the modalities of sexual selection.

In the case of a male-choice arrangement, two types of females were in contact with one type of males only. There was no need to raise the offspring of these females because they were dissected after the experiment and examined for the presence or absence of spermatozoa in their spermathecae.

2. Small populations. For some purposes, small populations of ten to twenty flies, or pairs in vials or milk bottles, replicated many times, are more convenient than large populations. This technique was used for male-choice (two types of females, one type of male), for female-choice (one type of female, two types of males), and for multiple-choice experiments. If no choice is offered, that is, if one type of female is in contact with one type of male, differences in sexual behaviour can be observed in many different combinations in successive experiments. This arrangement is useful for measuring female receptivity and male vigour.

3. Direct observation. Elens and Wattiaux (1964) developed an observation chamber for the direct observation of copulations during several hours, using a lens of low magnification. If the flies do not differ morphologically, they may be distinguished by clipping the apex of a wing in one type of flies. An analogous but very small device was employed for observing pairs of flies with a stereomicroscope.

Let us now consider the main modalities and factors of sexual selection as related to the maintenance of the heterogeneity of populations. The Hardy-Weinberg principle, fundamental in population genetics, is based on the random combination of female and male gametes. This is only possible if matings occur at random, that is, if there is complete panmixia. Early studies in *Drosophila melanogaster* on the hereditary transmission of morphological characters by Lutz (1911) showed that wild-type females and males have a large advantage in sexual selection when competing with flies having a mutant wing venation. Sturtevant (1915, 1921) made some observations on courtship and mating behaviour which led him to valuable conclusions. He understood the great utility the newly discovered mutants may have for this kind of behavioural studies and showed that the wing vibrations of the male play an important role in the excitation of the females. Extensive series of experiments with mutant types permit the conclusion that neither females, nor males, exercise a "choice" with respect to the mutant character, but that mutant types are usually less active than the wild-types. The courtship of mutant males is less vigorous and less persistent than that of wild-type males and they mate less often. They tend to mate more often with mutant than with wild-type females, the former being less inclined to move away.

7.3. Sexual selection in large experimental populations

A *Drosophila* population living in a population cage certainly does not closely represent the situation in a free-living natural population but, by crude approximation, it permits the study of factors which play a role under more complex natural conditions, even if a direct extrapolation of the experimental results to the situation in natural populations is not possible.

Nonrandom mating has been clearly demonstrated in population cages

TABLE 7.1

Sexual selection in large experimental populations of *D. melanogaster*.

Populations				Females			
				N fertilized by ♂			
		N	% fertiles	"cn"	"v"	"cn" and "v"	% ♀ fertilized by ♂ "cn"
♀ "v" ♂ "v" ♂ "cn"		74	94.59	66	2	2	95.71
		163	91.41	140	3	6	95.97
		168	90.47	139	11	2	92.10
		176	93.75	144	21	0	87.27
		253	89.72	214	7	6	95.59
		230	79.56	167	15	1	91.53
m		1.064	88.91	870	59	17	92.86 ± 1.40; $\chi^2 = 13.99$
							$P < 0.01$
♀ "cn" ♂ "cn" ♂ "v"		186	95.16	153	13	11	89.54
		173	56.07	94	2	1	97.42
		277	44.40	113	10	0	91.86
		139	52.52	66	5	2	91.78
		100	91.00	77	6	8	89.01
		246	79.26	172	15	8	90.25
		223	70.40	139	13	5	90.12
m		1.344	67.93	814	64	35	91.07 ± 1.08; $\chi^2 = 6.10$
							$P < 0.30$

(Boesiger, 1953). Two stocks were used: The first, newly derived from a natural population, and hence with a high degree of heterozygosity, was marked with the allele "cinnabar" (*cn*) affecting the eye-colour. The second, a highly inbred laboratory stock maintained for many years by brother-sister matings, was marked with the sex-linked allele "vermillion" (*v*) producing phenotypically about the same deviant eye-colour as does "*cn*". All populations consisted of one type of females, either "*cn*" or "*v*", and the two types of males in a 1:1 proportion. The offspring of each female revealed whether she was fertilized by a "*cn*" or a "*v*" male, or by both. Table 7.1 gives the results of 13 such experiments. There appeared to be a very large deviation from panmixia, about 92 % of "*v*" and of "*cn*" females being fertilized by "*cn*" males, despite the fact that the latter represented only 50 % of the males in the cage. Evidently, "*cn*" males have a strong advantage over "*v*" males in this case of sexual selection. Similar experiments with another vermillion and a wild-type stock showed also large deviations from panmixia. If the proportion of males in the population is high, the advantage of the more vigorous males is increased. Petit (1959) found in her population cages a similar selective advantage of wild-type males over mutant males marked with "Bar", independent of the genotype of the females. In agreement with Sturtevant's conclusions, these cases of strong sexual selection seem to be mainly, if not exclusively, a consequence of differences in the sexual vigour of males and the receptivity of females.

7.4. *Measurement of the vigour of males and the receptivity of females*

The differences in female receptivity and male vigour between the two stocks "v" and "cn" were measured in different ways. If pairs in the four possible combinations of the stocks are put in vials for 12 days, one can observe clear-cut differences in the percentages of fertile pairs. Under these circumstances there is no choice and no competition. The results given in Table 7.2 show that "v" females are more receptive and that "cn" males have a higher sexual vigour. The fourth column of Table 7.2 gives the corresponding figures extracted from the results of the experimental populations presented in Table 7.1. It is obvious that the differences in male vigour are much larger under conditions of competition in the population cage than in the pairs without competition. The fifth column of Table 7.2 shows the results for another measure of receptivity and vigour. Six females were put for 48 hr with one male and the number of fertilized females was determined afterwards. The fifth column gives the mean numbers which again reveal, under still other conditions, the higher receptivity of "v" females and the higher vigour of "cn" males.

TABLE 7.2

Receptivity of females and vigour of males from two stocks of D. melanogaster.

Pairs	N	% of fertile pairs	% of copulation in large populations	N fertilized ♀ out of 6
♀v × ♂v	200	61.0±3.4	6.3	3.43±0.28
♀v × ♂cn	302	80.2±2.3	82.6	5.09±0.21
♀cn × ♂v	200	54.0±3.5	6.1	3.31±0.28
♀cn × ♂cn	325	73.9±2.4	61.8	4.07±0.27

The two types of males, "v" and "cn", were also tested in populations consisting of six females and one male, using eight different types of females. Table 7.3 shows that "cn" males always fertilize a greater proportion of females than "v" males do. This experiment also yielded important differences in the receptivity of females.

TABLE 7.3

Differences in the receptivity of females and the vigour of males between different stocks of D. melanogaster.

Stocks of females	N of populations	Mean number of fertilized females by	
		♂ "cn"	♂ "v"
white	100	5.44	3.92
ebony	100	5.38	3.32
vermillion	220	5.09	3.43
+ Oregon R-C	220	4.35	3.27
forked	500	4.37	2.47
cinnabar	220	4.07	3.31
+ Perpignan	200	3.96	1.88
vestigial	300	3.37	2.28

In another set of experiments, the vigour of males from many stocks was measured in the same way in numerous replicated populations comprising six wild-type females, all from the same population, and one male (Boesiger, 1962). Some of the stocks were maintained for many years in the laboratory under conditions of inbreeding, the others were newly derived from natural populations. The mean percentage of fertilized females was 43.02 for the genetically more uniform males of the first group. The males with a higher degree of heterozygosity, derived from the natural populations, were more vigorous and fertilized 73.05 % of the females.

7.5. Behavioural differences between "v" and "cn" males

The direct observation of the behaviour of the four combinations of pairs with a stereomicroscope reveals several differences. Each pair was observed under standardized conditions of age, space, light, temperature, humidity, and hour. The "*cn*" males showed, as compared with the "*v*" males, a slightly shorter latency to start courtship and a considerably shorter latency to the beginning of actual copulation. The courtship of the "*cn*" males turned out to be much more persistent and intensive. This difference shows up in the mean number of copulation trials per male, which is much higher for "*cn*" males (see Table 7.4). The data also indicate a higher receptivity of the "*v*" females.

There also exists a difference in mean copulation duration for the four types of pairs, as presented in Table 7.4. The vigorous "*cn*" males had shorter copulation times with both types of females and the "*v*" females showed with both types of males a shorter copulation time. The female probably determines the duration of the copulation which seems to depend,

TABLE 7.4

Number of trials and duration of copulations of "*cn*" and "*v*" females and males.

Constitution of pairs		Copulations		Mean age of first copulation in hr
		Mean N of trials	Duration in min	
♀ *v*	♂ *v*	0.88	16.2	38 h 35 min
♀ *v*	♂ *cn*	3.93	13.7	34 h 51 min
♀ *cn*	♂ *v*	1.35	20.9	40 h 53 min
♀ *cn*	♂ *cn*	4.39	16.0	36 h 22 min

at least partially, on the quantity of transferred sperm. The number of offspring after one copulation is much lower for "*v*" males, indicating lower quantities or lesser activity of the sperm.

Another behavioural difference between the two stocks "*v*" and "*cn*" concerns the mean age of first copulation after hatching of pairs put in vials (Boesiger, 1962). The right-side column of Table 7.4 shows that, with both types of females, the "*cn*" males copulate earlier and that copulation takes place earlier for the more receptive "*v*" females.

7.6. *The genetical basis of the receptivity of females and the vigour of males*

The results of the following analysis have been established for a few stocks only. They cannot be generalized, but they probably apply for many cases and may be quite relevant to the genetical structure of populations.

Some of these experiments are summarized in Table 7.5. Line A presents the percentages of fertilized females in a large number of replicated small populations comprising six females and one male in the four combinations made with the original "*v*" and "*cn*" stocks. The higher receptivity of "*v*" females and the higher vigour of "*cn*" males are evident.

Individual females from the two stocks were crossed with one male from a wild-type stock. By sib-mating the F_1, three new "*v*" and three "*cn*" lines were produced. After this outcrossing, they became again homozygous for the marker loci "*v*" or "*cn*". Lines B, C and D of Table 7.5 present the results for three of the new stocks. Compared with A, there are large differences in female receptivity and male vigour, caused by the introduction of part of the wild-type genome. The results in lines B, C and D show a highly significant heterogeneity. The data of experiments B, C, and D still show a higher vigour of "*cn*" males, but not a higher receptivity of "*v*" females.

Lines E, F, and G of Table 7.5 present the results obtained after many generations of intercrossing the two stocks, leading to isogenicity of the "*v*" and the "*cn*" stocks, except for the two marker loci. Line E gives the percentages of fertilized females in the four mating combinations after ten generations. The differences between the vigour of "*cn*" and "*v*" males, as compared with the original stocks, are reduced but still existing. The four matings still yield significantly different results. Line F shows the results after 40 generations. There are not any significant differences in female

TABLE 7.5

Changes in female receptivity and male vigour of "v" and "cn" flies due to outcrossing.

State of stocks (3)	N of tested males	Fertilized females in % Females "cn" ♂ "cn"	Females "cn" ♂ "v"	Females "v" ♂ "cn"	Females "v" ♂ "v"	χ^2 (1)	P
A	440	67.88	55.15	84.94	57.12	163.84	0.001
B	444	71.66	65.91	76.33	55.25	75.17	0.001
C	440	86.81	60.60	70.60	50.30	218.32	0.001
D	440	68.78	50.15	63.62	64.54	54.63	0.001
$\overline{m}$	1.324	75.78	58.91	70.21	56.72		
χ^2 (2)		67.435	35.207	26.162	28.112		
P		0.001	0.001	0.001	0.001		
E	1.411	74.55	71.55	66.11	59.87	129.72	0.001
F	453	51.96	53.62	54.39	58.84	7.46	0.05
G	309	59.48	55.26	57.07	58.59	1.57	0.70

(1) These χ^2 are calculated from the counted numbers of fertilized and non-fertilized females and concern the four mating combinations.
(2) These χ^2 concern one type of mating combination for the three strains B, C, and D.
(3) See explanation in the text.

receptivity and male vigour between the two stocks; the four combinations show a reasonable homogeneity. The results in line G were obtained exclusively with wild-type females and males, taken from generation F_{10}. The data are very homogeneous and resemble those obtained with "v" and "cn" flies (line F). The experiments suggest that the alleles "v" and "cn" and their wild-type alleles do not affect female receptivity and male vigour, which are most probably determined by polygenes. The many generations required to achieve isogenicity which is needed for our result indicate that the number of responsible genes cannot be small.

It is also apparent that the male vigour, after isogenicity is reached, is not fixed at an intermediate level between the values of the two original stocks, but at the low level of the "v" males. The original "v" stock was highly inbred, but the "cn" stock had a high degree of heterozygosity. Possibly, high male vigour is the outcome of a large degree of heterozygosity so that inbreeding will reduce it. More direct evidence for the advantage of heterozygous males was furnished by a comparison between the percentages of fertilization by males of two highly inbred laboratory stocks marked, respectively, by the genes "forked" and "sepia" and hybrid males obtained from the two reciprocal crosses between these stocks. The results are presented in part A of Table 7.6. The hybrid males, in small populations of ten females and one male, fertilize two to three times more females than the males of the original inbred stocks. When two non-inbred stocks, newly bred from natural populations were crossed, the hybrid males did not show an increase in sexual vigour. Part B of Table 7.6 gives an example of this situation. Other evidence for the effect of the degree of heterozygosity on male sexual vigour was provided *a contrario* by a comparison between the percentages of fertilization by males of the two inbred isogenic stocks "v" and "cn" and hybrid males obtained from reciprocal crosses between the two isogenic stocks. Hybridization of coisogenic stocks cannot increase the degree of heterozygosity. As part C of Table 7.6 shows, there is indeed no difference in male vigour.

TABLE 7.6

Comparison between the sexual vigour of males from inbred stocks and hybrids.

	Stocks of the males	N tested	% fertilized females
A	forked, inbred laboratory stock	267	23.78
	sepia, inbred laboratory stock	302	28.31
	hybrids, sepia × forked	227	74.27
	hybrids, forked × sepia	237	78.43
B	Aloxe, natural population	296	60.50
	Montpellier, natural population	233	77.51
	hybrids, Aloxe × Montpellier	241	72.94
	hybrids, Montpellier × Aloxe	238	81.93
C	vermillion } isogenic stocks	173	22.72
	cinnabar }	173	17.46
	hybrids, $v \times cn$ } of the isogenic stocks	245	25.27
	hybrids, $cn \times v$ }	241	21.16

7.7. Advantage of inversion heterozygotes in sexual selection

Several comparisons between homo- and heterokaryotypes with respect to sexual selection have been made. In *Drosophila pseudoobscura*, Spiess (1970) found that males heterozygous for the gene arrangements in a chromosome tend to fertilize more females and to exhibit a shorter latency to copulation than homozygotes. For *Drosophila persimilis* Spiess (1970) showed that the relative advantage of inversion heterozygotes is greater if the flies are maintained in unfavourable temperature conditions.

An elegant demonstration of the advantage of inversion heterozygotes has been given by Brncic and Koref-Santibanez (1964) for *Drosophila pavani*. Observing the behaviour of pairs, they divided them into three classes according to: males having copulated within 30 minutes, males having displayed only courtship, and males without any sexual activity. The frequency of heterokaryotypes was highest in the first class and lowest in the third. Sperlich (1966) found in populations of *Drosophila subobscura* from the mediterranean region, with a higher degree of chromosomal polymorphism, higher frequencies of copulation than in northern populations containing a low number of structural heterozygotes.

In *Drosophila robusta*, Prakash (1969) did not find any superiority of heterokaryotypes over homokaryotypes when each chromosome was studied separately. But certain chromosomal associations clearly showed advantages of multiple heterokaryotypes. This is not surprising. Heterokaryotypes do not go with advantages in sexual and natural selection simply because the gene arrangements are different. They do so only if the heterokaryotype constitutes a favourable combination of gene blocks, or supergenes, which has been brought about by natural and sexual selection. For instance in *Drosophila persimilis*, Spiess and Langer (1964) found a considerable advantage in the mating speed of males with the homokaryotype WT/WT over the homokaryotype KL/KL, while the heterokaryotype WT/KL was intermediate. Thus, sexual selection may contribute much to the maintenance of genetical heterogeneity.

7.8. *Selective advantage of hybrids from crosses between wild-type stocks*

Another aspect of the advantage of heterozygotes has been studied in comparisons between several elements of the sexual behaviour of individuals from inbred lines and hybrids. Fulker (1966) observed in small populations of six virgin females and one male, kept together for 12 hr, the latency to the first copulation, the number of copulations per male, the number of fertile copulations, and the number of offspring. He compared the results for males of six inbred laboratory stocks and six types of hybrids obtained by crosses between stocks. The results are summarized in Table 7.7. In most cases the performances of the hybrid males are superior.

TABLE 7.7

Comparison between sexual performances of inbred and hybrid males of *D. melanogaster*.

	Genotype of males	Latency of first copulation	Observed number of copulations	Number of copulations resulting in fertilization	Number of offspring produced
Inbred lines	Edinburgh (Ed)	9.2	4.6	4.0	147
	6 CL	34.6	1.6	1.0	33
	Samarkand (S)	26.4	2.8	1.8	98
	Wellington (W)	16.4	4.2	3.6	236
	Oregon (Or)	18.2	2.8	2.4	158
	Florida (F)	14.8	3.6	2.8	118
Hybrids	F × Ed	3.6	6.2	5.4	302
	F × W	6.8	5.8	5.4	340
	Or × W	5.6	5.2	5.0	270
	6 CL × Ed	5.6	5.0	4.6	234
	6 CL × S	7.6	3.2	1.4	63
	Or × Ed	5.2	5.4	4.2	225

(After Fulker 1966)

7.9. Single-gene effects in sexual selection

Recessive mutations in the homozygous condition depress in most cases the general vigour and, consequently, exert also a deleterious effect on the sexual vigour of males. Several authors have presented data showing the inferiority, of mutant males, for example Lutz (1911) for abnormal wing venation Sturtevant (1915) for the loci white, vermillion, and forked, Reed and Reed (1950) for white, Petit (1951) for forked, Bastock (1956) for yellow, and Elens (1957) for ebony. Sometimes the effects of single genes are very conspicuous. The mean numbers of females fertilized by hybrid males, obtained from reciprocal crosses between two highly inbred laboratory strains of *Drosophila melanogaster* marked by the alleles "vestigial" and "yellow", were very different (Boesiger, 1962). When "vestigial" females were crossed with "yellow" males, the F_1 hybrid males (wild-type) fertilized 81.6 % of the females. The reciprocal cross produces males hemizygous for the sex-linked gene "yellow"; these males fertilized only 22.9 % of the females, despite the hybridization. The X-chromosomes of the males from the two reciprocal crosses are different ones, but it is reasonable to assume that the gene "yellow" itself is responsible for this large difference. Many other results, obtained with several yellow strains of different origin, have consistently revealed strong effects of "yellow" upon the sexual behaviour.

As Spiess (1970) pointed out, mating success is a very complex character and implies many factors, such as vigour, discrimination, preference, sex-drive, and sensitivity. One must add that there is no clear-cut difference between a specific sexual vigour and the general vigour of animals. This does not mean that there is a complete overlap, or that there is no genetically-based difference at all, but the interactions between the two kinds of vigour, which are both complex, do not permit a precise qualitative and quantitative determination of the roles of individual genes in sexual selection. The selective value is not an intrinsic and constant property of a gene. For natural selection as well as for sexual selection, one may determine the influence of a given gene under a given condition. This influence, improperly called "its selective value", depends on the genetic background, the frequency of this gene in the population, the genotypes of the other organisms in the ecological community, and very much also on the abiotic ecological conditions. For all these reasons, establishing any specific effects of a given gene on sexual selection and, *a fortiori*, on one of the elements of sexual selection, is difficult. The genetic study of the effects of "*v*" and "*cn*" on male vigour in sexual selection, as presented above, shows how fallacious the attribution of a given effect to a particular marker gene may be. Behavioural or other differences between two stocks marked by specific alleles cannot be attributed to these loci if the stocks are not coisogenic. Even if this elementary precaution has been taken, it is still possible that the observed effect is not due to the marker gene itself, but to a very closely linked gene.

7.9. *Selection for quantitative aspects of sexual behaviour*

Starting from a heterogeneous population of *Drosophila melanogaster*, Manning (1961) selected within parallel lines the first, respectively the last, ten pairs seen mating. In two lines he obtained a moderate increase over 25 generations of selection, and in two other lines subjected to the opposite selection pressure a decrease in mating speed. There were marked fluctuations over generations which may be attributed to unknown environmental factors. It would seem that the selection caused accumulations of alleles which affect mating speed: in the unselected control line about 60 % of the matings were observed during the first five minutes, in the selected lines the corresponding percentages were 10 % and 85 %, respectively. Hybrids between a fast and a slow selection line presented an intermediate mating speed. Further analysis by Manning (1961), Spiess (1970), and Kessler (1968) showed that the situation is a complex one and that several polygenic systems are involved. Not only polygenic interactions within the individual have to be considered, but also the reactions of the partner (see Kessler, 1968; *Drosophila pseudoobscura*). After twelve generations of selection for fast and slow mating speed, the mating indices turned out to be very different for different paired combinations of individuals from the control line and the two selected lines.

In the context of this chapter the following hypothesis may be advanced. Since selection yielded responses in both directions, and since the hybrids between the selected lines showed intermediate values, the unselected control lines and the original natural population were probably characterized by a high degree of heterozygosity for genetic factors responsible for mating speed. In other cases of selection for quantitative differences, it has been established that the populations returned more or less rapidly to the initial level of the trait selected after selection was abandoned. For instance, Kessler found in *D. pseudoobscura* a tendency in the line selected for high mating speed to return to the control level if selection is terminated. In a few cases (see, for example, Lerner, 1954, and Palenzona et al., 1972), it has been shown that the gain under selection was accompanied by a reduction in fecundity. Apparently, natural and sexual selection establish and control those gene pools which are well-adapted to the ecological conditions.

7.10. Conclusions

This chapter presents some data and arguments in favour of the hypothesis of efficient participation of sexual selection in the maintenance of the genetic heterogeneity of natural populations. Sexual selection doubtless has several other functions. It may lead to sexual isolation which, apart from or in combination with geographical isolation, is the main barrier preventing gene exchange between species. In cases of morphologically extremely similar species, as for example *Drosophila pseudoobscura* and *D. persimilis* or the *Anopheles gambiae* complex, or in the case of the *D. paulistorum* complex, sexual isolation reveals the existence of different species or incipient species. Some authors claim that intraspecific and sympatric sexual isolation between different genotypes might be considered as a mechanism for sympatric speciation.

Another problem is that of the differences observed in fecundity. Heterozygous males generally show a high fecundity. It has sometimes been said that male fecundity is practically unlimited because of the large number of spermatozoa. This is not so, not even in Drosophilids. One can measure the fecundity of males by counting the offspring obtained after one copulation with a female from a stock possessing a high egg-laying capacity (Boesiger, 1963). Table 7.8 shows the very large differences between the numbers of offspring of inbred and hybrid males, in addition to a less important difference in fecundity between inbred and hybrid females. The combined effects of high sexual vigour, high general vigour, and high fecundity of heterozygous males give them a great advantage over homozygotes and constitute a homeostatic mechanism for the maintenance of the heterogeneity of populations.

TABLE 7.8

Offspring produced by inbred and hybrid males and females of *D. melanogaster*.

| Parents | | Mean number | Coefficient of |
Females	Males	of offspring	variation
Aloxe	sepia, inbred	16.5 ± 3.6	148.9
Aloxe	forked, inbred	61.5 ± 14.6	153.3
Aloxe	hybrids, se/f	399.3 ± 18.1	30.3
Aloxe	hybrids, f/se	488.4 ± 16.4	21.5
sepia, inbred	hybrids, f/se	124.8 ± 22.7	110.8
forked, inbred	hybrids, f/se	159.4 ± 16.3	63.9
hybrids, f/se	hybrids, f/se	394.2 ± 13.4	21.2
hybrids, se/f	hybrids, se/f	405.6 ± 15.1	23.2

For females, the situation has been less well studied. Table 7.8 shows that hybrid females possess a higher fecundity than inbreds. Some of the results reported in this chapter indicate that homozygous, inbred females have a higher receptivity than heterozygotes. Spieth (1951) found in some races of

D. americana americana that if females have a low receptivity, if they do not easily mate, the males tend to be very active sexually. We found, especially for the stocks "vermillion" and "cinnabar", a relation between the degree of heterozygosity and the sexual behaviour of females and males. Heterozygous females have a low receptivity but a high fecundity. Since they probably mate in natural populations only once, it seems favourable that they do not waste their high genetical capacities by mating with a "sluggish", genetically less fit homozygous male. Such a heterozygous, very "resisting" female has to be stimulated intensively and persistently in order to accept copulation. Only a heterozygous male is able to do it. Since matings with homozygous males of low fecundity will not leave many offspring, natural selection and sexual selection must have favoured low receptivity in females and high sexual activity in males.

Finally, genetical analyses of a behaviour are usually very complex ones. A typological approach alone, looking for a direct and simple relation between a particular gene and the behaviour or element of behaviour, is in most cases useless. Let us hope that behaviour genetics will develop in the near future, not only because of its great intrinsic interest, but also because of its great significance for a better understanding of non-typological genetics and the complex interactions between gene pools and environments.

References

Bastock, M. (1956): A gene mutation which changes a behaviour pattern. *Evolution*, **10**, 421.

Boesiger, E. (1953): Activité sexuelle des femelles et des mâles de deux souches mutantes de *Drosophila melanogaster*. *C.R. Acad. Sci.*, **237**, 1180.

Boesiger, E. (1954): Sur la fertilité des mâles de deux souches mutantes de *Drosophila melanogaster*. *C.R. Acad. Sci.*, **239**, 1320.

Boesiger, E. (1958a): Influence de l'hétérosis sur la vigueur des mâles de *Drosophila melanogaster*. *C.R. Acad. Sci.*, **246**, 489.

Boesiger, E. (1958b): Déterminisme génétique de l'activité sexuelle des mâles de deux souches mutantes de *Drosophila melanogaster*. *C.R. Acad. Sci.*, **246**, 1304.

Boesiger, E. (1962): Sur le degré d'hétérozygotie des populations naturelles de *Drosophila melanogaster* et son maintien par la sélection sexuelle. *Bull. Biol. Fr. Belg.*, **96/1**, 3.

Boesiger, E. (1963): Comparaison du nombre de descendants engendrés par des mâles homozygotes et des mâles hétérozygotes de *Drosophila melanogaster*. *C.R. Acad. Sci.*, **257**, 531.

Boesiger, E. (1967): La signification évolutive de la sélection sexuelle chez les animaux. *Scientia*, **102**, 207.

Brncic, D. and Koref-Santibanez, S. (1964): Mating activity of homo- and heterokaryotypes in *Drosophila pavani*. *Genetics*, **49**, 585.
Dobzhansky, Th. and Boesiger, E. (1968): Essais sur l'Evolution. Masson, Paris.
Elens, A. A. (1957): Importance sélective des différences d'activité entre mâles ebony et sauvage dans les populations artificielles de *Drosophila melanogaster*. *Experientia*, **13**, 293.
Elens, A. A. and Wattiaux, J. (1964): Direct observation of sexual isolation. *Drosophila Inform. Serv.*, **39**, 118.
Fulker, D. (1966): Mating speed in male *Drosophila melanogaster*: A psychogenetic analysis. *Science*, **153**, 293.
Kessler, S. (1968): The genetics of *Drosophila* mating behaviour. I. Organization of mating speed in *Drosophila pseudoobscura*. *Anim. Behav.*, **16**, 485.
Lerner, M. (1954): Genetic Homeostasis. Oliver and Boyd, Edinburgh.
Lutz, F. E. (1911): Experiments with *Drosophila ampelophila* concerning evolution. The effect of sexual selection. *Publ. Carneg. Inst.*, **143**, 36 (lecture of 1909).
Manning, A. (1961): The effects of artificial selection for mating speed in *D. melanogaster*. *Anim. Behav.*, **9**, 82.
Merrell, D. J. (1949a): Mating between two strains of *Drosophila melanogaster*. *Evolution*, **3**, 266.
Morgan, T. H. (1909): Experimentelle Zoologie. Leipzig.
Morgan, T. H. (1932): The Scientific Basis of Evolution. London.
Palenzona, D., Rocchetta, G. and Jacuzzi, A. (1972): The relationship between fitness and response to selection in *Drosophila melanogaster*. *Theor. Appl. Genet.*, **42**, 65.
Petit, C. (1951): Le rôle de l'isolement sexuel dans l'évolution des populations de *Drosophila melanogaster*. *Bull. Biol. Fr. Belg.*, **85**, 392.
Petit, C. (1959): Le déterminisme génétique et psychophysiologique de la compétition sexuelle chez *Drosophila melanogaster*. *Bull. Biol. Fr. Belg.*, **92**, 248.
Prakash, S. (1969): Chromosome interaction affecting mating speed in *Drosophila robusta*. *Genetics*, **60**, 589.
Reed, S. C. and Reed, E. W. (1950): Natural selection in laboratory populations of *Drosophila*. II. Competition between a white-eye gene and its wild type allele. *Evolution*, **4**, 34.
Selous, E. (1929): Schaubalz und geschlechtliche Auslese beim Kampfläufer (*Philomachus pugnax*). *J. Ornith.*, **77**, 262.
Sperlich, D. (1966): Unterschiedliche Paarungsaktivität innerhalb und zwischen verschiedenen geographischen Stämmen von *Drosophila subobscura*. *Z. Vererbungsl.*, **98**, 10.
Spiess, E. B. (1970): Mating propensity and its genetic basis in *Drosophila*. In: Essays in Evolution and Genetics in Honor of Theodosius Dobzhansky. M. K. Hecht and W. C. Steere, eds. North-Holland Publ. Co., Amsterdam.
Spiess, E. B. and Langer, B. (1964): Mating speed control by gene arrangement carriers in *Drosophila persimilis*. *Evolution*, **18**, 430.
Spieth, H. T. (1951): Mating behavior and sexual isolation in the *Drosophila virilis* species group. *Behavior*, **3**, 105.
Sturtevant, A. H. (1915): Experiments on sex recognition and the problems of sexual selection in *Drosophila*. *J. Anim. Behav.*, **6**, 351.

CHAPTER 8

Variability in the aggressive behaviour of *Mus musculus domesticus*, its possible role in population structure

J. Busser, A. Zweep and G. A. van Oortmerssen

8.1. Introduction

Agonistic behaviour among conspecifics will shape the social structure of populations. In recent years, evidence has been obtained that the social structure of a population may affect its dynamics and, through this, its genetic composition. In microtine rodents living in favourable habitats, population size has often been found to fluctuate fairly regularly, but no direct environmental influences have been demonstrated. Chitty (1967, 1970) has suggested that such fluctuations might be caused by the presence of different genotypes, the fitness of which is determined by different levels of aggressiveness: the less aggressive types being adapted to a situation of low density and the aggressive types being adapted to high density. A high proportion of aggressive animals, in high-density populations, supposedly decreases the population size, so that with the passage of time the living-conditions for non-aggressive animals improve, leading in turn to an increase in numbers. In house mice, too, population numbers may fall and rise, sometimes showing outbreaks (Hall, 1927; Southern, 1954; Pearson, 1963).

Despite the extensive research on house mice, it has not yet been established whether all house mice show a demic population structure based on territoriality, or whether other social structures or variants of the demic structure exist. *Mus musculus* originates from Central Asia; from there the species has spread by different routes. Very likely, these sub-groups have diverged in evolution, so that now different subspecies can be distinguished (Schwarz and Schwarz, 1943; Zimmermann, 1950). Van Oortmerssen (1970) has demonstrated quantitative differences in behaviour between inbred strains of house mice which reflect behavioural adaptations to different habitats which are likely to have been present already in the feral populations from which the ancestors of the inbred strains have been taken. BALB mice appeared to be well-adapted to surface-living, showing territorial behaviour and demonstrating a sensitive aggression-flight balance. C57BL mice appeared to be more adapted to living in holes; they did not show territorial behaviour and their aggression-flight balance appeared to be relatively insensitive. Most research on feral house mice has probably dealt with *Mus musculus domesticus* (see, for instance, Anderson, 1970). However, studies like those of Eibl-Eibesfeldt (1950) and Newsome (1969a and b, 1970) may actually have been carried out with mice of another subspecies, since both authors, in contrast with reports on *M. m. domesticus*, describe large groups of mice that share all holes in a certain area. Newsome (1969b) has concluded that his mice were living in societies; Eibl-Eibesfeldt has called them "Groszfamilien". Crowcroft and Rowe (see Crowcroft, 1966) probably incorrectly claim that their family-groups confined in pens are similar to the "Groszfamilien" of Eibl-Eibesfeldt (see van Oorsmerssen, 1970). The findings indicate that the subspecies may differ considerably in agonistic behaviour, social structure, and genetic background.

The feral house mice of The Netherlands belong to the subspecies *Mus musculus domesticus* (Zimmermann, 1950). This subspecies is found in Western Europe, Canada, and the northern part of the U.S.A. In comparison to mice of other subspecies, they seem to be well-adapted to commensal life. As commensals, they have followed man all over the world in the last centuries. Especially along trade routes and in coastal regions in different parts of the world, one may therefore expect *M. m. domesticus* to be present. They may be found outdoors and indoors. In The Netherlands they inhabit all kinds of buildings and shelters the year around, but only during the summer are they regularly caught outdoors. Indoors they hold a monopolistic position, outdoors they must compete with other species such as *Microtus arvalis*, *Micromys minutus*, and *Apodemus sylvaticus*. Only the latter can occasionally be caught indoors. In the following we will deal mainly with *Mus musculus domesticus*.

These mice tend to display territorial behaviour, which leads to the formation of small groups of mice that live together and probably defend their territory against neighbouring groups as well as against intruders. However, the findings do not agree in all respects. It remains unclear how the territorial groups are composed, whether they constitute an isolated family-group (deme), or whether mice from elsewhere may be accepted, respectively implying strong inbreeding or a certain degree of outbreeding. Strong inbreeding as a result of isolation between family-groups has been proposed by Lewontin and Dunn (1960) as an explanation of the discrepancy between the expected and observed frequencies of *t*-alleles in natural populations of house mice. However, even in the case of very low interdemic migration rates (e.g. 3 %), one must assume an effective breeding size of the deme of less than 4 (Levin et al., 1969). This might mean that only the dominant male in the territorial group participates in the reproduction (DeFries and McClearn, 1972). In addition it is often assumed that the dominant status is normally passed on from father to son; this would enhance inbreeding considerably. A combination of this phenomenon with small size of the breeding unit would uphold the original hypothesis of Lewontin and Dunn. However, for neither of these assumptions has definite proof been obtained.

The dispersal ranges of house mice are rather small (up to 5 km). The home ranges are much smaller, usually not more than a few square meters. Small streams and ditches act as geographical barriers since house mice do not swim voluntarily. Conceivably, house mice living in different localities may be differentially adapted as a consequence of geographical isolation. Because of this we should be aware of possible differences between populations belonging to the same subspecies.

It will be clear that, in order to study the genetics of aggressive behaviour in houce mice, one should be informed about the role aggression plays in the social structure and dynamics of the mice used. We therefore made a demographic study into a local feral population and a number of laboratory populations. Results of this will be presented together with findings on the relations between attack latency, dominance, and reproductive success.

8.2. The "Kooipolder" population

In order to collect information on feral house mouse populations in our locality, one of us (Busser, unpublished results) has started a descriptive ecological study, using the catch-mark-release method in the Kooipolder. This polder (about 100 acres) is surrounded by canals (Fig. 8.1). It is used for small industries. In the middle of the polder a large aviary (10 × 15 m) is present. It is bordered by reed-beds on three sides and by a vegetable

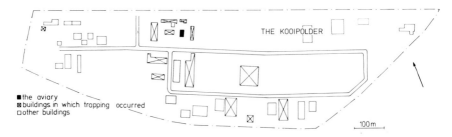

Fig. 8.1. The trapping area.

garden on the fourth. Because the group of mice in the aviary is probably representative for the entire Kooipolder population, it is described in more detail here. In the aviary and its immediate vicinity (about 0.6 acres), trapping was started in June 1970 by means of Longworth livetraps. Indoors, the traps stood 2 m apart, outside the aviary 10 m. They remained in one spot from the onset of the experiment. Except during trapping nights (two or three times a week), the mice could freely enter and leave the baited traps. A few direct observations at night indicated that probably all mice present were caught. The results were analyzed by a calendar of catches. These studies have not yet been completed.

Right from the beginning of the trapping experiment, a gradual increase in numbers occurred (Fig. 8.2), ranging from an average of 8 animals up to 26 animals ten months later (Phase I). At that time (May 1971), the aviary population collapsed and numbers dropped to the original level within three to four months (Phase II). After September 1971, the numbers continued to decrease more slowly for the next twelve months (Phase III). After that period until September 1973, only rarely a mouse was caught (Phase IV). The increase during phase I was caused by new litters being born throughout the winter up till April 1971. The young mice were caught at the age of three to four weeks, but most of them disappeared some time after the first catch, males on the average at 3 weeks and females at 10 weeks. Thus 6 weeks after the first catch, 89 % of the juvenile males (N = 113) and 65 % of the juvenile females (N = 94) had disappeared. Of the juveniles present for a longer period, the males on the average stayed 11 weeks, the females 23 weeks.

The trapping results furthermore revealed that the aviary contained two territories. The males restricted their activity ranges to their own territory, but the females were frequently caught in either one. Within 7 months, 34

Agression and population structure in mice

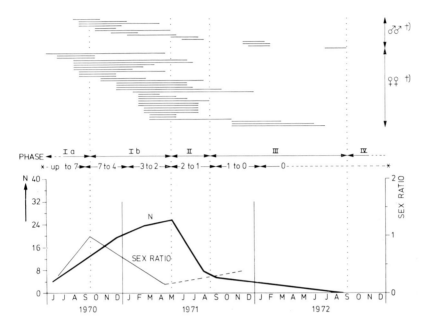

Fig. 8.2. The aviary population. N = Number of animals known to be present. * = Increase and decrease in number of young born per week over indicated period. † = Calender of catches for animals that stayed longer than six weeks. For further explanation, see text.

instances of females moving from one territory to the other were recorded. Occasionally mature animals intruded into the aviary, of which some had been caught elsewhere before. Of these intruders (3 males and 9 females), some stayed longer than one or two weeks, viz. two males stayed 8 and 34 weeks, respectively, and seven females stayed 5, 5, 8, 14, 21, 23, and 47 weeks, respectively. Another remarkable result are the changes in the sex ratio. During phase I it rose, at first, from 0.3 to 1 (Phase Ia, Fig. 8.2), but it went down to its original level by the time phase II began (Phase Ib, Fig. 8.2). Over phases II and III, the sex ratio appeared to be rather variable, its mean value slowly rose again to a value of about 1. During the increase phase to halfway the decrease phase, nearly all mice caught were in good physical condition. Later on we found a higher percentage of mice, especially young ones, which showed a suboptimal physical condition.

It seems clear, then, that these mice live in territories each containing one to four mature males, three or more females, and a number of young born in that group. Contrasting with the findings of Anderson (1965), these demes do not seem to be strongly isolated, since quite regularly females and sometimes also males will get the chance to intrude into the demes and settle.

Their numbers, as has also been shown by others, rise and fall in a way that reminds us of similar phenomena in field mice (Krebs et al., 1973). Demographic changes in other house mouse populations, however, were clearly related to the seasons (Berry, 1968; Newsome, 1969a, b). In the Kooipolder population this was certainly not the case, since the production of offspring went on throughout the winter of 1970–1971. This might have been due to favourable environmental conditions in the aviary at that time, but this cannot explain the sudden collapse of the population in the summer of 1971, when conditions were at least as good. It is possible that in the summer of 1971 many mice left the aviary to settle in the fields. If so, it remains unclear why the remaining portion did not grow in number again during the next winter, despite the presence of just matured males and females. Since no change in environmental conditions could be detected, one might ask whether intrinsic changes of the population were responsible. The changes in the sex ratio observed in the Kooipolder population may be considered as an indication that such intrinsic changes do occur. Changes in social behaviour which affect the social structure and composition of a population might exert similar effects on house mouse populations as those proposed by Chitty (1967) for field mice. It therefore seemed worthwhile to attempt to trace in our house mice phenotypes differing for aggressiveness, dominance, or both.

8.3. Dominant, subordinate, and tolerant males

In preliminary experiments, the behaviour of inbred male mice (CPB-s/Gro) in their home cage was studied in confrontations with another male. CPB-s mice tend to behave territorially, as previously shown by van Oortmerssen (1970). The results suggested that two types of males can be distinguished: an aggressive type that attacks soon after encountering its opponent, and a non-aggressive type that attacks after a delay or not at all. The latter shows frequently grooming or social grooming. The existence of these two types was confirmed provisionally in a subsequent test in which six males, three of each type, were confined in one cage. The following events were observed:

Fighting occurred between the mice of the aggressive type until one of them became dominant. The defeated animals huddled together in a place where they could avoid further attacks by the dominant male. They fled from every other mouse when outside their hiding-place. Males of the non-aggressive type never started a fight. Although these mice were also attacked by the dominant male, they seemed to be less disturbed by this than the defeated aggressive ones.

After the first male (1) had become dominant, we took him out. One day and a half later, a second aggressive-type male (2) achieved dominance. After this animal had been removed no signs of dominance were observed for several days, but gradually one of the non-aggressive males (3) developed some hostile behaviour towards the others. This male was taken out, leaving two non-aggressive males (4 and 5) and a defeated aggressive male (6). For the next few weeks no aggressiveness was seen between them, after which they were removed in the order indicated. All six were kept in isolation for a month or more. Then, they were put again into the same territory in reversed order, starting with 5 and 6. Male 6 became dominant over males 3, 4, and 5, but male 2 took over after his reintroduction. He remained the dominant one, since male 1 did not succeed in taking over.

These preliminary results justified a more extensive and intensive study concerning the existence of aggressive and non-aggressive types of males and their role in populations. This study was carried out by P. P. van der Molen (1973, unpublished) using the inbred strains CPB-s/Gro and C57BL/LiA/Gro, their F_1 hybrids, and feral *M. m. domesticus*. Of each group he observed a number of family populations (C57BL, 8; CPB-s, 12; F_1 hybrids, 10; feral mice, 2). A population consisted of four brothers and two sisters that stayed together from birth. At the age of 2 to 3 months, they were introduced into a large cage ($2 \times 1 \times 1$ m) in which a small elevated platform (20×20 cm) was present that could be reached only by climbing a rope. Three nest-boxes were present on the floor.

The results of this study confirmed our earlier impression. In all populations, except the C57BL, some males (aggressive males) became actively involved in fighting. Soon one male (a so-called α male) became dominant and chased the other males. The other aggressive males (ω males) retreated to the platform where they lived as outcasts. A few of these at first engaged in combat on the platform, but this stopped as soon as the physical condition of the mice deteriorated. After that, they peacefully huddled together and came down only for a quick meal, mostly at a time the dominant male was inactive.

In all F_1 and feral populations and in most CPB-s populations, males were present that never initiated a fight (β males). These were also subjected to attacks by the α male, some became wounded sometimes, but this did not induce them to retreat to the platform. They seemed to tolerate the aggressive behaviour of the α male. They apparently stimulated aggression in the α male to a lesser extent than the ω males did, for the number of attacks by the α male dropped considerably soon after the ω males had retreated; the β males could then more or less freely roam the cage. A short time after the dominant males had been taken out or had died, the dominant status was taken over by ω males in all instances, never by β males. This applied only if the α male had been the superior one for a relatively short time. If he had been dominant for a longer period, a β male sometimes assumed a slightly dominant status.

In the Kooipolder population, Busser could distinguish two groups of juvenile males. Males leaving the population within 6 weeks after the first catch may represent non-tolerant aggressive males (ω). The other males stayed much longer; they may be β males that tolerate aggressive behaviour. The findings for juvenile females were similar, but generally they seem to be more tolerant than males. Possibly, tolerant and non-tolerant types are found also among females. A female in oestrus present in the laboratory populations aroused all males. On some occasions even the ω males climbed down and faced the aggression of the α male. The β males as well as the α male, and on a few occasions even ω males, were seen mating. Judging from their behaviour and from vaginal plugs, both the α male and the β males achieved successful matings.

The C57BL populations revealed quite a different picture. Only in one population an aggressive dominant male was found; he was extremely aggressive, attacking not only males but also females. In the other C57BL populations all animals lived more or less peacefully together. Although some quarreling occurred, no animal was chased to the platform. This is in accordance with previous findings by van Oortmerssen (1970), who concluded that C57BL mice show a social structure different from that of territorial mice. Since the F_1 mice showed territorial behaviour, we may conclude that many genetic factors controlling behaviour that causes territoriality are dominant.

8.4. A possible role of tolerance in population dynamics

Van der Molen (personal communication) proposed an alternative hypothesis with regard to the role of agonistic behaviour in the regulation of the numbers of house mice. In growing populations of house mice of the territorial type, selection for tolerance might occur instead of selection for aggressiveness as proposed by Chitty (1967) for field mice. After founding a territory, a particular male may generate both aggressive and tolerant juveniles. As soon as they have become mature, the former will leave and settle elsewhere, the latter will either stay or leave much later. If these β animals sire offspring, as they seemed to do in our experimental populations, the proportion of young β males in the next generation is expected to increase if the two behavioural phenotypes indeed are genetically determined. This would make it possible for large numbers of animals to occupy a small area. Large numbers have been found for instance by Selander (1970) in barns and hen houses. His analyses of biochemical variation in such populations suggested the presence of a number of adjacent demes, each containing a large number of animals. We have observed β males, and even females, taking part in defending the territory. Thus, in stable habitats with firmly established territories, the territories could persist even after the original dominant male has died. Until now this situation has never been observed in the field, however.

When the fraction of tolerant mice in a particular habitat goes up, nomadic aggressive males may get a chance to settle between or even within established territories. Such invasions would lead to one of the following phenomena: Either an intruding α male is accepted by the resident tolerant inhabitants, which would merely result in a new genotype contributing to the reproduction within a deme, or the intruder chases away many residents, resulting in mass migration in that area. This would give the impression of a plague. If it were true that higher densities of populations are accompanied by selection for high aggressiveness, the occurrence of plagues would be highly improbable. The first reason is that long before densities sufficiently high for a plague to occur are reached, the increased aggressiveness will upset the social organization and young mice will not survive. Secondly, there are contradictory data on the physical condition of mice in plagues. Newsome (1969, p. 356) states that all his captured mice were low in body weight, reproductively inactive, covered with mites, and suffering from severe anaemia. He does not report any wounds. By contrast, Crowcroft (1966, p.

115), reporting on a plague in the same Australian district (South Hummocks near Port Wakefield), saw only mice that were well-fed and very active. On the basis of our hypothesis one should initially expect healthy well-fed mice in plagues and later mice in suboptimal physical condition, whereas with the hypothesis of increased aggressiveness one should expect only mice in poor physical condition and covered with the wounds received during many aggressive encounters.

8.5. Dominance and reproduction

The finding that subordinate males participate in reproduction contrasts with findings of DeFries and McClearn (1970). They tested the relation between social dominance and reproductive success in so-called triads, consisting of three cages connected to each other with Y-shaped plexiglass tubing. The animals could freely move from one cage to another. In addition to a number of females, two or three males from different strains (inbred and outbred) were put in. In nearly all triads one male became the dominant. Paternity was determined on the basis of coat colour. Among the 169 litters obtained from 75 triad experiments, in each of which a dominant male was present, 91–95 % (pooled results) were sired by the dominant male, the others were either fathered by subordinates or were mixed litters. From these experiments DeFries and McClearn concluded that in the field the dominant male has a much higher probability of contributing to the gene pool than a subordinate.

The experimental triad populations, however, do not seem in any way to be comparable with natural populations that inhabit a territory; the males in the triads did not form a natural territorial group. Clearly all subordinate males were wounded. From several studies on natural populations, including the one reported in this chapter, we know that two or more males may live together in one territory. Without confinement, as in the aviary population described above, most males and females leave, only a few stay and tolerate each other. In large confinements, like the pens of Crowcroft and Rowe, similar phenomena can be observed. Thus, the probability of experimentally combining tolerant males and females is very small. In some triads subordinate males managed to sire offspring (5.9 % of the litters). In these triads the males very likely tolerated each other, reflecting a situation that is found under natural conditions. DeFries and McClearn refute the objection that their triad populations are not representative by pointing out that wounds are also found under natural conditions. However, the examples given by them concern populations with a considerable amount of migration. Lidicker (1966) has shown that peaks in the number of wounds correspond closely to peaks in individual movement, i.e. when large numbers of mice intrude into existing territories. Rowe and Redfern (1969) have shown that such intrusions will build up the intraterritorial aggressiveness.

8.6. Attack latency and reproduction

The results of a selection experiment supported the presence of distinct behavioural phenotypes in our house mice. Male mice were selected by one of us (van Oortmerssen) for long and short attack latencies. This parameter was chosen because a male is most effective in defending his territory if he attacks the intruder right after meeting him. This selection experiment, in which CPB-s and feral mice were used, failed because the lines died out in the first and second generation.

In order to measure the attack latency in an environment that resembled the natural situation as closely as possible, special test cages ($120 \times 30 \times 30$ cm) were designed which contained three compartments divided by transparent sliding doors. The test males were kept in litter groups until maturity, they were then separated and provided with two females. At the age of 3 to 4 months, one male was put in the left compartment (A) of a test cage with nesting material. Prior to testing they lived for at least two days in this compartment. Once a day, the sliding door that gave access to the middle compartment (B) was opened for about 10 min to enable the test mouse to explore the area adjacent to its territory. On the successive test days a strange standard opponent was introduced into the third compartment (C). Compartment C is a small transparent box giving access to compartment B by a perforated transparent sliding door. Thus, a test mouse in B could see, hear, and smell the opponent. All opponents came from the CPB-s strain; they behaved in a standard way due to regular training for non-aggressiveness in a strange environment. After introduction of an opponent into C, the sliding door between A and B was opened. Two latencies were measured: firstly, the time to sniffing at the sliding door B-C, measured from the moment of entering B (meet latency), and, secondly, the time elapsing between the first contact with door B-C (at this moment the door was opened) and the first attack (attack latency). As soon as one of the mice attacked, the animals were separated to avoid a possible influence of combat on subsequent tests. Each animal was tested 5 times with a maximum score per test of 1800 seconds. Mean scores were calculated for each animal.

Only a slight association was found between meet latency and attack latency (Spearman's rank correlation coefficient, $r_s = 0.14$; $P < 0.01$). This confirmed our impression that house mice notice each other only at close quarters, which indicates that our measure of attack latency probably is a correct one. With respect to attack latency, two distinct categories of males were detected. Firstly, there were males that attacked in all 5 tests. These males, without exception, showed short latencies, the means never exceeding 400 sec. Secondly, there were males that once or more often failed to attack; they showed longer latencies, all means being higher than 400 sec. Some of these males never showed any attack at all. Males of the long latency class, unlike the other ones, displayed a variety of social behaviours in the test situation, ranging from frequent nosing and social grooming to attempted copulation, which depended on the docility of the opponent. The latter finding may indicate that these males were sexually motivated. Of CPB-s mice in the P-generation of selection, 47 males were tested, 24 of which turned out to belong to the short latency class and 23 to the long latency class. For 20 feral mice tested, these numbers were 11 and 9. For the inbreds as well as the feral mice, the breeding results of low and high latency groups appeared to be quite different. In the low latency class, males proved to be fertile in percentages of 83 and 81, respectively, but in the high latency class these percentages were 25 and 22.

Similar results were obtained by Zweep (unpublished data). He tested 32 feral mice for attack latency on their home ground in encounters during which they were allowed to fight to victory or defeat: 17 turned out to belong to the short latency class, 15 to the long latency class. In the short latency class 82 % of the males appeared to be fertile; in the long latency class 20 % were fertile, 40 % were fully infertile, and the remaining 40 % produced only one litter in their whole life. Some of these litters consisted of one or two pups only. All males had ample opportunity to prove their fertility. In the last experiment, low latency always led to victory; in the high latency class 70 % of the males were defeated sooner or later. Clearly, the results among the three groups of mice are highly consistent. Table 8.1 shows the pooled results. It can be concluded that the aggressive males in a population have a much better chance to produce offspring than the non-aggressive males and that the dominant male in a territorial breeding unit very likely is responsible for the production of most of the young. The question remains to what extent tolerant non-aggressive males take part in the reproduction in a natural population. From preliminary observations on laboratory populations we expect it to be very small indeed. Further research is needed to answer this.

TABLE 8.1

The relation between attack latency and reproductive success in feral and inbred male mice. LL = low latency; HL = high latency.

Strain	Number of ♂♂ tested	Number of LL	Number of HL	Reproductive success of LL ♂♂ N	%	Reproductive success of HL ♂♂ N	%	
CPB-s	47	24	23	20	83	6	25	*)
Feral	20	11	9	9	81	2	22	*)
Feral	32	17	15	14	82	3	20	o)
Total	99	52	47	43	83	11	23	

*) Territorial borderline encounters. Results obtained by van Oortmerssen.
o) Homeground encounters. Results obtained by Zweep.

8.7. Genetics of aggressive behaviour

It is obvious that the ideas advanced in this chapter will require further support. The hypothesis of two behavioural morphs playing an important role in the social structure and the genetic composition of territorial house mouse populations is amenable to further verification. It may provide us with a clue in our attempts to clarify the genetic aspects of aggressive behaviour in house mice. We need to know now whether the two phenotypes are genetically determined, or whether they are shaped by specific experiences, or both.

Selection for attack latency resulted in complete extinction of the high as well as the low latency lines in the second generation of selection. This phenomenon resembles that found in previous selection experiments with fraying, a nestbuilding behaviour character observed in CPB-s mice (van Oortmerssen, 1970). For fraying, it has been shown that a balanced polymorphism is responsible for the genetic variability in this character, involving variation in only one factor or gene block. Animals heterozygous for the factor for fraying appeared to be fully fertile, the homozygotes either were completely infertile or they produced infertile offspring. The similarity of effects in the two selection experiments support the idea that aggressive and non-aggressive types of males reflect a genetically determined balanced polymorphism. If this were true, we might perhaps consider the greater part of the animals with a short latency to be heterozygotes. In every generation, apart from heterozygotes, homozygotes will appear, demonstrating short and long latencies concomitant with different reproductive capacities. In that way, variability in attack latency would be maintained in the population. An important question is whether short attack latency is positively correlated with victory, and long latency with defeat. Some indication of this was found in the data of Zweep presented above.

Our results are at variance with those of Lagerspetz (1964; p. 51). In a selection experiment on aggression, using a complex rating method, she did neither find any differences with respect to reproductivity between aggressive and non-aggressive selection lines of house mice, nor did she find evidence for a balanced polymorphism. This might be due to differences in taking care or handling of the mice or to methodological differences. An alternative possibility is that our findings and those of Lagerspetz reflect the situation that, in house mice of different origins, different mechanisms exist which control social behaviour and population structure.

References

Anderson, P. K. (1965): The role of breeding structure in evolutionary processes of *Mus musculus* populations. In: Mutation in Population, Proceedings of the Symposium on the Mutational Process, Prague. R. Honcariv, ed. (Academia, Prague) pp. 17–21.

Anderson, P. K. (1970): Ecological structure and gene flow in small mammals. In: Variation in Mammalian Populations. R. J. Berry and H. N. Southern, eds. (Academic Press, London, New York) pp. 299–325.

Berry, R. J. (1968): Ecology of an island population of the house mouse. *J. Anim. Ecol.*, **37**, 445.

Chitty, D. (1967): The natural selection of self-regulatory behaviour in animal populations. *Proc. Ecol. Soc. Aust.*, **2**, 51.

Chitty, D. (1970): Variation and population density. In: Variation in Mammalian Populations. R. J. Berry and H. N. Southern, eds. (Academic Press, London, New York) pp. 327–333.

Crowcroft, P. (1966): Mice All Over. (Foulis, London).
DeFries, J. C. and G. E. McClearn (1972): Behavioural genetics and the fine structures of mouse populations. A study in micro-evolution. In: Evolutionary Biology, Vol. 5. Th. Dobzhansky, M. K. Hecht and Wm. C. Steere, eds. (Appleton-Century-Crofts, New York) pp. 281–291.
Eibl-Eibesfeldt, I. (1950): Beiträge zur Biologie der Haus- und der Ährenmaus nebst einigen Beobachtungen an anderen Nagern. Z. Tierpsychol., 7, 558.
Hall, E. R. (1927): An outbreak of house mice in Kern County, California. Univ. Calif. Publ. Zool., 30, 189.
Krebs, D. J., M. S. Gaines, B. L. Keller, J. H. Myers and R. H. Tamarin (1973): Population cycles in small rodents. Science, 179, 35.
Lagerspetz, K. (1964): Studies on the aggressive behaviour of mice. Ann. Acad. Sci. Fenn. B., 131, 1.
Levin, B. R., M. L. Petras and D. I. Rasmussen (1969): The effect of migration on the maintenance of a lethal polymorphism in the house mouse. Amer. Naturalist, 103, 705.
Lewontin, R. C. and L. C. Dunn (1960): The evolutionary dynamics of a polymorphism in the house mouse. Genetics, 45, 705.
Newsome, A. E. (1969a): A population study of house mice temporarily inhabiting a South Australian wheatfield. J. Anim. Ecol., 38, 341.
Newsome, A. E. (1969b): A population study of house mice permanently inhabiting a reed-bed in South Australia. J. Anim. Ecol., 38, 361.
Newsome, A. E. (1970): An experimental attempt to produce a mouse plague. J. Anim. Ecol., 39, 299.
Oortmerssen, G. A. van (1970): Biological significance, genetics and evolutionary origin of variability in behaviour within and between inbred strains of mice (Mus musculus). A behaviour genetic study. Behaviour, 38, 1.
Pearson, O. P. (1963): History of two local outbreaks of feral house mice. Ecology, 44, 540.
Schwarz, E. and H. K. Schwarz (1943): The wild and commensal stocks of the house mouse, Mus musculus Linnaeus. J. Mammal., 24, 59.
Selander, R. K. (1970): Behaviour and genetic variation in natural populations. Amer. Zoologist, 10, 53.
Southern, H. N. (1954): Control of rats and mice. Vol. 3. House mice. (Univ. Press, Oxford).
Zimmermann, K. (1950): Zur Kenntnis der mitteleuropäischen Hausmäuse. Zool. Jb. (Syst.), 78, 301.

PART II

Phenogenetic and regulatory aspects

CHAPTER 9

Activity and sexual behaviour in *Drosophila melanogaster*

B. Burnet and K. Connolly

> as a representative
> of the insect world
> i have often wondered
> on what man bases his claims
> to superiority
> everything he knows he has had
> to learn whereas we insects are born
> knowing everything we need to know
>
> don marquis

9.1. Introduction

The principal frame of reference in psychology is human behaviour and experience. Psychologists tend therefore to be interested in behaviour which has direct links to human concerns and human problems. From this it follows that on the whole psychologists will choose to work with higher species and in particular with mammals. This is amply borne out by even a cursory examination of the literature where it is clear that, apart from man himself, the favourite animal is the rat. Mammals exhibit a rich repertoire of behaviour and many of the characteristics of man, for example learning and problem solving, can be demonstrated to varying degrees. On the whole the farther an animal is removed from our own species the less confident we are in ascribing to it components of behaviour which we know to exist in man. Whilst this position is understandable and indeed reflects a most necessary caution it nevertheless lacks a carefully thought-out rationale. Dethier (1964) has challenged the justification for this and argued that an arbitrary selection of some hazy point on the phylogenetic scale beyond which common elements do not exist is open to question. From his own studies he has shown that motivation can usefully be studied in insects and perhaps even concepts such as mood are appropriate.

The insect brain differs greatly from the mammalian brain; indeed Vowles (1961) has put forward the hypothesis that the insect nervous system differs from the vertebrate at all functional levels. Since the cell bodies are small and lie at the periphery rather than being surrounded by dendrites, and because the receptive areas of dendrites are smaller, it has been argued that there are severe limitations on integration within the insect nervous system. These differences cannot be compensated for by increasing the number of cells since not only does overall brain size impose a limitation but the number of cells in sense organs and motor systems is also limited. Whilst these differences do not rule out a rich and varied behavioural repertoire they do suggest that the behaviour of insects is likely to be simpler and, perhaps more important, less variable.

Behaviour genetics has three major areas of concern: (1) Formal genetics is concerned with the extent to which variation in a behavioural phenotype is genetic in origin, and with mapping the location on chromosomes of genes which affect behaviour. (2) Population genetics deals with the factors influencing the allele frequencies in populations. This area is thus concerned with evolution and such questions as isolating mechanisms many of which are essentially behavioural in nature. (3) Physiological and developmental genetics seek to explore the pathways between the primary gene product and behaviour. Studies in this field thus involve gene-environment interaction at a physiological level and lead to explorations of biosynthetic pathways. *Drosophila* have been used in behavioural studies in each of these domains though our concern here is directed primarily towards causal explanations, hence our focus is in the third, physiological, area.

Behavioural characters are almost invariably more remote from the primary gene product than either physiological or biochemical characters. This has led some to the view that single-gene studies are of very limited utility and that the biometrical approach has greater power in this context. Whilst it is clear that biometrical models are necessary to explore some important questions, it is our view that the single-gene approach coupled in certain cases with selection studies provides a powerful tool in the search for causal explanations in behavioural biology. In essence one is subjecting a living system to controlled perturbations in order to examine their consequences on the functioning of the system and then make inferences regarding the manner of its organisation.

Behavioural studies with *Drosophila* have been concerned with fixed action patterns and not with instrumental learned responses. The suggestion that *Drosophila* may be capable of instrumental learning (Murphey, 1967) has not been confirmed (Yeatman and Hirsch, 1971; Murphey, 1973) and so far as we are aware there is, as yet, no convincing evidence of learning in this animal. This necessarily makes *Drosophila* an experimental organism of limited interest to those primarily concerned with learning ability. Recently, however, Benzer (1973) has reported that under appropriate conditions conditioned avoidance responses have been demonstrated in *Drosophila*. Publication of the detailed observations will be awaited with interest. It may well be that past failure to demonstrate learning ability in *Drosophila* reflects a lack of insight into what constitutes an appropriate learning situation for a fly. The absence of demonstrable learning ability is for some the source of serious misgivings about the value of *Drosophila* studies in behavioural genetics. It, however, has its advantages because, on the whole, the fixed action patterns which make up the behaviour of these flies are stable. Given that for any genetic analysis it is necessary to objectively determine units for measuring the phenotype, it will be easier to do this when one is dealing with fixed action patterns rather than behaviours which are in a state of flux as a consequence of the animal's learning. This is not to argue that previous experience is an unimportant consideration in attempts to understand *Drosophila* behaviour nor that learning cannot be objectively measured. It is simply that in the search for appropriate units a relatively stable phenotype has great advantages.

None of the foregoing should be taken to imply that the behaviour of *Drosophila* is simple and fixed in any gross sense. The sexual behaviour of *Drosophila*, which is described below, involves an elaborate sequence of action patterns and the interaction between males and females is often highly complex. The subtlety of this communication process is made apparent by experiments which cause distortions to be introduced. By comparison with the behaviour of mammals, that of *Drosophila* is a good deal more stable and less influenced by early experience and learning. From this standpoint it has many advantages when seeking genetic and physiological correlates and antecedents.

Given that our objectives are to discern genetic components of behaviour and to explore the development of behavioural phenotypes our task is one of identifying the particular gene or genes in question, then one of determining the site of their influence on behaviour. In the case of a single gene mutation which is linked with a change in behaviour of a specific kind it is difficult to say with any certainty what the primary focus of the genetic alteration is. The site at which the mutant gene exerts its primary effect may be far removed from the affected organ but in order to trace the path from gene to behaviour it is necessary to find the primary focus at which the gene acts and exerts its effect in developing behaviour. Hotta and Benzer (1970) have made use of genetic mosaicism to 'dissect' the nervous system of *Drosophila*. Using a stock carrying a ring-X chromosome it is possible to produce individuals in which a portion of the body is mutant male whilst the rest is female. In such sex mosaics the division line between normal and mutant tissues can occur in various orientations. In their studies on abnormalities of visual function in flies, Hotta and Benzer (1970) found that mutant eyes always functioned abnormally irrespective of the amount of normal tissue elsewhere, indicating that the primary source of visual deficits in these animals is in the eye itself. In a more recent paper Hotta and Benzer (1972) have extended Sturtevant's concept of a fate map to develop a technique for constructing maps of internal behavioural foci with reference to external body landmarks. Because of a correspondence between the position of cells on the surface of the *Drosophila* blastula and body parts which they will eventually form, it is possible to build up a fate map of the blastula and locate on this map mutant behavioural foci. From the position of a focus it is possible to infer what the corresponding adult tissue will be, thus some mutations can be shown to result in abnormal behaviour because of defects in sense organs, in the nervous system, or because of abnormalities in effector mechanisms. These techniques offer a means of unravelling the immensely complex and beautifully integrated adult organism. The progression is thus from the one-dimensional gene to the two-dimensional blastoderm, thence to the three-dimensional organism and finally to the fourth dimension of time in which the organism's behaviour occurs.

In our view *Drosophila* is a singularly valuable organism in which to investigate the genetic basis of behaviour. What follows is an account of researches in our laboratory over the past few years into activity and sexual behaviour of the fly.

9.2. Selection for changes in locomotor activity

Most healthy animals engage in a variety of motor activities some of which have no obvious or immediate significance. Many of these behaviours appear to be independent of external stimulation or any obvious internal need state. Thus when specific behaviours are observed, identified, and logged separately there remains a residue which may be appropriately labelled 'general activity' or more appropriately spontaneous locomotor activity. Even when an animal's major physiological needs are satisfied it will rarely remain quiescent. Further there are marked differences between animals in the level of activity which they show.

There are difficulties of definition with respect to the term spontaneous activity. Hinde (1966) uses the term spontaneity to refer to changes in the output of a system without any corresponding change in the input. However, in so doing he points out that the distinction between spontaneous and environmentally elicited activity is one of degree. This is because any activity which appears to be spontaneous may be the end product of a long series of physiological events which were originally initiated by an external stimulus.

Beginning from a heterogeneous stock of Pacific strain wild-type *Drosophila melanogaster*, Connolly (1966, 1968) selected for increased (high active) and decreased (low active) levels of locomotor activity defined as distance covered per unit time. An open-field apparatus made from a perspex box ($10 \times 10 \times 0.5$ cm) with the lid marked off in cm squares was used to measure activity. A single fly was introduced into the apparatus, allowed some time to recover from the disturbance caused by the transfer, and the number of squares crossed in one minute was recorded. The response to selection over some 115 generations is shown in Fig. 9.1. The unselected control population maintains an essentially stable level rarely varying outside the limits 20–40, though there are inter-generation fluctuations which probably represent for the most part minor environmental variations. By about generation 25 of the selection the mean score of the inactive line was only 2.5, leaving little margin for further change. Beyond generation 33 the open-field apparatus was an insensitive measure since a large proportion of the animals scored zero. Between generations 25 and 60 the active line showed little or no systematic change, though the inter-generation fluctuations were increased somewhat. From generation 60 onwards the active line showed a substantial increase in the level of locomotor activity. This accelerated response to selection may well be due to recombination between linked genes controlling high activity which will in turn lead to the formation of "extreme" gametes which, being favoured by selection, increase in frequency.

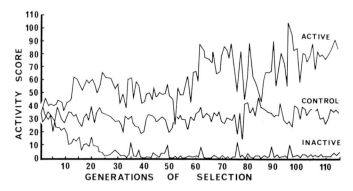

Fig. 9.1. Response to selection for adult locomotor activity.

Relaxed selection lines were set up at generation 94 of selection and maintained for 20 generations. Over 20 generations of relaxed selection there was no indication of any systematic regression of the selected lines to the level of the unselected control population. The absence of any drift towards the activity level of the controls provides some evidence for the stability of the gene pools of both the experimental populations. The heritability of locomotor activity was estimated from a mid-parent-offspring regression which gave a value of $h^2 = 0.51 \pm 0.10$.

Not all locomotor activity is spontaneous. Connolly (1967) distinguished between activity and reactivity and examined the selected lines in this respect. Reactivity is not limited to responses to other animals but may also occur in relation to inanimate features of the environment. On the basis of the test employed, the active and inactive lines showed no differences in reactivity. Connolly thus concluded that activity and reactivity are under the control of different sets of genes, and further that the major phenotypic difference between the two selected lines was one of activity level.

Although studies on activity in *Drosophila* have hitherto focused exclusively on adult behaviour, patterns of larval activity are also of considerable interest. In contrast to the aerial life of the adult fly, the larva has a semi-aquatic life style. On a suitable food medium it moves about slowly, whilst feeding in the substrate. The larva feeds by shovelling with the mouth hooks

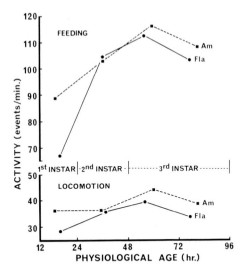

Fig. 9.2. Feeding and locomotor activity in larvae at different ages in the Amherst (Am) and Florida (Fla) inbred wild-type strains. Larval age is measured in hours from the time of eclosion from the egg (After Sewell, 1973).

which are actuated by the cephalopharyngeal sclerites and this is accompanied by a sucking action of the pharynx. The rate of larval feeding is easily monitored under a dissecting microscope by placing the larva in a drop of fresh 2 per cent aqueous yeast solution. After a short settling-down period the number of cephalopharyngeal retractions per minute of continuous feeding can be recorded with a hand tally counter. Alternatively, larvae can be observed feeding in situ under axenic conditions in a translucent culture medium such as Sang's Medium C (Sang, 1956). Locomotor or ambulatory activity may be scored by placing the larva in an open-field situation on a plain agar substrate in a Petri dish and recording the number of waves of sequential contraction which pass along the body segments in locomotion.

Under optimal environmental conditions larvae feed continuously with short pauses at the ecdysis between instars. Fig. 9.2 shows that the feeding rate is age-related. It is initially low in first instar larvae rising steadily to a maximum during the first half of the third instar. The feeding rate falls rapidly towards the end of the third instar when the larvae show reversal of their geotactic response and move out of the substrate to pupate. The rate of locomotor activity remains relatively constant throughout the larval period.

Larval feeding rate responds readily to selection. The response to bidirectional selection in 72-hour-old larvae from a heterogeneous base population is illustrated in Fig. 9.3. There was a steady change in feeding rate in both directions in each pair of selected lines. The realised heritabilities based on the first 20 generations of response, averaged for the two replicate lines, were 0.18 in the high direction and 0.12 in the low direction.

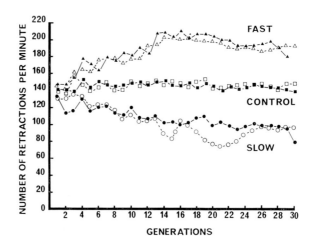

Fig. 9.3. Response to selection for larval feeding rate measured in third instar larvae aged 72 hr after eclosion from the egg (After Sewell, 1973).

The correlated responses in larval locomotor activity in the lines selected for feeding activity is shown in Fig. 9.4. There was no systematic change in larval locomotor activity correlated with increase in feeding rate. Faster feeders did not move about more actively than larvae in the control lines from the base population. Progress towards slow feeding rate was however accompanied by a reduction in locomotor activity. The asymmetry of the correlated response to selection suggests that reduced feeding rate has, in part at least, been achieved by a general reduction in larval activity which could, for example, be due to the recruitment of deleterious alleles causing a depression in the basal metabolic rate. The absence of any significant correlated response in locomotor activity with increasing feeding rate is more meaningful because it is an indication of the discrete nature of the different constellations of gene frequencies controlling the two activities. This point can be pursued further by asking whether there is any correlation of locomotor activity in the adult with either of the measures of larval activity. The larvae from the high and low adult activity selected lines illustrated in Fig. 9.1 did not differ significantly from each other or from the control population with respect either to feeding rate or locomotor activity. Conversely, there was no significant correlation between larval feeding rate and the locomotor activity of the adult flies from the selected lines illustrated in Fig. 9.3. These results based on three non-homologous measures therefore provide no evidence for a general component underlying the levels of activity for the different behaviours, and consequently give no support for a concept of "general activity".

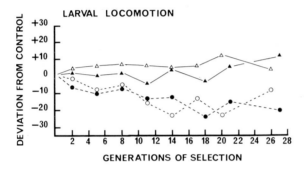

Fig. 9.4. Larval locomotor activity in the lines selected for feeding rate. Locomotor activity is expressed as the deviation in retractions per min from the intra-experimental control base population score for each generation (After Sewell, 1973).

9.3. Biochemical substrate of activity

Investigations into the genetic basis of behaviour pose questions regarding the physiological and biochemical pathways linking the genome with behavioural phenotypes. Rick et al. (1967) have related GABA (gamma-aminobutyric acid) production to the behaviour of rats selectively bred for high and low emotional reactivity and suggested that the increased level of arousal shown by the reactive strain was related to reduced production of GABA within the cerebral cortex. Since one of the parameters used to measure emotional reactivity in the rat is locomotor activity, and further because the evidence for GABA as an inhibitory transmitter in nervous tissue is strong, Tunnicliff et al. (1969) assayed GABA/glutamate ratios in head tissue of flies from the active, inactive, and control lines. No systematic or significant differences were found. However, an assay of the biogenic amines, serotinin, noradrenaline, and dopamine in the same populations revealed an interesting pattern of differences between the lines in respect of noradrenaline and dopamine (Table 9.1). On the basis of these findings Tunnicliff et al. hypothesised that the level of spontaneous locomotor activity is mediated at the biochemical level by the balance existing between these two compounds. This in turn suggests that the additive effects of selection have been on systems controlling these two amines. Such a buffered system, being dependent on changes in the levels of two substances, has considerable adaptive advantage in maintaining an optimum level of phenotypic expression.

The hypothesis relating the balance of noradrenaline and dopamine to spontaneous locomotor activity was investigated by Connolly et al. (1971) who experimentally manipulated dopamine level. By rearing flies on a medium containing various concentrations of gamma-hydroxybutyric acid the level of dopamine was increased in the high activity line and a corresponding drop in activity was observed, thus providing support for the hypothesis.

TABLE 9.1

Concentrations of biogenic amines expressed in μg per g body weight for adult females of the active, control, and inactive lines (from Tunnicliff et al., Comp. Biochem. Physiol., 1969).

Population	Serotonin	Noradrenaline	Dopamine
Active	0.30±0.14	1.36±0.19	0.44±0.03
Control	0.35±0.04	0.93±0.20	0.49±0.11
Inactive	0.32±0.06	0.62±0.15	0.87±0.28

9.4. Courtship and mating behaviour

The males of many animal species display to the females immediately prior to mating. Such displays may involve several forms of behaviour. The animals may assume particular postures, carry out certain patterns of movement, or emit specific calls. This process of behavioural display is called courtship. In essence courtship is a process of communication; information is transmitted between the male and female of the same or closely related species. The process serves a number of functions, dependent to some extent on the species involved. It enables the identification of a conspecific and conversely aids in discriminating against other species thus serving to maintain sexual isolation. In many species courtship also serves to signal whether a potential mate is in a suitable physiological state for mating. Any behavioural process involving more than one animal is inevitably complex since it entails an interaction and progressive modification of behaviour as information is transmitted. In courtship the male and female each have a repertoire of acts the utilisation of which is not independent of the behaviour of the other. There consequently exists a close interdependence between the potential mates, each influencing the behaviour of the other.

Species where the behaviour of the two sexes interacts sequentially in time provide rich but complex material for behavioural analysis. Courtship may thus be investigated on different functional levels; (1) the response repertoire of male and female should be described and analysed, (2) typical patterns of interaction between the sexes may be sought, (3) the consequential effects of the other's behaviour, which itself occurs in a dynamic equilibrium, can be studied. Thus threshold phenomena and motivational factors will interact with signals from the potential mate to further complicate the process.

The study of courtship behaviour in Drosophilids was begun by Sturtevant (1915) who described the process in *Drosophila melanogaster*. Subsequently mating behaviour in *Drosophila* has been studied by a number of persons and the species-typic patterns have been described, in varying degrees of completeness, for about 200 species. Much of the experimental work involving courtship in *Drosophila* has been geared to examining evolutionary aspects and consequences.

The complex patterns of mating behaviour shown by *Drosophila* involve a series of acts which can be observed and identified as units or elements of behaviour. Once the qualitative description of the behaviour of each sex has been accomplished, quantitative descriptions of patterns of interaction can be sought. For any given species the elements or units of courtship are typically performed many times although there is considerable quantitative and qualitative variation within a species. The behaviour of the species with which we have worked, *Drosophila melanogaster*, has been described in considerable detail by Bastock and Manning (1955) who paid most attention to the elements of behaviour which make up the male's repertoire.

Various authorities (Sturtevant, 1915; Bastock and Manning, 1955; Spieth, 1952) have stated that when mature males of *D. melanogaster* are introduced

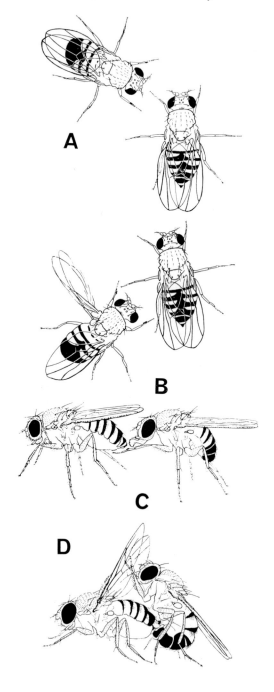

Activity and sexual behaviour in Drosophila 213

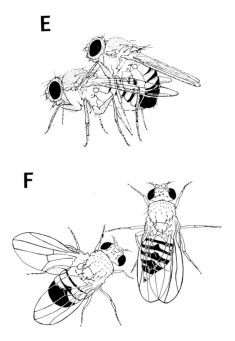

Fig. 9.5. Courtship behaviour in *Drosophila melanogaster*. A, orientation of the male to the female. B, wing vibration by the male. C, the male licks the female's genitalia with his proboscis. D, mounting by the male with genital contact. E, flies in copulation. F, a rejection response by the female to courtship. The female turns her abdomen towards the male and extends her ovipositor.

into the presence of females courtship is initiated by the male tapping the body of the female with his forelegs. In the course of observing a very large number of courtships we have rarely witnessed the male tapping the female, courtship beginning more usually without any physical contact between the sexes. The initial component of courtship was called by Bastock and Manning (op. cit.) orientation and involves the male in approaching the female usually to within 2 or 3 mm. The male faces the female usually to one side and initially slightly behind her (Fig. 9.5A). If she moves off he follows maintaining his position behind and to one side of her. If the female stands motionless he may circle round her whilst remaining close and facing her. Superimposed upon the orientation component is vibration. Whilst orientated to the female the male extends the wing closest to her head out through 90° from his body and vibrates it rapidly up and down before returning it to the resting position. In the process of this wing display the male lowers the trailing edge of the wing below that of the leading edge such that the blade of the wing is tilted towards the female (Fig. 9.5B). Bouts of vibration last usually for a few seconds (though during longer courtships trains of intermingled brief and long bouts are frequent) and are often repeated many times. If the female remains stationary the male will circle round whilst vibrating until he positions himself directly behind her with his head close to the tip of her abdomen. In this position he lowers his abdomen, usually vibrates again, and with a lunging motion of his extended proboscis licks the female's genital region. In this third, licking, component of courtship the labial lobes of the proboscis are opened and the forelegs extended under the female's abdomen (Fig. 9.5C). The licking is followed usually by the male curling his abdomen round and under his body whilst rearing upward with his head and thorax to achieve genital contact. The movement is accompanied by an extension upward and outward of the forelegs above the female's abdomen so positioning them as to spread the female's wings (Fig. 9.5D). If the female permits intromission the male forces her relaxed wings apart with his forelegs and head, grasps her abdomen with his middle legs and the dorsal surface of her wings with the tarsal claws on his forelegs (Fig. 9.5E). In this position copulation takes place. During copula the pair usually remain motionless though the female may vibrate her wings intermittently and occasionally kick with her hind legs. Copulation usually lasts for approximately 15–20 minutes though variations due to genotypic differences have been reported (MacBean and Parsons, 1966). At the completion of copulation the male disengages his genitalia and dismounts. Occasionally genital disengagement is incomplete and the male pushes upwards and backwards and in so doing elongates his abdomen, turns through 180°, and withdraws his penis.

The structure of the male's behaviour during courtship is of course much influenced by the female's responses. If the female is very active and continually moving in the observation cell, bouts of vibration tend to be short and licking does not occur frequently. Similarly the male may make several attempts at copulation before succeeding. These attempts vary in their completeness from what appears initially as a successful mount to a lunge at the female by a male with his abdomen curled right under towards his thorax. We have also observed on many occasions what appears to be an intention movement on the part of the male prior to licking the female's genitalia. Here the male with his abdomen lowered slightly as in Fig. 9.5C will bob his abdomen quickly but slightly up and down, the orientation of the abdomen in respect of the thorax remaining the same.

If a male attempts to copulate with an unreceptive female who does not spread her vaginal plates, thus preventing intromission, he will rarely attempt to complete the mount but instead drop back into his original position behind her and resume courtship. The whole sequence of vibration, circling the female, and licking proceeding over again. Whilst circling the female the male may exhibit a display termed scissoring with one or both wings. The wings are scissored in and out some 70° from the resting position but not vibrated. Scissoring often grades into vibration (Manning, 1959). In some instances where the female of a pair appears relatively unreceptive the male may break off courtship for several minutes and either move independently of the female's path, stand motionless, or quite often preen. After a rest or some interpolated activity he then resumes courtship.

A range of responses on the part of the female towards the courting male have been described and collectively labelled rejection responses (Spieth, 1952). These include "decamping" (running or flying away from the male if space permits), kicking rearwards at the male with her hindlegs, flicking of her wings very briefly and vigorously, depressing her abdomen down towards the substrate, or extruding the ovipositor and angling her body towards the male so as to direct the ovipositor towards him (Fig. 9.5F). Connolly and Cook (1973) have examined the development and efficacy of these rejection responses in the Pacific strain of *D. melanogaster*. They report that the typical pattern displayed by young virgin females (females probably not sufficiently physiologically mature enough for mating) is flicking, whilst mature fertilised females engage in extrusion. In the case of the Pacific strain of *melanogaster* none of these responses inhibited the male's courtship though extrusion did prevent intromission from taking place. Males were observed to attempt copulation with a female whose ovipositor was extruded. In contrast other inbred strains of *melanogaster* respond differently to extrusion. Connolly et al. (1974) observed that extrusion by the female resulted in the male breaking off courtship in the Novosibirsk strain.

Differences in courtship behaviour amongst inbred lines of the same species have so far received little attention but marked differences between species, even closely related species, have been described (Spieth, 1952). Striking differences have been described in the Hawaiian drosophilids by Spieth (1966). In *Drosophila picicornis*, for example, the male positions himself in front of the female, if she attempts to avoid him he arcs back and forth to prevent her moving. When she stands motionless he will strike his proboscis down against the substrate whilst periodically opening and closing the labellar lobes. This action is followed by a further complex series of movements involving the forelegs and antennae. From the courting stance the male vibrates his forelegs and both wings alternately. The female of this species also signals acceptance (not known in *melanogaster*) by assuming a particular posture where she spreads her wings and extrudes her ovipositor. This leads to a further sequence of behaviour on the part of the male prior to copulation. Other Hawaiian species exhibit lek behaviour in which the male must attract the female to a lek or mating site.

Patterns of courtship, as has been stated above, involve communication between the male and female, signals are transmitted and received by each member of the pair The nature of these signals, how they are transmitted and received, present important problems in elucidating behavioural mechanisms. Courtship is also of obvious and great evolutionary significance since it is very largely through behavioural differences that sexual isolation is maintained (Mayr, 1963).

9.5. Genetic variation in courtship and mating speed

Rapid and efficient mating is an important component of fitness in *Drosophila* because it determines which individuals will transmit their genes to the succeeding generation. As Spiess (1970) has pointed out, natural selection will tend to focus on different aspects of sexual behaviour in males and females. In males it will tend to promote persistent courtship activity to ensure that females are inseminated, whereas in females high receptivity must be compromised with the need to discriminate between conspecific males and males of different species. Mating speed is consequently a resultant of interaction between these partially conflicting requirements. Mature virgin 3–4-day-old adult flies from a vigorous wild stock normally copulate within 2–5 min after mixing the sexes, but there is a considerable amount of variation between strains. Several authors have provided evidence that a substantial proportion of the variation in mating speed is genetic in origin (reviewed by Spiess, 1970, and Parsons, 1973).

The most rigorous analysis so far undertaken using quantitative genetic procedures is that of Fulker (1966) for male mating speed. Five males from each of six inbred strains and from each of a sample of six F_1 progenies of crosses between the lines were used. Each of the sixty males from the twelve genotype groups was then paired with six virgin females – one female being drawn from each of the six inbred lines to provide a standard test situation for the males. Female contributions to strain differences in mating speed were consequently held constant in the analysis. The flies were observed for a 12-hour period after which each female was placed in a separate culture and scored for yield of progeny produced. Fulker examined the relationship between time to first copulation, number of copulations, number of copulations resulting in fertilisation, and the total number of offspring resulting from each copulation. Product-moment inter-correlations of measures showed that males which mate rapidly on the first occasion also copulate more frequently and successfully, and leave the most offspring. The inter-correlations suggested that the number of copulations resulting in fertilisation could be used as a convenient measure of mating speed since the measures were practically interchangeable ($r = 0.96$). The procedure would have the advantage of avoiding constant observation over the 12-hour period. This fact was exploited in the performance of a replicated diallel cross between the six inbred lines in all 36 possible ways to yield progeny from within each of the inbred lines together with the 15 F_1 crosses and their reciprocals. Five males from each group of offspring were tested with six females each drawn from the inbred lines as before. Analysis of the diallel cross yielded a heritability (narrow sense) of 0.36. There was a relatively high level of heterosis in the crosses, and strong directional dominance for high frequency of mating. Directional dominance for fast mating speed in *D. pseudoobscura* was also reported by Kessler (1969). Since the additive genetic effects accounted for only approximately half the total genetic contribution to the character, response to selection would be expected to be slow and mainly in the direction of reduced mating speed. This prediction agrees with the results obtained by Manning (1963) who selected for speed of mating in males and observed a response for low mating speed but not for high mating speed. Selection for mating speed applied to both sexes simultaneously was effective in both directions but gains were greatest in the direction of slow mating speed (Manning, 1961) and involved changes affecting the behaviour of males and females. Fulker (1966) believes that the genetic architecture for genes controlling mating speed in his lines argues for a history of strong natural selection for maximum rather than intermediate or slow mating speed.

Courtship behaviour represents an ongoing and dynamic interaction between individuals of opposite sex and involves a complex series of discrete but inter-connected behavioural elements. Differences in mating speed could arise as a function of differences in sexual response thresholds, in either sex, or be due to variation in the intensity, or quality, of courtship stimuli offered by the male. Apart from the work of Manning (1961, 1963) relatively little attention has been directed towards variation in the individual elements of courtship behaviour which could account for differences in mating speed between inbred or selected lines. Speed of mating is made up of two gross components (1) courtship *latency*, and (2) *duration* of courtship. These can most conveniently be measured by placing a single pair of virgin flies in a circular mating chamber of opaque perspex, or other suitable material, with a clear perspex or glass lid (19 mm diameter and 7 mm deep). This confines the flies within a limited space, so that their behaviour can be continuously observed under a binocular microscope, whilst allowing reasonable freedom of movement. It is important that the flies are transferred to the chamber without anaesthesia because this markedly affects their behaviour. Latency is the time elapsing between the introduction of the pair into the cell and the beginning of courtship, defined conservatively as the first bout of wing vibration by the male. Differences in courtship latency may under appropriate circumstances be a reflection of differences in the sexual response threshold between males. Duration of courtship is measured from the beginning of courtship until copulation occurs. Courtship duration is a function both of the sexual response threshold of the female and of the quality and intensity of courtship stimuli provided by the male.

Connolly et al. (1974) found significant differences in the latency and duration of courtship in inbred wild-type strains differing in mating speed. Courtship latency differs markedly between males of different genotypes and depends on female state. Three day old male flies from two isogenic wild-type strains, Novosibirsk and Formosa, were tested in a single-pair mating situation with females from each of the two strains. The females were: (1) 6-hour-old virgins (collected without etherisation), (2) mature 3-day-old virgins, and (3) 3-day-old fertilised females. Fifty single-pair matings of each type were observed and the number of pairs to initiate courtship within 10 min recorded in Table 9.2. Novosibirsk males actively court mature virgin females of both strains but they ignore young virgin females altogether. Formosa males in contrast actively court young females more readily than mature females. Studies over a longer period show that the Formosa males have a longer latency with mature than with young females, whereas very few of the Novosibirsk males will commence courting the immature females and so have "infinite latency". The behavioural factors underlying these differences in male interaction with virgin female state are presently unknown.

TABLE 9.2

Number of males initiating courtship (defined as the first bout of wing vibration) within a 10 min period with females of different states. Further details are given in the text.

Males	Formosa females		
	Young	Mature	Fertilised
Formosa	32	22	4
Novosibirsk	0	47	13
Males	Novosibirsk females		
	Young	Mature	Fertilised
Formosa	47	25	3
Novosibirsk	2	50	8

(Data from Eastwood, unpublished).

Males of both genotype groups (Table 9.2) show a marked reluctance to court fertilised females, and this phenomenon is brought more sharply into focus in the data summarised in Fig. 9.6 which compares the reaction of Novosibirsk and Pacific males to virgin and fertilised females. Pacific males differ from Novosibirsk and Formosa males in that they actively court fertilised females so that the differences in reaction to mature female state is genotype dependent. The proportion of time spent by the males in performing each of the principal courtship elements is shown in Fig. 9.6. The male vibrates his wings whilst he is orientated to the female, so that the wing vibration element is superimposed over the orientation element. The Figure shows the total proportion of time during which the male was orientated to the female whether or not he was vibrating, so that the two components are measured independently. The measure is consequently different from that used by Bastock and Manning (1955) which has been criticised by Cane (1966).

Novosibirsk and Pacific males show marked differences in the proportions of time spent in the various courtship elements. With mature virgin females Novosibirsk males spend significantly less time in orientation and licking and make fewer attempts to copulate. The difference between the courtship performance of males of the two strains is even more striking when they are paired with fertilised females. The Pacific males court these as actively as virgins and, apart from a lower rate of licking, the proportion of time spent performing the different courtship elements is the same with both types of female. Novosibirsk males in contrast either ignore fertilised females, or commence by orientating and subsequently break off courtship, although a few males occasionally persist in courting and attempt copulation.

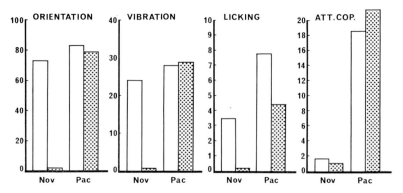

Fig. 9.6. Elements of male courtship behaviour towards virgin (white columns) and fertilised (stippled columns) females in the Novosibirsk and Pacific inbred wild-type strains. The mean percentage of courtship time spent in orientation, vibration, and licking by the male was measured in single-pair matings. Attempted copulation was measured as the mean number of events per male. Sixty pairs of flies of each kind were each observed from the commencement of courtship by the male until copulation occurred, or for a standard period of 30 min (After Connolly et al., 1974).

Females of both strains show a varied repertoire of rejection responses towards courting males (Fig. 9.7), and there is a marked change in the frequency of these responses caused by impregnation which we shall have occasion to discuss in greater detail presently. Virgin females show high rates of kicking, together with wing flicking, curling and jumping, whereas fertilised females make these movements much less frequently and use genital extrusion as their characteristic response. There are significant strain differences with respect to the rates at which these rejection responses are given. Pacific virgin females show higher rates of kicking, and Pacific fertilised females exhibit significantly higher rates for all the rejection responses with the exception of fending.

Novosibirsk flies have a rapid and efficient courtship so that some 50 per cent of pairs copulate within a minute or so from the time that the sexes are brought together. The males are readily accepted by the females and consequently do not need to make repeated attempts to copulate. Pacific flies in contrast are slow to mate, and a period of approximately 10 min is required before 50 per cent of the pairs have copulated. The Pacific males provide a high intensity courtship which is adequately stimulating since, as Connolly et al. (1974) have shown, they exhibit rapid mating speed with females of other strains, whilst Pacific females are equally slow to mate with males of other genotype groups. The reason for the slow mating speed within the Pacific strain is that Pacific females have a high threshold of sexual receptivity and require a greater input of courtship stimulation before they will accept.

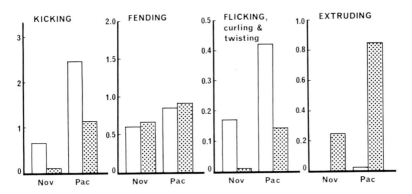

Fig. 9.7. The rejection responses made by virgin (white columns) and fertilised (stippled columns) females to male courtship in the Novosibirsk and Pacific inbred wild-type strains. Kicking, fending, flicking curling and jumping, and extruding, measured in a single-pair mating situation, are each expressed as the mean number of responses per min for each type of female (After Connolly et al., 1974).

The difference in reaction to fertilised females on the part of males of the two strains lies in the fact that extrusion by the female inhibits courtship by Novisibirsk males whereas this rejection movement has no such effect on Pacific males which continue to offer an intense and persistent courtship to the female. It seems probable that the absence of response by the Pacific male to rejection signals by the female represents an adaptation to the female's high acceptance threshold.

Although certain major gene mutations are known to have effects on orientation or vibration, we have as yet little insight into the genetic basis of normal variation in the separate behavioural elements of courtship for either sex. Eastwood (1974) has made a start here by constructing all possible combinations of chromosomes from two strains with fast and slow mating speed, and finds that differences in several of the elements of male courtship behaviour are due to discrete chromosome effects with little evidence of interaction. For example, the X-chromosome appears to have a major effect in determining the amount of wing vibration. This approach could be usefully employed to examine the genetic basis of the difference in reaction of males to fertilised females described above.

9.6. Visual stimuli

The role played by visual stimuli in courtship and mating in *Drosophila* has received a good deal of attention in the literature particularly with regard to ethological isolation between species. Burnet and Connolly (1973) list four principal ways in which visual factors may influence courtship behaviour: (1) Differences in phototaxis and responsiveness to light may lead to assortative mating through differential association of flies in light gradients. (2) Males may perceive females visually and therefore need light for efficient courtship. (3) The females may employ visual cues for species recognition, and require visual stimulation to become sexually excited by the courting male. (4) Light may have a general effect on the activity of the flies which could in turn have consequences for their readiness to court and to copulate. Grossfield (1971) has offered a classification of *Drosophila* species into a system of three classes with respect to the effects of light on their mating behaviour. Class I comprises species which mate equally well in light and in darkness. In class II courtship is inhibited by darkness. Species in class III are those in which mating is blocked by darkness. *D. melanogaster* is known to be capable of mating in darkness (Spieth and Hsu, 1950) and this has led Grossfield (1971) to place this species in his class I light-independent species.

Hardeland (1971) and Hardeland and Stange (1971) have reported that *D. melanogaster* males show more courtship activity in darkness than in the light. They consequently conclude that visual stimuli play a minor role in the mating behaviour of this species. This is surprising, because although flies can mate in darkness they take longer to do so than in light. If mating is more efficient in light the expectation would be that natural selection in *D. melanogaster* would have favoured flies which preferentially court in the light. In these experiments courtship events were recorded for mixed groups of both sexes consisting of six males and six females observed under red light which is effective darkness for the flies. Courtship events defined as "the reaction of one male to one female" were recorded for 10 minute intervals over a period of 24 hours. Now, under the conditions of this experiment the females used are unlikely to have remained virgin. As we have seen, in light the males of some strains of *D. melanogaster* do not persistently court fertilised females, because their courtship is inhibited by rejection responses made by such females. The predominant rejection response made by fertilised females is extrusion and if, as seems possible, this acts as a visual stimulus then males courting in darkness will fail to perceive it and continue to court whereas under lighted conditions their courtship activity should be inhibited. Hardeland (1971) notes that courtship in darkness seldom leads to copulation, as would be expected since fertilised females are markedly unreceptive to male courtship approaches. The diurnal rhythm of courtship activity may thus depend on the perception of inhibitory stimuli from the environment rather than upon an endogenous physiological rhythm. Hardeland (1971) finds that phase shifts in the light/dark cycle result in appropriate shifts in the courtship rhythm which shows that the rhythmicity is entrained in the light/dark cycle. Under continuous light, courtship activity is suppressed (Hardeland and Stange, 1971), which is what we expect if inhibitory visual stimuli from the females are important. Consequently, Hardeland's conclusion that visual stimuli are of relatively minor importance in the mating behaviour of *D. melanogaster* does not necessarily follow from his results.

Earlier evidence pointing to the importance of vision in mating behaviour was provided by Geer and Green (1962) working with the eye pigment mutants *white* and *white-apricot*. In the light, the allele for lighter pigmentation was associated with a reduction in mating success whereas in complete darkness mating was at random. Mutants such as *white* which lack the screening pigment normally present in the primary and secondary pigment cells of the compound eye show greatly attenuated visual acuity and are effectively blind to movement in the external environment. They are consequently unable to perceive visual stimuli during courtship.

The dark red eye colour of wild-type flies is due to the presence of a mixture of red pterin and brown ommochrome pigments produced by independent biosynthetic pathways (Ziegler, 1961). When these pathways are simultaneously blocked by two independently acting mutant alleles *vermilion* (*v*) and *brown* (*bw*) the result is a phenotypically white eye similar to that in the sex-linked mutant *white*. Double mutant *v;bw* flies are effectively blind like *white* mutants, but flies with either the brown or the red pigment singly have normal visual acuity not significantly different from wild-type. The block in synthesis of ommochromes from tryptophan caused by the *v* mutation can be bypassed by dietary provision of kynurenine, one of the missing intermediates in the pathway. This leads to an improvement in visual acuity which is dependent upon the density of brown pigment synthesised (Burnet et al., 1968). Environmental compensation for a missing or defective enzyme, in this case tryptophan pyrrolase, due to the presence of the *v* allele results in a phenocopy of the action of the wild-type allele. Connolly et al. (1969) found that the white eyed mutant *v;bw* flies were at a disadvantage in competitive mating tests. When synthesis of ommochrome pigment was restored in the mutants by feeding kynurenine, their mating success was indistinguishable from that of the *bw* control flies. Competitive mating tests using flies possessing varying amounts of the red eye pigment, caused by mutant genes acting sequentially in pterin biosynthesis, also revealed a significant dependence of mating performance on eye pigment density.

Burnet and Connolly (1973) observed cage populations maintained either in continuous light or in continuous darkness in the same constant-temperature room. Under a system of random mating, each population contained two phenotype groups: white eyed *v;bw*, and red eyed *v;+/+* and *v;+/bw*. In continuous darkness the frequency of white eyed flies showed no significant directional change over a period of 10 generations, indicating that the white eyed flies were not at any marked selective disadvantage under those conditions. The situation in the populations maintained in constant light was quite different. The frequency of white eyed flies declined steadily and substantially over the period of observation, indicating that they were at a strong competitive disadvantage caused by the deficit in mating ability under these conditions.

Female flies do not appear to show any obvious dependence on visual cues during courtship since Connolly et al. (1969) found no significant difference in mating success between females with pigmented and non-pigmented eyes. Detailed analysis of male courtship behaviour showed however that the reduced mating success of white eyed *v;bw* males is due to their inability to establish and maintain contact with the female during courtship. It is the orientation component which is primarily affected, but given sufficient time the males can succeed in copulating with the female. Visual cues are certainly important in the male's perception of the female's body orientation. Spieth (1966) found male flies court and frequently attempt to copulate with the head end of a stationary decapitated female. When the head is glued back into position they no longer make this mistake. In other species vision is of more critical importance. *D. subobscura* males for example will begin to court decapitated females but do not copulate because these females fail to give the necessary visual sign-stimuli for the male to complete his courtship. Visual information, whilst not essential for the initiation of courtship nor for the attainment of copulation in *D. melanogaster*, is important to the extent that flies which are unable to use visual cues have a less efficient courtship, and consequently a lower fitness, than those which are able to use them. For this reason the species might more appropriately be included as a facultative dark-mating species in class II of Grossfield's classification.

9.7. Auditory stimuli

The wing vibration element is of particular importance in the courtship of *Drosophila*. During the wing display described above the male provides patterned auditory stimuli directed towards the female. Comparative studies on sound production within the genus have been made by Ewing and Bennet-Clark (1968) who have shown that the sounds produced by vibration, which they have called the courtship song, consist of a train of sound pulses. In *D. melanogaster* each pulse consists of a single up and down movement of the wing. The wing mechanism involved in the production of the song is described by Bennet-Clark and Ewing (1968) who found that the sounds produced in courtship are similar to a component of the flight tone. The acoustics of the song are discussed by Bennet-Clark (1971). A male fly standing on a flat substrate produces his courtship wing beat at 45° to the substrate so that the sound output is directed over an arc of 90° in the direction of the female. The male normally remains orientated towards the female whilst he is vibrating so that the sounds he is producing are directed maximally towards her and are not received by other flies in the immediate vicinity. The song consists of trains of single-cycle pulses of sound 3 msec in length separated by intervals of 34 msec, the frequency of sound within pulses is 324 Hz (Bennet-Clark and Ewing, 1969; Ewing and Bennet-Clark, 1969).

The courtship song serves two important functions. It acts as a species-specific signalling code whereby the female is able to distinguish between courtship of a conspecific male and that of a male of a different species, and is consequently important in the maintenance of sexual isolation between species. By artificially simulating songs with different characteristics, Bennet-Clark and Ewing (1969) were able to show that pulse interval rather than pulse length is the critical parameter determining the species-specific nature of the song.

The song is also one of the elements which sexually stimulates the female to accept copulation. Males which have had their wings removed are not readily accepted by virgin females and consequently show greatly reduced mating speed. When groups of virgin females were exposed to a 5 min period of prestimulation with a simulated courtship song prior to mixing with wingless males they showed an increase in receptivity (Bennet-Clark, Ewing and Manning, 1973). Shorter periods of prestimulation, and the use of pulse intervals of vibration of half or twice the natural interval had no effect on female receptivity. The females appear to summate wing vibration over time as was suggested earlier by Bastock and Manning (1955) and Manning (1967) and the effects of these stimuli evidently perseverate for several minutes in the absence of continuing stimulation.

When males are artificially deprived of their wings, a certain proportion do succeed in stimulating the females sufficiently to mate with them but only after a lengthy period of courtship. Cook (1973a and b) obtained a significant increase in the receptivity of females to wingless males in response to selection and this was accompanied by a reduction in the period of courtship time required by the males to achieve copulation. The changes which he observed involved an adaptation in the female's courtship processing rather than an improvement in the male's courtship ability. The wings of *vestigial* mutants are reduced in size to stumps. Some laboratory stocks of *vestigial* mutants are nevertheless remarkably efficient in their mating behaviour. The *vestigial* females show an increased receptivity to courtship by *vestigial* males, but the courtship behaviour of the males is also changed. In a competitive mating situation involving *vestigial* males and males of unrelated wild-type strains, we have found that the mutant males are equal and in some instances superior in mating ability. The reason for this appears to be that the *vestigial* males maintain a persistent close-contact courtship of the female, with a high frequency of licking and attempts to copulate. This results in the amusing situation that whilst the wild-type male "warms up" the female with wing vibration stimuli it is the *vestigial* male which successfully copulates with her!

For the courtship song to be effective in species recognition, leading to the maintenance of sexual isolation between sibling species, there must be limits to the amount of phenotypic variation within species which can be tolerated. The observations of Bennet-Clark, Ewing and Manning (1973) already show that wide deviations in the sounds produced by males are likely to be maladaptive, but we have as yet no information about the intraspecific limits of tolerance to variation in the characteristics of the wing vibration code or its genetic architecture in *D. melanogaster*. It is known from the work of Ewing (1969) that some of the differences in the courtship songs of the sibling species *D. pseudoobscura* and *D. persimilis* are controlled by genes located on the X-chromosome. A promising avenue of approach in this direction will be the analysis of the acoustic characteristics of wing vibration signals produced by males which court persistently but which are nevertheless refused by receptive virgin females.

Manning (1967) investigated the effect of removing all, or different parts, of the antennae on female sexual receptivity and found that the arista and Johnston's organ are both necessary for perception of the wing vibration stimuli of a courting male. The arista is a feather-like structure formed from the modified terminal segments of the antenna (Fig. 9.8). It acts like a sail, or lever arm, actuating the bulbous flagellum which stimulates chordotonal sensillae located in Johnston's organ.

Burnet, Connolly and Dennis (1971), investigating the processing of auditory information by female flies, used combinations of mutant genes which cause stepwise reduction in the size of the arista. The morphological effect of these mutants is illustrated in Fig. 9.8. Reduction of the effective surface area of the arista lowers the sexual receptivity of virgin females due to inadequate perception of wing vibration stimuli from the courting males. Such females vigorously reject the males' advances by kicking. Wing vibration is thought to induce female receptivity through a process involving summation of excitatory stimuli such that the female accepts only when some necessary threshold is reached, and in addition some kind of specific identification process is also operating to determine the response of the female.

Individuals which are genetic mosaics in which the head end is male

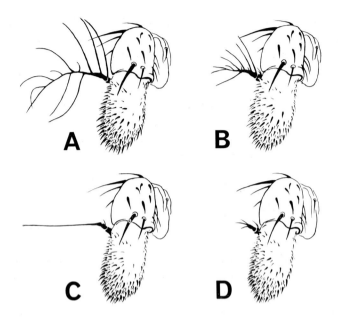

Fig. 9.8. The structure of the arista in wild-type and mutant flies. A, wild-type. B, the *aristaless* mutant. C, the *thread* mutant. D, the double mutant *aristaless;thread*.

whilst the rest of the animal is female are reported to court normal females and perform wing vibration (Benzer, 1973). Hollingsworth (1955) reports a similar situation in *D. subobscura*. The wings are vibrated by motor nerve impulses from the thoracic ganglion so that a female thoracic ganglion is capable of producing the courtship song typical of the male under the direction of a male brain. Evidently the thoracic ganglion in a female must be preprogrammed with a copy of the male wing vibration code even though not normally required to produce it. Ewing (1969) has suggested the idea that there is a reference oscillator, genetically determined, in the neural circuitry which is used by females as a comparator for afferent signals. and which males use as a comparator for efferent signals. The auditory stimuli received by the female from the male could be transduced and the neural representation matched against the correct copy of the signal that is genetically determined in the female. This interesting notion leads to a more detailed interpretation of the behavioural effects in the antennal mutants. Reduction in the effective size of the arista will cause decreased stimulation of Johnston's organ so that the signal transduced will be weakened especially in *al;th* mutants. Consequently it will easily be distorted by other extraneous auditory signals and by random activity within the nervous system, resulting in a low signal-to-noise ratio. Consequently the female is unable to obtain a correct match on comparison with the standard and so persistently rejects the male. Rejection may of course occur in other situations even when the match is correct, because the female is influenced by hormonal and other physiological factors. The fact that some females do eventually accept the male after prolonged courtship suggests that some alternative and less effective mechanism may also be active in influencing their receptivity, but another possibility is that the female, rather than accepting the male after identification, is in fact "taken by force".

A perhaps unexpected finding was that $al;th$ mutant males with morphologically the most abnormal aristae showed reduced competitive mating ability, indicating that some component of the mutant males' courtship is inadequately stimulating to the females. Earlier studies by Begg and Packman (1951) had shown that flies homozygous for the mutant gene *antennaless*, which exhibit bilateral absence of antennae, also show reduced mating ability. That the courtship of the arista mutants should be inadequate was surprising because these males spent a high proportion of their courtship time in wing vibration and actually had significantly longer bout lengths of wing vibration than wild-type males. The possibility suggested by Burnet et al. (1971) to account for the deficit in mating ability in these mutants is that auditory feedback via Johnston's organ, together with feedback from the wings themselves, is necessary for control of the onset and termination of wing vibration and also for the maintenance of the specific interpulse interval which is used by the females for species discrimination. Information feedback from the antennae is known to be important for the control of wing movements during flight. In *Calliphora*, information about air currents is transmitted via Johnston's organ by movements of the pedicel-flagellum joint, and is used to control the wing beat and antennal positioning reactions (Burkhardt and Gewecke, 1965). This evidence for the involvement of peripheral feedback loops involving the antennae in the control and coordination of flight gives at least indirect support for the suggestion that some similar feedback system may operate in the control of wing vibration in the courtship of *D. melanogaster*.

9.8. Physiological factors influencing courtship and mating

Mayr (1950) believed that the reason why the antennae are essential for the sexual receptivity of the female is that they are responsible for the perception of olfactory stimuli from the male. Subsequently, Ewing and Manning (1963) found that olfactory stimuli are unlikely to be important in the sexual stimulation of females. The converse may not be true since Shorey and Bartell (1970) have reported that courtship activity of males may be influenced by a volatile female sex pheromone. Indeed very little is presently known about the physiological factors which underly differences in male sexual activity. In *D. melanogaster* and in *D. subobscura* males with normal external genitalia and accessory glands, but which lack functional testes or sperm, are apparently quite normal in their ability to initiate and complete courtship and mating (David, 1963; Maynard Smith, 1956), so that male sex-drive is not obviously conditional upon fertility.

The sexual receptivity of virgin females is age related. Females are unreceptive on the day of emergence from the pupa but all become fully receptive within 48 hours of emergence. Receptivity declines in aged females some 20–30 days after emergence. Manning (1967) distinguishes two separate processes controlling the sexual behaviour of female *Drosophila*. The first he calls switch-on and switch-off of receptivity which determines whether or not a female is accessible to stimulation by male courtship. The second process is courtship summation. Switch-on evidently involves some kind of physiological threshold since in an individual female the change occurs rapidly in an all or none fashion from the non-receptive to the receptive state. By allowing virgin females to be courted to acceptance, but preventing actual insemination so that their courtship duration could be sequentially retested, Cook (1973c) was able to show that once a female is sexually mature her courtship processing system is set to require a constant amount of courtship for acceptance to occur. That is, acceptance by the female depends upon summation of courtship stimuli up to a fixed threshold.

Growth and maturation of the ovaries take place rapidly during the first 48 hours after emergence so that the switch-on of sexual receptivity occurs as the female becomes fully fertile. Although sexual receptivity and fertility are temporarily related, switch-on does not depend upon fertility. In *female sterile* mutants the ovaries remain in a rudimentary condition caused by a breakdown in the normal processes of ovogenesis due to abnormalities of oogonial and nurse cell differentiation (King, 1970). Females homozygous for this mutation nevertheless become sexually receptive at the same time as wild-type females (Burnet et al., 1973). Maynard Smith (1958) reports that *D. subobscura* females which lack ovaries are also normally receptive. Maturation of the ovaries is stimulated by the juvenile hormone which is produced by the corpora allata of the adult female following a period of inactivity during metamorphosis. In *Drosophila* the corpora allata together with the corpora cardiaca form part of the ring gland complex. Manning (1967) found that females which had received an implant of corpus allatum with associated complex of corpus cardiacum and hypo-cerebral ganglion on the day before emergence from the pupa showed switch-on of sexual receptivity on day 1, that is 24 hours before his control females. Females showing precocious development of sexual receptivity also had larger ovaries than the control flies. Manning (1967) considers that switch-on is a direct consequence of the release of juvenile hormone from the corpora allata, although the mode of action of the hormone, directly or indirectly, on the neural systems involved is presently unknown. According to Highnam and Hill (1969), the corpus allatum hormone can be regarded as a gonadotrophic hormone directly influencing uptake by the oocyte during vitellogenesis of materials made available by neurosecretory hormones. Juvenile hormone is also known from work on other insects to modify protein synthesis to the production of vitellogenic proteins for the developing ovary. Thus the parallel development of sexual receptivity and fertility would seem to be due to the dependence of both systems on the action of the juvenile hormone (Fig. 9.9).

A possibility arising from the action of juvenile hormone on the switch-on of female sexual receptivity is that the hormone might also have some influence on the setting of the threshold of acceptance. A test of this notion would obviously involve direct assay of the level of circulating hormone in females from strains such as Novosibirsk and Pacific which differ markedly in their receptivity. Because of its involvement as a gonadotrophic hormone, virgin females with a high rate of egg production might be those with a high titre of circulating juvenile hormone. Cook (1973c) found a negative correlation between fecundity and courtship duration for Pacific females, which provides provisional support for the idea. Further work in this direction, together with studies using the technique of corpus allatum transplantation which Manning (1968) has used so successfully, should be interesting, but present evidence is of itself insufficient to upset the assumption that the factors which control the setting of the acceptance threshold are primarily neural rather than hormonal.

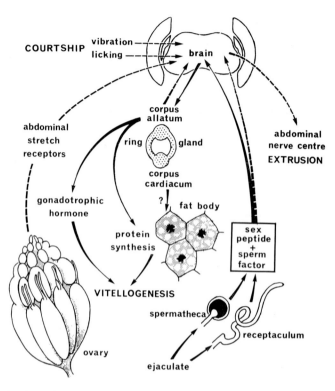

Fig. 9.9. A schematic representation of some of the effects of fertilisation in relation to the physiological factors important in determining the behaviour of the fertilised female fly.

Impregnation of the female is followed by switch-off in her receptivity to subsequent courtship for a period which is determined by the fertility of the mating. Impregnation is also accompanied by an increase in egg production which lasts for the period over which the female carries live sperm. As the sperm supply stored in the spermathecae becomes exhausted, after some 8–10 days, egg production falls back to the virgin level and the female regains her receptivity to courtship and mating. Manning (1967) described two effects of mating on the switch-off of sexual receptivity, a short-term copulation effect and a sperm effect. The short-term copulation effect was observed in virgin females mated to sterile males of a special stock which produce non-motile sperm: "sterile-mated" females showed switch-off in their sexual receptivity lasting some 48 hours, after which they returned to the receptive condition, and there is a temporary increase in egg production by sterile-mated females of approximately the same duration as the non-receptive period (David, 1963; Cook, 1970). These observations suggest that the continuing presence of live motile sperm is in some way essential for maintaining the non-receptive state in mated females.

The suggestion, originally made by Kummer (1960), that a secretion from the male paragonial glands stimulates the increase in oviposition rate has been confirmed by transplanting male paragonial glands into females (Garcia-Bellido, 1964; Merle, 1969). The effective agent extracted in pure form by Chen and Bühler (1970) is a peptide (sex-peptide) which is not present in virgin females but appears in fertilised females at low concentration following impregnation. Burnet et al. (1973a) found that virgin females receiving paragonial gland implants into the abdominal haemocoele show switch-off in their sexual receptivity accompanied by an increase in the rate of oviposition. Both effects were transient and did not last beyond 72 hours after gland implantation, compared with 6–10 days in normally fertilised females. By no means all females receiving implanted paragonia show the induced increase in oviposition rate. There is a marked heterogeneity of response caused by complete inhibition of oviposition in a proportion of females due, apparently, to the effects of surgical interference. This heterogeneity of response would be obscured when measurements of oviposition rate are based on groups of females as used by Merle (1969) rather than on individual female records. Virgin females which received an implant of a portion of the testis did not differ significantly from unoperated virgin control females in their sexual receptivity and showed no increase in rate of oviposition. The simplest interpretation of these observations is that switch-off in sexual receptivity like the induced increase in oviposition rate in females is caused by the presence in the paragonial secretion of a pheromone which may be identical with sex-peptide, although this has yet to be proved by injection of the purified substance isolated by Chen and Bühler (1970).

The action of paragonial gland secretion on mating refusal and oviposition is known to occur in other Diptera including *Aedes* (Fuchs et al., 1969), and studies on mating refusal in intact and decapitated *Musca* females by Leopold et al. (1971) point to the brain as the receptor site for the active substance. The action of the pheromone in *Drosophila* may well be to stimulate an increase in neurosecretion and/or the activity of the corpora allata leading to increased production of gonadotrophic hormone (Fig. 9.9). The duration of the period of mating refusal induced by paragonial gland implants in *Drosophila* (48–72 hr) is considerably shorter than that induced by fertilisation but is closely similar to the short-term mating effect observed by Manning (1967) and Cook (1970) in "sterile-mated" females, and so provides additional evidence for a sperm effect. The fact that testis implants into the body cavity have no significant effect either on receptivity or egg production indicates that the sperm factor must be mediated through the spermatheca or receptaculum seminis of the female, or else it may only be active in the presence of sex-peptide. The sperm factor may be chemical, or a direct neural inhibitory effect depending on the presence of sperm in the seminal receptacles. Manning (1967) observed that in fertilised females the return to the receptive state takes place some 24 hours after the seminal receptacles are virtually empty. In *D. subobscura* mating refusal persists after exhaustion of the female's sperm supply (Maynard Smith, 1956). These observations seem more to suggest the attenuation of some chemical influence rather than release of neural inhibition.

Mating refusal in fertilised females is accompanied by a change in their repertoire of courtship rejection responses. The characteristic rejection movements made by virgin females are, as we have seen, fending and kicking. After impregnation the rates of kicking and fending are lower and genital extrusion, much less frequently observed in virgin females, is the modal response to courtship. Host females with testis implants had high rates of kicking and fending compared with females with paragonial gland implants which showed reduced rates of kicking and fending appropriate to the fertilised state (Burnet et al., 1973). There was no significant difference in rate of extrusion between females with paragonial and testis implants, however.

The act of extrusion involves movements resembling an exaggerated version of oviposition and the same or similar abdominal muscle groups are probably involved in both activities. These muscles are known to have polyterminal polyneuronal innervation so that their motor activity could be under the control of different neural centres. Grossfield and Sakri (1972) have shown that decapitated females continue to lay eggs. Although such headless females do not show extrusion they kick vigorously in response to contact by a courting male (Spieth, 1966), so that the extrusion response is evidently controlled by the brain. It is possible that the pheromone present in the ejaculate acts upon receptor sites in the brain in such a way as to cause changes in the expression of motor patterns under localised control of the abdominal ganglia forming part of the thoracic ganglion centre. The presence of sex-peptide may be a sufficient condition for inhibiting kicking and fending, but the expression of extrusion may depend upon some additional factor. Grossfield and Sakri (1972) have suggested that the degree of distension of the ovaries may provide informational input to the thoracic ganglion centre influencing post-reflex oviposition. The rate of extrusion could, in a similar way, be influenced by informational inputs to the brain or thoracic ganglion centre from abdominal stretch receptors or from the ovary (Fig. 9.9). Prima facie evidence for this idea is that fertilised females which are homozygous for the mutation *female sterile* and which consequently do not have fully developed ovaries show a significantly lower rate of extrusion than their fertilised normal female sibs. In the light of this finding it is perhaps possible to explain why there was no significant difference in rate of extrusion between groups of virgin females receiving paragonial or testis implants. As we have already mentioned, the surgical interference involved in injecting the implants caused a cessation of oviposition in a large proportion of female hosts in both operated groups which is indicative of physiological changes affecting the ovary which may have interfered with the extrusion response. An obvious prediction on the basis of these suggestions which requires experimental testing is that fertilised females which have been kept on a protein-free diet, which prevents normal growth and maturation of the ovaries, should have a lower rate of extrusion than their sisters which have been fed on a normal diet and so have fully developed mature ovaries.

The mating refusal response can be overcome by forced matings which occur when females are confined at high population density in the presence of a large number of males. In a female which has been multiply inseminated, the sperm is stored and utilised on a "last in – first out" basis so that the offspring which she produces are fathered by the last male which copulated with her. Males which produce a pheromone in their ejaculate, which induces refusal of subsequent mating by the female, will clearly have an advantage in ensuring that their genes contribute to the gene pool of the succeeding generation. Duration of copulation in *D. melanogaster* is male-determined (MacBean and Parsons, 1967), and would itself be expected to be a component of fitness because those males will be at an advantage which remain in copulation for a sufficient period beyond sperm transfer to ensure that the pheromone reaches its site of action to switch-off the female's receptivity and so make sure that their sperm are used for fertilisation. This expectation is supported by the observation that in crosses between inbred lines there is substantial dominance for long copulation duration (Eastwood, unpublished data). In the lines studied by MacBean and Parsons (1967) short copulation duration was apparently associated with a reduction in the pretransmission period before sperm transfer had taken place. It would be interesting to know whether selection for a reduction in the period of copulation following sperm transfer is associated with a correlated increase in the frequency of second mating by females.

Pleiotropy

A single major gene mutation may have effects on various aspects of the phenotype. This phenomenon is known as pleiotropy, and consideration of the relationship between different phenes may give a useful indication of the primary site of action of the gene locus concerned. Wilcock (1969) has criticised the value of studies on the pleiotropic effects of major gene mutations affecting behaviour and regards many of the cases described in the published literature as examples of what he describes as trivial pleiotropy which have contributed little of interest to psychology. Whilst Wilcock is right to stress the value of the biometrical approach to the study of behaviour it remains true that advances in several important areas of biochemistry, not least the pathways and regulation of intermediary metabolism, have come from studies on the effects of major gene mutations, and there is no reason to suppose that the approach will be less fruitful in the study of behaviour, and its underlying biochemical basis. The *white* mutant has pleiotropic effects on the synthesis of eye pigments and upon visual acuity. Lack of visual acuity follows from the absence of eye pigmentation because visual discrimination by the compound eye depends upon optical separation of the individual ommatidia. A further pleiotropic effect of the mutant allele is that it reduces the efficiency of courtship in males, thereby reducing their mating success in a competitive situation. These phenes can be grouped into a causal sequence of effects. One can of course dismiss this as an example of "trivial" pleiotropy simply because it requires no great feat of intellect to deduce the order of cause and effect. However, since the flies can mate perfectly well in darkness it is not necessarily obvious why impaired visual acuity should reduce their mating performance, so that analysis of courtship of the mutant males provides information on the necessity for visual cues and the relative importance of orientation in relation to the other elements of male courtship.

In other instances the causal connection between different phenes is by no means obvious. It is worthwhile to look at a specific mutant in some detail in order to show how consideration of the separate effects on the phenotype may suggest guide lines for the design of more detailed investigations into the nature of the pathways through which the action of the gene on behaviour may be mediated. We have chosen the *yellow* mutant, which provided one of the first examples of a single-gene mutation affecting the behaviour of *Drosophila*.

Sturtevant (1915) first noticed that *yellow* mutant male flies are usually unsuccessful relative to wild-type males in a competitive mating situation. It seemed possible that this could be because the behaviour of the mutant males is altered in some way which makes them less stimulating to the females. Bastock (1956) showed that although *yellow* males do eventually mate they take longer to do so. The *yellow* males were slower to begin courtship, and once courtship had begun, they had to continue to court for a longer period to achieve copulation with the female. The difference in behaviour evidently depended on the genotype of the male because wild-type males were equally successful with *yellow* and wild-type females, whereas *yellow* males showed the same disadvantage with each kind of female. Bastock (1956) compared the courtship behaviour of wild-type and *yellow* males and found that the mutant males vibrate their wings for shorter periods and have longer intervals between bouts of vibration than wild-type males. The deficiency in mutant male courtship behaviour may be a consequence of differences in the kinds of stimuli provided by the females to wild-type and mutant males. The females might, for example, react against the changed appearance or smell of the mutant males, but this possibility seems unlikely because the courtship records for the females gave no convincing evidence of differences in number of rejection responses during courtship by mutant and wild-type males. The difference in vibration bout length between *yellow* and wild *melanogaster* males was also apparent in their courtship with *D. simulans* females which rejected both equally (Bastock, 1956).

A significant finding is that locomotor activity of *yellow* mutant males is lower than that of wild-type males. A specific *yellow* allele which we are studying was introduced into the genetic background of an isogenic wild-type stock by repeated backcrossing over many generations. The locomotor activity of *yellow* mutant males, and their wild-type male sibs grown under identical environmental conditions, was tested in the open-field situation in an activity chamber. The results illustrated in Fig. 9.10 show that the activity of the mutants is consistently and significantly lower than that of wild-type males, indicating that the *yellow* allele has effects on locomotor activity as well as on sexual behaviour.

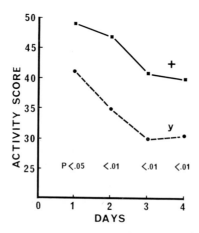

Fig. 9.10. Locomotor activity of wild-type and *yellow* adult male flies. Activity is expressed as the mean score per min for 25 males of each genotype group, measured individually in an open-field situation on each of four successive days after eclosion from the pupa (After Wilson, unpublished).

We have found that the time taken for wild-type males to copulate was approximately 2 min (Burnet et al., 1973b). During this period *yellow* mutants showed longer intervals between bouts of orientation than the wild-type males. That is they broke off courtship for longer intervals. The *yellow* males consequently showed longer intervals between bouts of wing vibration and for this reason engage in a less intense courtship of the female than do wild-type males. As shown by Bastock (1956) the mean bout length of vibration in *yellow* males is significantly shorter than in wild-type. That is they provide less wing vibration stimuli during bouts of courtship. However, it is difficult to see this in itself as the primary reason for the reduced courtship success of *yellow* males. If the antennae are removed from females they cannot perceive wing vibration stimuli, yet *yellow* males are still less successful. Presumably this is because contact stimuli required to bring the female to acceptance are given at longer intervals due to longer duration of breaks between bouts of orientation by *yellow* males. The deficit in mutant courtship seems to be consistent with a generally lower level of sexual motivation in *yellow* males rather than with a specific deficiency in the wing vibration element, although the possibility that there are also differences in the quality of the wing vibration signals offered by *yellow* and wild-type males merits investigation.

Wild-type flies show a characteristic distribution of black pigmentation on the abdominal tergites and bristles, but in *yellow* mutants these areas remain light in colour. Mutant alleles at the *yellow* locus are by no means identical in their effects on the phenotype (Green, 1961; Lindsley and Grell, 1968). Green has also found (personal communication) that not all *yellow* mutants studied by him show a deficit in mating success. Our observations confirm the earlier findings of Bastock (1956) that residual genetic variation can modify the effects of the *yellow* allele on mating success. We have observed that the *yellow* allele which we are studying causes a deficit in male courtship success in different inbred wild-type backgrounds but the competitive disadvantage shown by the mutant males varies, from relatively mild to severe, depending on the particular genetic background with which one is dealing.

The importance of the *yellow* mutant is that we have here a gene with pleiotropic effects on body colour and on behaviour. The black pigmentation of *D. melanogaster* is mainly an indole melanin (Fattoruso et al., 1965) formed from the amino acid tyrosine, an essential nutritional requirement of the fly. A possible connection between the pigmentary and behavioural effects of the *yellow* allele is suggested by the fact that this amino acid is the precursor for three important biochemical pathways illustrated in Fig. 9.11.

Tyrosine is converted by hydroxylation to dopa (3,4-dihydroxyphenylalanine) which may be converted to dopa quinone, a precursor of melanin. Dopa is also a precursor for dopamine and noradrenaline which are known to be present in *Drosophila* brain (Tunnicliff et al., 1969), and consequently provides a common link between pigmentation and the synthesis of catecholamines which may subserve a function related to neural transmission. Tyrosine is also a key substance for the hardening and darkening of the cuticle (Karlson and Sekeris, 1964; Lunan and Mitchell, 1969). The mechanisms regulating these pathways are of considerable complexity and not fully understood in *Drosophila*. Monophenolase (tyrosinase-A_1) oxidises

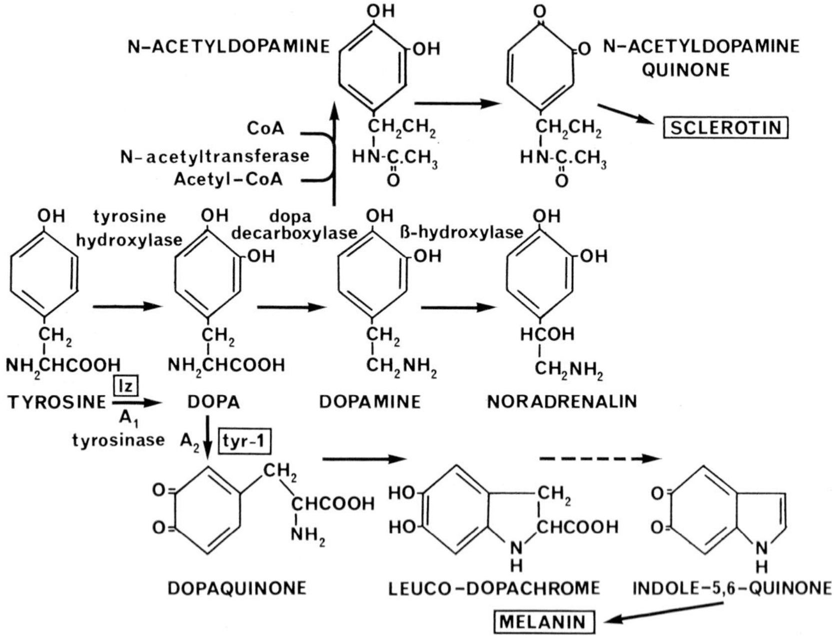

Fig. 9.11. Pathways for biosynthesis of melanin, sclerotin and catecholamines from tyrosine.

tyrosine to dopa, and both monophenolase and diphenolase (tyrosinase-A_2) have been shown to be able to oxidise dopa to dopaquinone which spontaneously polymerises to the pigment melanin (Mitchell and Weber, 1965). *Drosophila* phenol oxidases require activation in extracts. Mitchell and Weber (1965) and Mitchell et al. (1967) have shown that activation involves at least five different protein components: A_1, A_2, A_3, P, and S. Aggregation of these different protein subunits may yield at least seven different enzyme components at different stages of development. Activity of both monophenolase and diphenolase is affected by mutation at the *lozenge* locus on the X-chromosome (Peeples et al., 1969), and a locus on the second chromosome, designated *tyr-1* (synonym α_1) may be a structural gene controlling diphenolase activity (Lewis and Lewis, 1963).

An especially important period in the synthesis of components for phenol oxidase production is in the last half of the third larval instar. In wild-type individuals phenol oxidase activities reach a maximum at puparium formation, reducing rapidly over the next 20 hours to be followed by a smaller peak in activity between 55 and 72 hours. Mitchell et al. (1967) found that although the *yellow* mutant has the normal potential for enzyme activity in late third instar it has little at puparium formation. It appears that the loss in phenolase activity which normally occurs at puparium formation takes place too soon. The potential for phenolase activity in the mutant rises again towards the end of the pupal period and is actually higher than in wild-type in the adult fly at emergence. The *yellow* gene may, as Mitchell et al. (1967) suggest, act as regulator for phenol oxidase activity. Interallelic complementation at the *yellow* locus (Green, 1961) seems to indicate that the gene is responsible for the synthesis of a protein which is at least a dimer, and it is also possible that this is, or regulates the availability of, one of the subunit components of phenolase. Unfortunately it is not yet known whether monophenolase (tyrosinase-A_1) is also involved in the pathway leading to synthesis of dopamine present in the brain or whether a separate tyrosine hydroxylase subserves this function. One consequence of this might be that the *yellow* gene in some way has effects on the activity of two separate enzymes. Instances of this kind are of course already known. For example, lactose synthetase has two different subunits. The A subunit has enzyme activity with lactosamines, while in combination with the B subunit forms lactose synthetase (Brew, 1970), the structural gene for the A subunit thus controls the activity of two different enzymes. The *yellow* locus might control the activity of monophenolase and tyrosine hydroxylase because it is the structural gene for, or regulates the availability of, a subunit component common to both enzymes.

The possibility that the behavioural abnormalites of the *yellow* mutant are referable to some as yet unidentified change in the metabolism of catecholamines implies that assay of brain catecholamine levels in adult mutants would be worthwhile. As we have already seen, Tunnicliff et al. (1969) suggest that changes in spontaneous locomotor activity may be mediated through changes in the balance of noradrenaline and dopamine, and the fact that *yellow* mutants also show a reduced level of locomotor activity may be an additional indication that the *yellow* gene has effects on endogenous levels of these compounds. Konopka (1972) reports that dopamine in *yellow* adults is within 8 per cent of normal, but it is not stated whether the mutant males showed a deficit in mating success, so that it is difficult to assess the significance of this finding.

Phenocopies of the pigmentary deficiency of the *yellow* mutant were obtained by feeding larvae on a medium containing α-dimethyl tyrosine (3-

methyl, α-methyl tyrosine methyl ester HCl) by Burnet et al. (1973b). Unfortunately the yellow bodied adults produced in this way were unsuitable for behavioural tests and died within 2–3 days after emergence, but normal wild-type males which had been fed on adult food medium containing α-dimethyl tyrosine were significantly less successful in mating compared to control males. Like *yellow* mutants these males showed a significant increase in the mean duration of breaks between bouts of orientation and vibration.

Two other pigmentary mutants, *ebony* and *tan*, show changes in courtship behaviour (Crossley and Zuill, 1970) associated with altered levels of dopamine (Hodgetts and Konopka, 1973; Konopka, 1972). It is by no means clear, however, whether the deficit in mating speed in these mutants is a secondary consequence of the male's inability to respond to visual cues in courtship, or whether their abnormal electro-retinogram responses (Hotta and Benzer, 1969) and their reduced courtship success are both caused by altered catecholamine balance.

A potentially interesting avenue of approach which may, incidentally, afford more insight into the action of the *yellow* gene is screening for mutations affecting specific enzymes in catecholamine biosynthesis. The presence in brain and thoracic ganglion in *D. melanogaster* of dopa decarboxylase has been demonstrated by Dewhurst et al. (1972). As shown in Fig. 9.11 this enzyme is responsible for the synthesis of dopamine. It is inhibited by α-methyl dopa, and Sherald and Wright (1972) made use of this fact to isolate mutants with increased activity for dopa decarboxylase. Sparrow and Wright (1974) have isolated mutant alleles hypersensitive to the compound and which map to a locus on the second chromosome. The mutant alleles designated *l(2)amd* are recessive lethals causing cuticular abnormalities in homozygous larvae.

It seems not at all unlikely that the dopa decarboxylases active in brain and in cuticular tanning are distinct tissue-specific isoenzymes. Consequently the best strategy for identifying mutations affecting the brain enzyme would seem to be to make the initial screening by feeding the inhibitor to adult flies rather than larvae. Such mutants could be of great potential value in the study of *Drosophila* behaviour. More particularly, some of the behavioural consequences of changes in the activity of dopa decarboxylase may be similar to those found in *yellow* flies if this mutant gene controls one of the preceding reactions in the synthesis of dopamine.

9.9. Activity in relation to courtship and mating behaviour

A means of examining the relationship between different behaviours is provided by populations which differ markedly in respect of a given behaviour. It is thus of interest to consider whether differences in a basic character such as activity level affect other behavioural parameters, notably sexual behaviour. Manning (1961) selected lines for fast and slow mating speed, selection being made on both sexes simultaneously. Using a mass mating test Manning (1961) found that after some 7 generations of selection the mean time to copulation for his slow line was 80 minutes whilst that of the fast line was 3 minutes. Also he reported differences in locomotor activity, the activity level of the slow mating line being much higher than that of the fast line. On the basis of these findings he concluded that the differences in activity were as important as any direct changes in sexual behaviour produced by the selection procedure. He argued that the level of activity, as such, was not so important as the speed at which flies could switch from "activity responses" to "sexual responses". An alternative method of investigating the interaction of these behavioural systems is provided by measuring the speed of mating in lines which differ substantially in their activity level. If Manning's interpretation were correct, flies from the inactive population should mate more rapidly than flies from the active line.

The mass mating technique has been described in detail by Manning (1961). This was used with flies from the adult activity selection lines illustrated in Fig. 9.1. Fifty pairs of virgin males and females aged for 3 days following emergence were introduced into a half-pint milk bottle and pairs removed as they copulated. A note was made of the number of pairs copulating within successive 5-min periods of the 30-min test. The first test was made at generation 15 of selection, the results of which are presented in Fig. 9.12A. It is clear that there are no differences between the experimental and control lines. This method of measuring speed of mating is most sensitive in the early stages, any differences therefore are more likely to be apparent in the first few minutes of the test. Accordingly when the test was repeated on generation 31 of the selected lines the number of pairs mating in each successive minute was recorded. Still no differences between the lines were apparent. The difference in activity between the two experimental lines at this point in the selection was 42, greater than that found by Manning. On the basis of these results it would appear that the relationship which Manning observed between speed of mating and activity level was merely a correlated change rather than evidence for an interaction between the two behavioural systems.

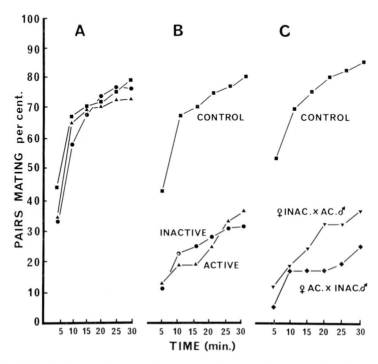

Fig. 9.12. Mating speed measured in a mass mating situation, expressed as the cumulative percentages of matings against time, for the lines selected for adult locomotor activity. A, mating speed at generation 15. B, mating speed at generation 63. C, mating speed within the control line and for both reciprocal pairings of active and inactive flies. Control (squares), active (circles), inactive (triangles), inactive female × active male (inverted triangles), active female × inactive male (diamonds).

Selection was continued and a further measure of mating speed was made at generation 63 of selection. At this point mating speed had decreased so markedly in the experimental lines that it was necessary to record the numbers copulating within 5-min intervals, rather than each minute (Fig. 9.12B). A further test of the direct importance of activity is provided by mating inactive females with active males and vice versa. Bastock (1956) reported that inactive females mated more readily than active females, one might therefore expect that females from the low activity line when mated with males from the high activity line would show a speed of mating similar to that of the controls. Reciprocal mating tests were therefore made on flies from generation 64 of selection. The results are shown in Fig 9.12C: females from the low activity line paired with males from the high line are slightly better than those in the reciprocal pairing but remain substantially slower than the controls.

The mass mating test, whilst useful for establishing any differences in overall speed of mating, is essentially crude and it does not provide a measure precise enough to indicate the nature of any behavioural changes which may have taken place. The systematic removal of pairs as they copulate reduces the population density as the test proceeds and inevitably reduces the probability of contacts between males and females. Speed of mating was studied using a single-pair test in order to control for possible artifacts in the mass mating procedure.

A single male and female, virgin and aged for 3 days after emergence, were used to measure courtship latency and duration of courtship. The results of the experiment are summarised in Table 9.3. The significant increase in the duration of courtship in the active line over that of the control population lends some support to Manning's earlier observation.

TABLE 9.3

Results from single-pair matings in active, inactive, and control lines; time in sec. Data based on 40 matings for each population.

	Population			Difference			
	Control	Active	Inactive	C/A t	P	C/I t	P
Mean courtship latency	66.0	68.4	30.0	0.77	n.s.	3.56	0.01
Mean duration of courtship	45.6	112.2	44.6	2.9	0.01	0.07	n.s.

The observed differences in mating speed could have arisen because the courtship behaviour of males from the high activity line may be in some way less stimulating than that of males from the other two populations, or because the threshold of receptivity of females from the high activity line may be raised. In order to investigate the first of these possibilities a further series of single-pair matings was tested and the individual bouts of the male's three principal courtship responses (orientation, vibration, and licking) were examined in detail. The major findings of this experiment are summarised in Table 9.4. These data provide no support for the hypothesis that the courtship behaviour of the males from the active line is in any obvious way less stimulating. Mean bout lengths for orientation and vibration are greater than in the controls, indicating that the courtship of the males from the high activity line is vigorous. Further, no differences were found in the frequency with which bouts of varying lengths occurred.

TABLE 9.4

Mean bout lengths in seconds of components of courtship behaviour in males of active, inactive, and control lines.

Parameter	Population						Difference			
	Active		Control		Inactive		C/A		C/I	
	Mean	SD	Mean	SD	Mean	SD	t	P	t	P
Orientation	7.5	3.9	3.8	2.1	4.8	4.2	4.4	0.001	1.2	n.s.
Vibration	2.1	0.5	1.6	0.3	2.0	0.6	4.2	0.001	3.6	0.01
Licking	0.7	0.2	0.7	0.3	0.9	0.3		n.s.		n.s.

The alternative hypothesis, to explain the changes in speed of mating as a consequence of a change in the threshold of receptivity in the active females, was examined by making reciprocal single-pair matings between the two selected lines. If the decrease in speed of mating in the active line were due to a change in receptivity on the part of those females then courtships between active females and inactive males should take as long as those between active males and active females. On the other hand, pairings of active males with inactive females should give courtship times of the same order as those found in the control population. The results of this experiment are summarised in Table 9.5. The mean courtship duration of the inactive females paired with active males is of about the same order as that recorded for the control population, whereas the courtship duration for the reciprocal pairing is enormously increased. Further, only 46 per cent of the active females mated within a one-hour test period. The duration of courtship for the pairing of active females with inactive males is very much longer than within line courtship times for the active line. This may well be due, as Bastock (1956) suggested, to active females receiving less stimulation. Males from the high activity line, because of their increased activity, are able to maintain contact with the female and also able to sustain a high level of courtship. The increase in mean bout length of orientation and vibration observed in the case of the active males may well be a result of secondary selection for vigorous courtship since only the most persistent of these males will leave descendants.

TABLE 9.5

Results from reciprocal pairings between active and inactive lines. Times are in sec except where otherwise stated. Tests extended over 60 min.

	Inactive female × Active male	Active female × Inactive male
Mean courtship latency	55.9	37.6
Mean duration of courtship	73.7	19.0 min
Per cent mating	100	46

Confirmation of the hypothesis of change in receptivity was obtained from a further experiment in which the frequency of rejection responses given by females from the three populations was measured. Data were obtained from a further set of single-pair matings where the courtship lasted in excess of 4 min. This period was chosen because the frequency of rejection responses is necessarily correlated with the duration of courtship. Observations were terminated after 8 min. The procedure thus amounted to observations on relatively unreceptive females. The results are summarised in Table 9.6, the mean number of rejection responses per group being; inactive 4.3, control 5.4, active 11.0. As is evident from Table 9.6, the most striking difference is in the frequency of extrusion where the active females produce far more responses. Connolly and Cook (1973) have shown that of the responses listed by Spieth (1952) extrusion is the most efficacious.

In a further and later study on mating speed Manning (1968) reported a similar finding. Working with a population of *Drosophila simulans* artificially selected for slow mating, Manning found that the females of this stock performed vigorous repelling movements to courting males, particularly extrusion. Although these results do not provide conclusive proof for the hypothesis concerning a shift in threshold of receptivity, they do lend strong support to the notion that the effects on sexual behaviour in the high activity line are mediated in this way.

TABLE 9.6

Total number of various rejection responses observed in 20 courtships in each of the three populations. Observations lasted between 4 and 8 min, but only a small proportion of courtships were completed when observations ended after 8 min.

Population	Kicking	Fending	Fluttering	Curling	Extrusion	Decamping	Total
Inactive	20	39	16	6	2	3	86
Control	50	23	25	3	3	4	108
Active	66	26	73	3	38	14	220

9.10. Incipient sexual isolation

Behavioural characters influence not only the evolutionary processes which lead to adaptation (Connolly, 1971) but also those processes leading to speciation. The essence of speciation is the production of two well-integrated gene pools from a single founder population. Species are themselves kept intact by what Dobzhansky (1937) termed isolating mechanisms. The mechanisms which isolate one species from another are the most important attributes which a species possesses since they are by definition the ultimate species criteria.

Mayr (1963) classified isolating mechanisms into two types, pre- and post-mating mechanisms. Of these, pre-mating mechanisms are the most important because they avoid the wastage of reproductive potential which has particularly serious consequences in the case of the female. Of the pre-mating mechanisms Mayr argues that the most important class is what he termed "ethological isolating mechanisms", these are mediated by differences in behaviour. Behavioural mechanisms of this kind are largely concerned with the production and reception of stimuli by the sex partners (for example, the different auditory patterns produced by the wing vibration of males of different *Drosophila* species), though behavioural differences in habitat selection may also lead to isolation (Pittendrigh, 1958).

Isolating mechanisms are by no means all-or-none phenomena. In closely related species the same basic pattern of courtship occurs. There are, however, quantitative differences which are sufficient to prevent the successful synchronisation of readiness to mate on the part of the male and female. Quantitative changes in certain aspects of the male's courtship pattern give rise to changes in mating success which become the starting point for divergence (Mainardi and Mainardi, 1966). After reviewing ethological isolation Mayr (1963) suggests that it arises, "... as a result of differences in activity rhythms, levels of activity and other generalised factors which influence the coordination of the drives of males and females".

Since our high and low activity lines differed markedly and also showed certain differences in mating behaviour, it is possible that random mating between the two lines may not occur. Any preponderance of homogamic over heterogamic mating would be evidence of incipient sexual isolation between the two lines and as such would be of interest in considering the origin of isolating mechanisms. Several techniques have been used to measure sexual isolation of which the multiple-choice test most closely approximates the situation likely to be found in nature. This method involves confining males and females of the different types together in equal numbers and counting the frequency of homogamic and heterogamic matings.

At generations 110 and 111 the cultures of the active and inactive lines were expanded to produce large numbers of flies of the same age. Virgin males and females of the two lines were aged 3 days from eclosion prior to being placed 10 males and 10 females of each line in the perspex chambers used to measure activity. In order to distinguish the two populations, the animals were marked on the thorax with a tiny spot of red or green paint. This marking was done 48 hours prior to testing and each animal was examined individually to ensure that the treatment had in no way damaged the wings or primary sense organs. As pairs copulated, the strain of male and female was noted and their position in the chamber marked. When *melanogaster* copulate they remain motionless for the duration of the copulation. Observations were maintained over 120 min. The test was repeated 12 times on two successive generations, 111 and 112. In all, 880 flies were involved. At generation 112 the unselected control population was expanded and progeny collected. On day 3 their activity was measured and they were sorted into two groups, those scoring more than 50 and those scoring less than 20. These high- and low-scoring flies were then marked and used as a control against which the data from the selected lines could be compared. In addition they provide a check on whether selection in a single generation gives rise to flies significantly different in their sexual preferences.

The results from the multiple-choice mating tests are given in Table 9.7. The proportion of homogamic matings is significantly greater than heterogamic matings in both generations ($\chi^2 = 25.4$; $P < 0.001$ and $\chi^2 = 20.8$;

$P < 0.001$, respectively). Various indices of sexual isolation have been developed (Parsons, 1967), of these the most appropriate for multiple-choice tests is that devised by Malogolowkin-Cohen, Simmons and Levene (1965). This joint isolation index is given by the equation

$$\frac{x_{1,1} + x_{2,2} - x_{1,2} - x_{2,1}}{N}$$

where $x_{1,1}$ and $x_{2,2}$ are the numbers of homogamic matings of strain 1 and 2, and $x_{2,1}$ and $x_{1,2}$ are the respective heterogamic matings. N is the sum of these, the total number of matings. The data obtained from the multiple-choice tests give indices of isolation of 0.43 and 0.39 for generations 111 and 112, respectively. The results from the phenotypically extreme animals taken from the control population show that mating is random, the value of χ^2 is non-significant.

Of the 24 separate multiple-choice tests carried out over 2 generations, only one showed an excess of heterogamic over homogamic matings. Whilst the indices of isolation are not large they nevertheless provide sound evidence of incipient sexual isolation. The control tests fail to suggest any tendency to assortative mating in flies chosen from the phenotypic extremes of the unselected line.

Whilst there have been experimental studies demonstrating that sexual isolation between related species may be changed, Koopman (1956) strengthened the isolation existing between *D. pseudoobscura* and *D. persimilis* by selecting against hybrids, and Kessler (1966) working with the same species selected successfully for and against isolation; there is little experimental data on the emergence of two populations from a common parental stock. Broadly there are two theories concerning the origin of isolating mechanisms. Dobzhansky (1937) suggested that they were ad hoc contrivances of natural selection when hybridisation became disadvantageous. This hypothesis is based upon the observation that hybrids between two species frequently have a lower fitness than either of the parental groups. The alternative hypothesis suggests that isolating mechanisms arise as an incidental by-product of genetic divergence in geographically isolated populations. The two theories are not mutually exclusive and any attempt to represent them as such leads to confusion. For natural selection to promote sexual isolation, the genetic raw materials must be present in the population but it is questionable whether these alone would lead to strong isolation unless favoured by natural selection.

TABLE 9.7

Results of multiple-choice mating tests between active (A) and inactive (I) lines and between phenotypic extremes from a control unselected population. H, high-scoring flies and L, low-scoring flies.

Test Number	Generation 111 of selection				Generation 112 of selection				Controls from generation 113			
	A×A	I×I	A×I	I×A	A×A	I×I	A×I	I×A	H×H	L×L	H×L	L×H
1	5	5	0	0	8	1	1	2	4	2	2	5
2	2	6	3	3	8	2	2	0	4	7	2	6
3	2	3	3	4	4	2	3	1	3	6	4	6
4	8	4	2	1	5	0	1	1	6	0	5	3
5	9	1	2	1	3	2	2	1	3	5	6	3
6	1	8	1	6	6	2	1	0	4	3	5	4
7	4	3	2	3	10	0	2	1	4	6	4	5
8	5	3	4	0	3	2	1	3	3	4	3	4
9	8	1	3	0	10	4	1	2	5	1	6	0
10	5	6	4	0	0	10	2	2	4	0	4	2
11	10	1	1	2	1	7	4	4	2	3	4	2
12	9	0	0	0	0	7	5	1	8	4	3	4
Total	68	41	25	20	58	39	25	18	50	41	48	44

The key to speciation is the splitting of an integrated population into two isolated ones. The essential demonstration that a single wild-type population can be converted into two populations mutually isolated in conditions which they must maintain was first made by Thoday and Gibson (1962). Disruptive selection for high and low chaeta number resulted in populations showing very little hybridisation. Del Solar (1966) investigated mating preferences in lines of *D. pseudoobscura* selectively bred for positive and negative geotaxis and phototaxis. Sexual isolation, slight but statistically significant, was found to have arisen as a by-product of selection. Our results presented above are similar; from a single base population divergent in respect of a given behavioural phenotype mating preferences were established in populations isolated for a little over 100 generations.

9.11. Conclusions

Some of the investigations reported and discussed in this chapter offer prima facie evidence of a causal association between activity and sexual behaviour in *Drosophila melanogaster*. The *yellow* mutant has pleiotropic effects on both these systems and there is the implication that a change in one may be contingent upon a change in the other. Further detailed studies are required to examine this possibility and the speculations entered into here outline the need for biochemical investigations linked to behavioural experiments. The evidence derived from directional selection for locomotor activity suggests that changes in mating speed are accidentally correlated with, and an indirect consequence of, changes in activity levels, but the indications are that normal variation in the two behavioural systems has a separate underlying genetic basis.

One consequence of changes in the gene pool of populations selected for differences in activity has been the development of incipient sexual isolation between them. Whether isolation follows from incidental changes in gene frequencies affecting characters other than activity, or whether it is causally related to the difference in activity itself will require more detailed investigation.

Any analysis of the organisation of behaviour in the fly requires to be undertaken at several levels and the approach to the functional biology of the whole organism which this entails is not only powerful but also offers some possibility of exposing the exquisite integration of living systems.

Acknowledgements

We are grateful to Dr. John Sparrow for permission to cite work in press, to Dr. David Sewell for permission to use material from his Ph.D. thesis, and to Miss Lynn Eastwood, Mr. Ronnie Wilson, Mrs. Jan Relton, and Mrs. Audrey Rixham for help in preparation of the manuscript.

References

Bastock, M. (1956): A gene mutation which changes a behavior pattern. *Evolution*, **10**, 421.
Bastock, M. and Manning, A. (1955): The courtship of *Drosophila melanogaster*. *Behaviour*, **8**, 85.
Begg, M. and Packman, E. (1951): Antennae and mating behaviour in *Drosophila melanogaster*. *Nature*, **168**, 953.
Bennet-Clark, H. C. (1971): Acoustics of insect song. *Nature*, **234**, 255.
Bennet-Clark, H. C. and Ewing, A. (1968): The wing mechanism involved in the courtship of *Drosophila*. *J. exp. Biol.*, **49**, 117.
Bennet-Clark, H. C. and Ewing, A. (1969): Pulse interval as a critical parameter in the courtship song of *Drosophila melanogaster*. *Anim. Behav.*, **17**, 755.
Bennet-Clark, H. C., Ewing, A. and Manning, A. (1973): The persistence of courtship stimulation in *Drosophila melanogaster*. *Behav. Biol.*, **8**, 763.
Benzer, S. (1973): Genetic dissection of behavior. *Sci. Amer.*, **229**, 24.
Brew, K. (1970): *Essays in Biochemistry*, **6**, 93.
Burkhardt, D. and Gewecke, M. (1965): Mechanoreception in Arthropoda: the chain from stimulus to behavioural pattern. *Cold Spring Harb. Symp. quant. Biol.*, **30**, 601.
Burnet, B., Connolly, K. and Beck, J. (1968): Phenogenetic studies on visual acuity in *Drosophila*. *J. Insect Physiol.*, **14**, 855.
Burnet, B., Connolly, K. and Dennis, L. (1971): The function and processing of auditory information in the courtship behaviour of *Drosophila melanogaster*. *Anim. Behav.*, **19**, 409.
Burnet, B. and Connolly, K. (1973): The visual component in the courtship of *Drosophila melanogaster*. *Experientia*, **29**, 488.
Burnet, B., Connolly, K., Kearney, M. and Cook, R. (1973a): Effects of male paragonial gland secretion on sexual receptivity and courtship behaviour of female *Drosophila melanogaster*. *J. Insect Physiol.*, **19**, 2421.
Burnet, B., Connolly, K. and Harrison, B. (1973b): Phenocopies of pigmentary and behavioral effects of the yellow mutant in *Drosophila* induced by α-dimethyltyrosine. *Science*, **181**, 1059.
Cane, V. (1967): Some ways of describing behaviour. In: Current problems in Animal Behaviour. Thorpe, W. H. and Zangwill, O. L., eds. (Cambridge University Press, Cambridge).
Chen, P. S. and Bühler, R. (1970): Paragonial substance (sex peptide) and other free

ninhydrin-positive components in male and female adults of *Drosophila melanogaster*. *J. Insect Physiol.*, **16**, 615.

Connolly, K. J. (1966): Locomotor activity in *Drosophila*. II. Selection for active and inactive strains. *Anim. Behav.*, **14**, 444.

Connolly, K. J. (1969): Locomotor activity in *Drosophila*. III. A distinction between activity and reactivity. *Anim. Behav.*, **15**, 149.

Connolly, K. J. (1968): Report on a behavioural selection experiment. *Psychol. Rep.*, **23**, 625.

Connolly, K. (1971): The evolution and ontogeny of behaviour. *Bull. Brit. Psychol. Soc.*, **24**, 93.

Connolly, K., Burnet, B., Kearney, M. and Eastwood, L. (1974): Mating speed and courtship behaviour of inbred strains of *Drosophila melanogaster*. *Behaviour*, **48**, 61.

Connolly, K., Burnet, B. and Sewell, D. (1969): Selective mating and eye pigmentation: An analysis of the visual component in the courtship behavior of *Drosophila melanogaster*. *Evolution*, **23**, 548.

Connolly, K. and Cook, R. (1973): Rejection responses by female *Drosophila melanogaster*: their ontogeny, causality and effects upon the behaviour of the courting male. *Behaviour*, **44**, 142.

Connolly, K., Tunnicliff, G. and Rick, J. T. (1971): The effects of gamma-hydroxybutyric acid on spontaneous locomotor activity and dopamine level in a selected strain of *Drosophila melanogaster*. *Comp. Biochem. Physiol.*, **40B**, 321.

Cook, R. (1970): Control of fecundity in *Drosophila*. *Dros. Inf. Serv.*, **45**, 128.

Cook, R. (1973a): Courtship processing in *Drosophila melanogaster*. I. Selection for receptivity to wingless males. *Anim. Behav.*, **21**, 338.

Cook, R. (1973b): Courtship processing in *Drosophila melanogaster*. II. An adaptation to selection for receptivity to wingless males. *Anim. Behav.*, **21**, 349.

Cook, R. (1973c): Physiological factors in the courtship processing of *Drosophila melanogaster*. *J. Insect Physiol.*, **19**, 397.

Crossley, S. and Zuill, E. (1970): Courtship behaviour of some *Drosophila melanogaster* mutants. *Nature*, **225**, 1064.

David, J. (1963): Influence de la fécondation de la femelle sur le nombre et la taille des oeufs pondus. Etude chez *Drosophila melanogaster*. *J. Insect Physiol.*, **9**, 13.

Dethier, V. G. (1964): Microscopic brains. *Science*, **143**, 1138.

Dewhurst, S. A., Croker, S. G., Ikeda, K. and McCaman, R. E. (1972): Metabolism of biogenic amines in *Drosophila* nervous tissue. *Comp. Biochem. Physiol.*, **43B**, 975.

Dobzhansky, Th. (1937): Genetics and the Origin of Species. (Columbia University Press, New York).

Eastwood, L. (1974): Unpublished Ph.D. thesis. University of Sheffield.

Ewing, A. (1969): The genetic basis of sound production in *Drosophila pseudoobscura* and *Drosophila persimilis*. *Anim. Behav.*, **17**, 555.

Ewing, A. and Bennet-Clark, H. C. (1968): The courtship songs of *Drosophila*. *Behaviour*, **31**, 288.

Ewing, A. and Manning, A. (1963): The effect of exogenous scent on the mating of *Drosophila melanogaster*. *Anim. Behav.*, **11**, 596.

Fattoruso, E., Piatelli, M. and Nicolaus, R. A. (1965): Su alcune melanine naturali. *Rc. Accad. Sci. fis. mat.*, Napoli, **32**, 200.

Fuchs, M. S., Craig, G. B. and Despommier, D. D. (1969): The protein nature of the substance inducing female monogamy in *Aedes aegypti*. *J. Insect Physiol.*, **15**, 701.

Fulker, D. W. (1966): Mating speed in male *Drosophila melanogaster*: A psychogenetic analysis. *Science*, **153**, 203.

Garcia-Bellido, V. A. (1964): Das Sekret der Paragonien als Stimulus der Fekundität bei Weibchen von *Drosophila melanogaster*. *Z. Naturforschung*, **196**, 491.

Geer, B. W. and Green, M. M. (1962): Genotype, phenotype and mating behaviour of *Drosophila melanogaster*. *Amer. Naturalist*, **96**, 175.

Green, M. M. (1961): Complementation at the yellow locus in *Drosophila melanogaster*. *Genetics*, **46**, 1385.

Grossfield, J. (1971): Geographic distribution and light-dependent behavior in *Drosophila*. *Proc. Nat. Acad. Sci. USA*, **68**, 2669.

Grossfield, J. and Sakri, B. (1972): Divergence in the neural control of oviposition in *Drosophila*. *J. Insect Physiol.*, **18**, 237.

Hardeland, R. (1971): Lighting conditions and mating behaviour in *Drosophila*. *Amer. Naturalist*, **105**, 198.

Hardeland, R. and Stange, G. (1971): Einflüsse von Geschlecht und Alter auf die lokomotorische Aktivität von *Drosophila*. *J. Insect Physiol.*, **17**, 427.

Highnam, K. and Hill, L. (1969): The Comparative Endocrinology of the Invertebrates. (Arnold, London).

Hinde, R. A. (1966): Animal Behaviour. (McGraw-Hill, London).

Hodgetts, R. B. and Konopka, R. J. (1973): Tyrosine and catecholamine metabolism in wild-type *Drosophila melanogaster* and a mutant, *ebony*. *J. Insect Physiol.*, **19**, 1211.

Hollingsworth, M. J. (1955): A gynandromorph segregating for autosomal mutants in *Drosophila subobscura*. *J. Genet.*, **53**, 131.

Hotta, Y. and Benzer, S. (1969): Abnormal electroretinograms in visual mutants of *Drosophila*. *Nature*, **222**, 354.

Hotta, Y. and Benzer, S. (1970): Genetic dissection of the *Drosophila* nervous system by means of mosaics. *Proc. Nat. Acad. Sci. USA*, **67**, 1156.

Hotta, Y. and Benzer, S. (1972): Mapping of behaviour in *Drosophila* mosaics. *Nature*, **240**, 527.

Karlson, P. and Sekeris, C. E. (1964): Biochemistry of insect metamorphosis. In: Comparative Biochemistry, Vol. 6. M. Florkin and H. S. Mason, eds. (Academic Press, New York).

Kessler, S. (1966): Selection for and against ethological isolation between *Drosophila pseudoobscura* and *Drosophila persimilis*. *Evolution*, **20**, 634.

King, R. C. (1970): Ovarian Development in *Drosophila melanogaster*. (Academic Press, London).

Konopka, R. J. (1972): Abnormal concentrations of dopamine in a *Drosophila* mutant. *Nature*, **239**, 281.

Koopman, K. F. (1956): Natural selection for reproductive isolation between *Drosophila pseudoobscura* and *Drosophila persimilis*. *Evolution*, **4**, 135.

Kummer, H. (1960): Experimentelle Untersuchungen zur Wirkung von Fortpflanzungsfaktoren auf die Lebensdauer von *Drosophila melanogaster*. Weibchen. *Z. Vergl. Physiol.*, **43**, 642.

Leopold, R. A., Terranova, A. C. and Swilley, E. M. (1971): Mating refusal in *Musca*

domestica: effects of repeated mating and decerebration upon frequency and duration of copulation. *J. exp. Zool.*, **176**, 353.
Lewis, H. W. and Lewis, H. S. (1963): Genetic regulation of dopa oxidase activity in *Drosophila*. *Ann. N.Y. Acad. Sci.*, **100**, 827.
Lindsley, D. L. and Grell, E. H. (1968): Genetic Variations of *Drosophila melanogaster*. *Carnegie Inst. Wash. Pub.*, **627**.
Lunan, K. D. and Mitchell, H. K. (1969): The metabolism of tyrosine-O-phosphate in *Drosophila*. *Arch. Biochem. Biophys.*, **132**, 450.
MacBean, I. T. and Parsons, P. A. (1966): The genotypic control of the duration of copulation in *Drosophila melanogaster*. *Experientia*, **22**, 101.
MacBean, I. T. and Parsons, P. A. (1967): Directional selection for duration of copulation in *Drosophila melanogaster*. *Genetics*, **56**, 233.
Mainardi, D. and Mainardi, M. (1966): Sexual selection in *Drosophila melanogaster*. The interaction between preferential courtship of males and differential receptivity of females. *Att. Soc. Italiana Sci. Nat. M. Civico Storia Nat. Milano*, **105**, 283.
Malogolowkin-Cohen, Ch., Simmons, S. and Levene, H. (1965): A study of sexual isolation between certain strains of *Drosophila paulistorum*. *Evolution*, **19**, 95.
Manning, A. (1959): The sexual behaviour of two sibling *Drosophila* species. *Behaviour*, **15**, 123.
Manning, A. (1961): The effects of artificial selection for mating speed in *Drosophila melanogaster*. *Anim. Behav.*, **9**, 82.
Manning, A. (1963): Selection for mating speed in *Drosophila melanogaster* based on the behaviour of one sex. *Anim. Behav.*, **11**, 116.
Manning, A. (1967): The control of sexual receptivity in female *Drosophila*. *Anim. Behav.*, **15**, 239.
Manning, A. (1968): The effects of artificial selection for slow mating speed in *Drosophila simulans*. I. The behavioural changes. *Anim. Behav.*, **16**, 108.
Maynard Smith, J. (1956): Fertility, mating behaviour and sexual secretion in *Drosophila subobscura*. *J. Genet.*, **54**, 261.
Maynard Smith, J. (1958): The effects of temperature and of egg-laying on the longevity of *Drosophila subobscura*. *J. exp. Biol.*, **35**, 832.
Mayr, E. (1950): The role of antennae in the mating behavior of female *Drosophila*. *Evolution*, **4**, 149.
Mayr, E. (1963): Animal Species and Evolution. (Belknap Press Harvard, Cambridge, Mass.).
Merle, J. (1969): Fonctionnement ovarien et réceptivité sexuelle de *Drosophila melanogaster* après implantation de fragments de l'appareil genitale male. *J. Insect Physiol.*, **14**, 1159.
Mitchel, H. K. and Weber, U. M. (1965): *Drosophila* phenol oxidases. *Science*, **148**, 964.
Mitchel, H. K., Weber, U. M. and Schaar, G. (1967): Phenol oxidase characteristics in mutants of *Drosophila melanogaster*. *Genetics*, **57**, 357.
Murphey, R. M. (1967): Instrumental conditioning of the fruit fly, *Drosophila melanogaster*. *Anim. Behav.*, **15**, 153.
Murphey, R. M. (1973): Spartial discrimination performance of *Drosophila melanogaster*: test-retest assessments and a reinterpretation. *Anim. Behav.*, **21**, 687.

Parsons, P. A. (1973): Behavioural and Ecological Genetics. (Oxford University Press, London).
Peeples, E. E., Geisler, A., Whitcraft, C. J. and Oliver, C. P. (1969): Comparative studies of phenol oxidase activity during pupal development of three lozenge mutants (lz^s, lz, lz^k) of Drosophila melanogaster. Genetics, **62**, 161.
Pittendrigh, C. (1958): Adaptation, natural selection and behavior. In: Behavior and Evolution. Roe, A. and Simpson, G. G., eds. (Yale University Press, New Haven).
Rick, J. T., Huggins, A. K. and Kerkut, G. (1967): The comparative production of gamma-aminobutyric acid in the Maudsley reactive and non-reactive strains of rat. Comp. Biochem. Physiol., **20**, 1009.
Sang, J. H. (1956): The quantitative nutritional requirements of Drosophila melanogaster. J. exp. Biol., **33**, 45.
Sewell, D. (1973): Larval feeding activity in Drosophila melanogaster. Unpublished Ph.D. thesis. University of Sheffield.
Sherald, A. F. and Wright, T. R. F. (1972): α-Methyl dopa as a screening agent for mutants of Drosophila melanogaster with elevated dopa decarboxylase activity. Genetics, **71**, 58.
Shorey, H. H. and Bartell, R. J. (1970): Role of a volatile sex pheromone in stimulating male courtship behaviour in Drosophila melanogaster. Anim. Behav., **18**, 159.
Solar, E. D. (1966): Sexual isolation caused by selection for and against positive and negative phototaxis and geotaxis in Drosophila pseudoobscura. Proc. Nat. Acad. Sci. USA, **56**, 484.
Sparrow, J. C. and Wright, T. R. F. (1974): The selection for mutants in Drosophila melanogaster hypersensitive to α-methyl dopa, a dopa decarboxylase inhibitor. Mol. Gen. Genet. (In press).
Spiess, E. B. (1970): Mating propensity and its genetic basis in Drosophila. In: Essays in Evolution and Genetics in Honor of Theodosius Dobzhansky. Hecht, M. K. and Steere, W. C., eds. (North Holland Publ. Co., Amsterdam).
Spieth, H. T. (1952): Mating behavior within the genus Drosophila (Diptera). Bull. Am. Mus. Nat. Hist., **99**, 401.
Spieth, H. T. (1966): Drosophilid mating behaviour: The behaviour of decapitated. females. Anim. Behav., **14**, 226.
Spieth, H. T. and Hsu, T. C. (1950): The influence of light on the mating behaviour of seven species of the Drosophila melanogaster species group. Evolution, **4**, 316.
Sturtevant, A. H. (1915): Experiments in sexual recognition and the problem of sexual selection in Drosophila. J. Anim. Behav., **5**, 351.
Thoday, J. M. and Gibson, J. B. (1962): Isolation by disruptive selection. Nature, **193**, 1164.
Tunnicliff, G., Connolly, K. and Rick, J. T. (1969): Locomotor activity in Drosophila. V. A comparative biochemical study of selectively bred populations. Comp. Biochem. Physiol., **29**, 1239.
Vowles, D. M. (1961): Neural mechanisms in insect behaviour. In: Current problems in Animal Behaviour. Thorpe, W. H. and Zangwill, O. L., eds. (Cambridge University Press, Cambridge).
Wilcock, J. (1969): Gene action and behavior: An evaluation of major gene pleiotropism. Psychol. Bull., **72**, 1.

Yeatman, F. R. and Hirsch, J. (1971): Attempted replication of, and selective breeding for, instrumental conditioning of *Drosophila melanogaster*. *Anim. Behav.*, **19**, 454.

Ziegler, I. (1961): Genetic aspects of ommochrome and pterin pigments. *Adv. Genetics*, **10**, 349.

CHAPTER 10

Genes affecting behaviour and the inner ear in the mouse

M. S. Deol

10.1. Introduction

This chapter is principally concerned with a large group of mouse mutants in which the behaviour and the inner ear are abnormal. It also includes mutants in which the inner ear is affected but no abnormalities of behaviour have so far been described, because it seems probable that the behaviour would, upon investigation, prove to be not quite normal. In addition, there is an account of audiogenic seizures, although there is only one unconfirmed report that they are associated with changes in the inner ear.

All these mutants are perhaps best described as neurological rather than behavioural in the conventional sense of the word, for the abnormalities of behaviour that they display are evidently pathological traits. They are, however, of considerable interest to the student of behaviour, and attention will be drawn to this aspect towards the end of the chapter.

10.2. Mutants with shaker-waltzer syndrome

Shaker-waltzer syndrome is the name given to a complex pattern of behaviour which is characterized by a tendency to run in circles, jerking movements of the head, hyperactivity, and abnormal responses to changes in position. These features may be accompanied by deafness, inability to swim, and tilting of the head. Although there are often clear and consistent differences in behaviour between different mutants, and even between different animals carrying the same allele, the similarities are so striking that the mutants form a readily identifiable group.

The shaker-waltzer syndrome can be caused by any one of about thirty different genes scattered over nearly three-quarters of the chromosomal complement of the mouse. This number would undoubtedly be much larger, were it not for the fact that new mutations of this kind are often not reported on account of the formidable task of establishing their relationship with or independence of known genes. Its magnitude is a reflection of the complexity of the mechanism or mechanisms underlying the syndrome as well as of the ease with which such mutations are detected. Some of these genes are recessive, while others are dominant. Some have no other known effects, while others are clearly pleiotropic, affecting structures such as the skeleton and the eye or traits such as coat colour in addition to behaviour. Some genes show full penetrance and expressivity, while the expression of others is greatly influenced by the genetic background. There is a large body of literature on this subject, and it is neither possible nor necessary to support all statements with references. For comprehensive lists of references the reader is referred to Grüneberg (1952), Sidman et al. (1965), Green (1966), and Deol (1968).

Perhaps the best way of presenting these mutants would be to give a detailed account of one of them, and then briefly mention some of the more interesting deviations from this pattern observed in others. The mutant waltzer (symbol v) may be taken as the prototype, for not only can it claim the longest history, but it became the subject of genetical studies before the re-discovery of Mendel's work. At that time these animals were known as dancing mice or waltzing mice or Japanese waltzers. This last name referred to their origin, for they were introduced into Europe from Japan. However, the Japanese themselves are reported to call them Nanking mice, for they are said to have really originated in China, where references to them go back to the year 80 B.C.

10.3. Behaviour

The behaviour of waltzer (v/v) mice does not appear to have been studied on a homogeneous genetic background, but differences between various stocks are relatively small. The tendency to run in circles is nearly always strong. It is particularly so when the animals are excited, such as after having been handled or placed in a new environment. A little prodding is often enough to elicit a response. This tendency appears to be more pronounced when the container is round. Circling is not continuous, but interruptions are not lengthy. The maximum number of revolutions without interruption that has been recorded is 407. A particular animal may show a marked preference for circling in one direction, but it will not do so exclusively. If two such mice are placed in one container they may or may not circle in the same direction. Obstacles in the way, such as a normal (non-waltzer) mouse, seem to have little or no effect. Some animals prefer to run in very tight circles, appearing to pivot on one fore-foot, so that counting the revolutions without the aid of instruments is difficult, but in general the circles are moderately wide.

Waltzer mice are markedly hyperactive, even when they are not actually circling. Their movements appear to be undirected, not resembling the exploratory behaviour of normal mice. The jerking movements of the head occur in the vertical plane. The head may be held for an instant in the lifted position, as if the animal were sniffing. These movements are virtually continuous. Their rhythm is more uniform than that of the circling movements, but they also are much more pronounced when the animal is excited or disturbed. Deafness is total and appears to be present from the beginning (normal mice begin to hear at the age of about 12–13 days). It is, however, possible that some hearing is present during the third week or so, at least on some genetic backgrounds. This may be so little as to be detectable by very sensitive tests only.

Waltzer mice cannot swim on the surface, but go under and gyrate violently in all directions, drowning unless rescued. Their righting reflexes on solid surfaces are, however, quite good. If they are held by the tail they make violent, uncoordinated movements, jerking the body about, whereas normal mice arch their backs and stretch out their forefeet, as if trying to reach a stable surface below. It has also been reported that they cannot walk in a straight line or cross a narrow bridge. They do not jump, and are liable to fall off the edge of a table. They do not become dizzy when rotated at speed in a cyclostat, but react to electrical stimulation of the brain in the normal manner. They make poor mothers, partly, no doubt, on account of hyperactivity. Whether their maternal motivations are also affected is not clear.

The syndrome is most marked in the young adult, and tends to weaken with age. There seems to be some evidence that females are more severely affected than males. Most of the overt symptoms can be suppressed by some voluntary mechanism, because they disappear during feeding and drinking and probably also during copulation. There seem to be no unusual movements during sleep.

Some other members of this group of mutants, such as jerker (je), closely resemble the prototype, while others differ from it in various ways. Circling, for instance, is much less common in varitint-waddler (Va), and when it does occur the speed is slow and the circles wider. The jerking movements may occur in the horizontal plane, as in fidget (fi) mice, which make weaving motions with their heads, or they may occur in both horizontal and vertical planes, as in Nijmegen waltzer (nv). They may also be much less marked or be intermittent only. The head may be held tilted to one side at a variable angle, as in kinky (Fu^{ki}) mice. The sense of hearing may not be affected at all, as in fidget or Nijmegen waltzer, or if affected, the onset of deafness may be delayed, as in shaker-1 ($sh-1$) mice, which can hear fairly well during the third and fourth weeks of life, or some slight residual hearing may be retained for a long time, as in pirouette (pi) mice. Hyperactivity may not be very pronounced, as in varitint-waddler, and swimming ability may be unaffected, as in Nijmegen waltzer. Mutant females may even make good mothers, as in shaker-1.

10.4. Pathology

The inner ear has been found to be affected in all mutants with the shaker-waltzer syndrome. On the basis of its pathology, these mutants can be roughly divided into two groups: those with degenerative abnormalities, and those with morphogenetic abnormalities. In the former group the development and fine differentiation of the inner ear appear to proceed normally until about the time the animals should begin to hear, but then various structures, particularly the neural elements, begin to degenerate. In the latter group the development takes a wrong turn at a much earlier stage, so that the gross structure of the organ is never normal. There are further important differences between mutants within these groups. Taking the degenerative group first, in Snell's waltzer (sv) all six regions of neural epithelium – the organ of Corti, the two maculae, and the three cristae – degenerate, while in shaker-2 ($sh-2$) only the organ of Corti and the macula of the saccule do so. In varitint-waddler the organ of Corti, the macula of the saccule, and the three cristae are affected, but not the macula of the utricle, and in jerker the organ of Corti as well as both the maculae are affected, but not the cristae. (It is to be emphasized that this account is based on observations with the light microscope only, and it is possible that electron microscopy would reveal abnormalities in structures that are now regarded as normal.)

In the morphogenetic group of mutants, the abnormalities may appear at the earliest stage of the development of the inner ear, as in kreisler (kr), a little later on, as in dancer (Dc), or considerably later on, as in shaker-with-syndactylism (sy). The extent of the abnormalities is also highly variable. In kreisler the entire inner ear is affected, often in such a way that it bears no resemblance whatever to the normal organ, while in dancer only the utricle and the superior and lateral semi-circular canals are affected. In fidget, the abnormalities are confined to the semi-circular canals, all three of which are abnormal, while in waltzer-type (Wt) they are limited to the lateral and posterior canals, and in Nijmegen waltzer to only the lateral canal.

Finally, in the mutant rotating (rg) both types of abnormalities, degenerative and morphogenetic, occur together; there is a degeneration of the macula of the saccule, and a severe constriction, frequently amounting to the closure of the lumen, in each of the canals.

The causes of degeneration in the inner ear are very poorly understood. It is, however, known that a humoral metabolic disorder is not involved, at any rate in shaker-1. The evidence comes from female mice that are homozygous for shaker-1 and heterozygous for Cattanach's translocation, an X-autosome translocation carrying the normal allele at this locus (Deol and Green, 1969). The normal allele becomes inactive in about half the cells in a random manner in accordance with the Lyon hypothesis and a patchwork of normal and abnormal cells is observed in the organ of Corti. This would not be the case if a generalized metabolic factor had been in operation. There is also some evidence that the primary fault might lie in the efferent fibres innervating the hair cells (Kikuchi and Hilding, 1965), the cell bodies of these fibres being in the brain. The causes of the abnormalities in the morphogenetic group are better understood. In many cases the primary defect seems to lie in the neural tube in the region where the inner ear is formed or in the ganglion cells from which its nerves arise, and the consequent disturbance of the inductive influence of these structures on the morphogenesis of the inner ear is believed to be responsible for the abnormalities observed.

10.5. Causal relationships

The relationship between the abnormalities of behaviour and of the inner ear is not as simple as it may seem at first sight. Since the inner ear is the seat of the sense of equilibrium, and this sense appears to be disturbed in these mutants, there is a tendency to assume that defects of the inner ear are responsible for the abnormal behaviour. This assumption is untenable for many reasons. First, extirpation of the whole or of a part of the inner ear on one or both sides does not produce this syndrome. Secondly, waltzing behaviour can be produced artificially by means of certain chemicals, and no abnormalities are observed in the inner ear in these mice (Goldin et al., 1948). Thirdly, gross differences in the pathology of the inner ear are found between mutants with indistinguishable behaviour. Fourthly, a study of the mutant kinky, in which the behaviour pattern varies greatly from animal to animal on certain genetic backgrounds, showed no consistent association between any particular defect of the inner ear and any specific peculiarity of behaviour, although it was clear that defects of the inner ear were in general much severer in animals with a more pronounced expression of the syndrome (Deol, 1966). This led to the conclusion that the abnormal behaviour is probably caused by some unidentified abnormality of the central nervous system, perhaps located in the brain stem or the cerebellum. The abnormalities of the inner ear may then be accounted for on the basis of inductive and functional relationships between the neural tube, the acoustic ganglion, and the otic vesicle during development, and perhaps later as well. This view does not conflict with the finding that the behaviour is generally, although not invariably, more severely affected in animals with severer defects of the inner ear. It is supported by studies on the mutants kreisler, dreher and dancer, in which abnormalities of the neural tube and the acoustic ganglion have been discovered, and on the artificially produced waltzing mice referred to above, in which widespread lesions have been observed in the cerebellum and the brain stem, but not, significantly, in the inner ear. It is also consistent with the fact that this syndrome can be caused by such a large number of genes at independent loci, for the physical basis of the abnormalities postulated here is not likely to be simple, or the chain of events leading to them short.

Studies similar to those on kinky mice described above were recently carried out on the mutant Nijmegen waltzer (nv), and these led to the same conclusions (Deol, unpublished). This mutant was discovered and described

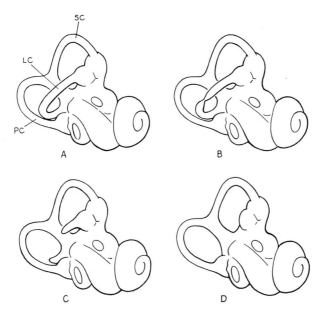

Fig. 10.1. Drawings of the right inner ear of normal and *nv/nv* mice. A) normal. B) *nv/nv*, lateral semi-circular canal constricted. C) *nv/nv*, lateral canal incomplete. D) *nv/nv*, lateral canal missing. LC = lateral canal; PC = posterior canal; SC = superior canal.

by van Abeelen and van der Kroon (1967). Affected animals tend to run in circles and shake their heads in both horizontal and vertical planes. As far as circling is concerned, the gene has incomplete penetrance and variable expressivity, many homozygotes appearing to have normal behaviour. Histological examination of 15 affected animals showed that the lateral semi-circular canal is either constricted in the middle or incomplete or altogether lacking (Fig. 10.1). Its crista is present, but it is generally somewhat reduced, and the spatial relationship between the cristae of the lateral and superior canals and the macula of the utricle may not be quite normal. To find out whether there were any consistent differences between homozygotes with normal and circling behaviour, the lateral semi-circular canal was examined in 25 *nv/nv* mice with normal behaviour and 30 *nv/nv* mice with affected behaviour, ranging in age from 25 to 55 days. In order to minimize differences due to the genetic background both types of animals were taken from the same matings. Observations were made on cleared heads, prepared by following the standard method for staining with alizarin, but without using

TABLE 10.1

The lateral semi-circular canal in *nv/nv* mice with normal and circling behaviour.

Lateral canal (Left/Right)	Normal	Circling
+/+	14	0
(+)/+	0	2
+/(+)	0	1
(+)/(+)	1	0
−/+	4	2
+/−	2	3
−/(+)	0	2
(+)/−	0	2
−/−	4	18
Total	25	30

+ = normal; (+) = constricted; − = incomplete or missing

the stain. The results are set out in Table 10.1. They essentially follow the same pattern as in kinky mice: abnormalities of the lateral canal are much more striking in animals with circling behaviour, but the affected behaviour cannot be ascribed to these abnormalities alone because they were also present in a number of mice with normal behaviour.

10.6. Mutants with uncomplicated deafness

There are two genes that cause total deafness without producing any of the locomotor disturbances associated with the shaker-waltzer syndrome. One of these, deaf (v^{df}), is an allele of waltzer. Its effects on the inner ear seem to be confined, as would be expected, to the organ of Corti, which begins to degenerate at about the same time and in the same manner as in the degenerative group of shaker-waltzer mutants. Deafness is present from the beginning, although some slight response to sounds may be obtainable in some animals during the third week or so. The other gene, deafness (*dn*), also causes abnormalities of the degenerative type, but in this case the macula of the saccule is involved to some extent in addition to the organ of Corti. Tossing movements of the head are occasionally observed in *dn/dn* mice, but they are generally so faint that only experienced observers can distinguish them from the normal sniffing movements. No residual hearing at any stage has been reported in this mutant.

Genes causing a pronounced degree of white spotting, such as Mi^{wh}, W^v and s^l, generally also lead to an acute loss of hearing. The loss is in most cases not total, for the organ of Corti is not affected in its entirety. The abnormalities of the inner ear follow a pattern quite different from anything observed in shaker-waltzer mice. The behaviour of affected animals is commonly supposed to be quite normal, but in view of suspected abnormalities of the neural crest in these animals it is probably not so (Deol, 1970a).

10.7. Audiogenic seizures

Some mice respond to certain types of sounds, such as the ringing of an ordinary door bell, by going into convulsions, which may end in death. These convulsions are called audiogenic seizures (detailed references in Fuller and Wimer, 1966). The susceptibility to seizures undoubtedly has a strong genetic element in it, for certain strains (e.g. DBA/2 and A) are much more susceptible than others (e.g. C57BL/6) and random-bred Swiss albino mice respond to selection for high and low susceptibility, but its genetic basis has long been a subject of controversy. This is to a great extent due to scoring difficulties, for the pattern of response is highly variable. The seizures cannot easily be put into quantitative terms for analysis purposes, for they differ greatly in intensity, speed of induction, liability to being fatal, and many other features. The trait was thought to be polygenic in origin, but Collins and Fuller (1968) have argued that in the strain DBA/2J proneness to initial seizures, as opposed to those in sensitized animals, is determined by a single pair of alleles. It is possible that there is more than one variety of the trait, each one caused by a different gene, as is the case with the shaker-waltzer syndrome. There is an obvious non-genetic element in the origin of seizures as well, for there can be striking differences in the behaviour of animals from the same inbred strain.

In its typical form, a seizure follows the following course. Within a very short time of the initiation of the stimulus the susceptible animal begins to show signs of agitation. If the stimulus is stopped at this stage nothing more may happen, but if it is continued, the animal starts walking rapidly, then running about, and finally has a convulsive fit, when it falls over to one side and starts kicking rhythmically. The convulsion may be mild and take the form of only a standing spasm, but it may be so severe that the trunk muscles get fixed in the inspiratory position. Such severe seizures usually result in death, although artifical respiration may sometimes succeed in reviving the animal.

Susceptibility to seizures is affected by a number of factors, some of which appear to affect different strains in different ways. The most effective range of sound frequency is 12000–25000 cycles per second, and the most effective range of intensity 86–104 db. A drop in body temperature decreases susceptibility, but prestimulation by sub-convulsive doses of sound decreases it in some strains and increases it in others. Perhaps the most important single factor affecting susceptibility is age, there being a period when the response is at its most intense and can be induced in the largest proportion of animals. In the DBA/2 strain this period lies between 20 to 35 days, there being a rapid decline in susceptibility after that, while in the A strain the susceptibility, although less severe, remains fairly constant up to the age of 8 months. In all strains old mice are more resistant to seizures, and when attacks do occur, they are mild. It appears from these observations that susceptibility may be a function of the state of "maturation" of the nervous system. The effects of propylthiouracil and triiodothyroxine point towards the same conclusion: propylthiouracil, which is known to retard the development of the nervous system, decreases susceptibility and the risk of death from seizure, and triiodothyroxine, which presumably has the opposite effect on the nervous system, advances the age at which susceptibility first appears.

The physiological basis of seizures remains a mystery. A number of suggestions have been made, but there is little or no confirmatory evidence. Clues may lie in the effects of various treatments on susceptibility or metabolic differences between susceptible and resistant strains or similarities between audiogenic seizures and those of other types. Treatments with insulin or glutamic acid reduces susceptibility, but castration increases it. Tranquilizers in general reduce it. No differences have been found between susceptible and resistant strains in the rate of glycolysis and the activities of cytochrome oxidase, malic dehydrogenase, succinic dehydrogenase, and alkaline phosphatase, but a reduction in the activity of adenosine triphosphatase has been observed in a susceptible strain. As this reduction was significant only during the period of high susceptibility, it has been suggested that the rate of oxidative phosphorylation may be a factor in the origin of these seizures. Audiogenic seizures bear some similarities to those induced by electric shocks or pentylenetetrazol, but there are also some clear differences. Susceptible mice have a lower threshold for response to electric shocks than resistant ones, and the same applies to treatment with convulsant drugs, of which in general smaller doses are required for susceptible animals. In searching for clues to the physiological basis of seizures it should not be forgotten that one may well be dealing with more than one genetic entity.

No changes associated with susceptibility to seizures have been observed in the central nervous system, but there is an unconfirmed report by Darrouzzet (in discussion following Deol, 1970b) that abnormalities of the hair cells in the organ of Corti occur. These were identifiable only with the electron microscope, but were present from the day of birth. He also found degeneration of afferent fibres, which might simply be the consequence of the hair cell abnormalities, but efferent fibres appeared to be unaffected. He does not mention which strain was used, but these observations, if confirmed, would undoubtedly provide an important clue.

10.8. Wider considerations

The value of these mutants to the student of behaviour is obvious. They not only provide a rich source of material for studies of a purely behavioural or neurological kind, but promise to bring the two closer by revealing correlations between neurological traits and peculiarities of behaviour. The gross anatomy of the mouse brain is now well understood, and with the immense advances made in neurological science in recent years and the vast array of techniques now available, the problem of identifying anatomical, physiological or biochemical abnormalities in the central nervous system no longer appears as formidable as it did a few years ago. Investigations of this kind would doubtless benefit from Ungerstedt's (1971) finding that vigorous circling behaviour can be induced in the rat by treatment with amphetamine following damage to the nigrostriatal dopamine system.

Attempts should be made to find out whether any of the symptoms of the shaker-waltzer syndrome can be suppressed or altered by means of drugs, preferably those with known mechanisms of action, or by electrical stimulation of various parts of the central nervous system. It would also be interesting to see to what extent the normal inhibition of the symptoms during feeding and drinking can be interfered with artificially. Purely behavioural studies may be aimed at finding out to what extent the circling behaviour is consequent on hyperactivity and being reared in the confined space of an ordinary mouse cage. Mutant mice raised in very large pens may well show different patterns of behaviour. Observations on segregating litters may show to what extent normal animals are affected by being reared with mutant ones, and vice versa. Effects of hereditary uncomplicated deafness on behaviour should also be carefully determined. The general impression is that such animals are much more docile.

Studies on the expression of these genes in feral mice may also prove rewarding. Deafness, for instance, may well alter the behaviour of feral mice, mitigating the "wildness". The shaker-waltzer syndrome may also have quite a different manifestation on the "wild" genetic background. It will probably not be necessary to carry out too many backcrosses for these purposes, because "wildness" is quite evident even in the F_i offspring of crosses between feral and laboratory mice.

As the shaker-waltzer syndrome seems to occur only in rodents, it might be thought that these genes are peculiar to this order of mammals and that the conclusions drawn from studies on these mutants have only a limited appliation. This view would almost certainly be erroneous. It is improbable, on general evolutionary grounds, that such a large number of genetic loci could be limited to a single order. It is much more likely that these loci are widespread, and have similar rates of mutation in other species of mammals, but that the changes in behaviour caused by them are different. Their effects on hearing, however, would be expected to be more constant, and it is possible that some types of apparently uncomplicated deafness in man are determined by such loci (Deol, 1968). In any case, even in the mouse genes producing the full syndrome and those causing uncomplicated deafness are not entirely unrelated, as is evident from the alleles waltzer and deaf.

In the same manner, conclusions drawn from studies on mice with audiogenic seizures may also have wider application. The phenomenon has also been reported in the rabbit, and the probability is strong that a systematic search would reveal its occurrence in many more species of mammals. It is possible that the severe discomfort felt by some people when subjected to grating or screeching sounds is in some way related to it.

References

Abeelen J. H. F. van and P. H. W. van der Kroon (1967): Nijmegen waltzer – a new neurological mutant in the mouse. *Genet. Res., Camb.*, **10**, 117.

Collins R. S. and J. L. Fuller (1968): Audiogenic seizure prone (asp): A gene affecting behavior in linkage group VIII of the mouse. *Science*, **162**, 1137.

Deol, M. S. (1966): The probable mode of gene action in the circling mutants of the mouse. *Genet. Res., Camb.*, **7**, 363.

Deol, M. S. (1968): Inherited diseases of the inner ear in man in the light of studies on the mouse. *J. med. Genet.*, **5**, 137.

Deol, M. S. (1970a): The relationship between abnormalities of pigmentation and of the inner ear. *Proc. roy. Soc. Lond. B*, **175**, 201.

Deol, M. S. (1970b): Mouse mutants with inner ear defects and their value in biomedical research. In: Les Mutants Pathologiques chez l'Animal. Sabourdy, M., ed. (Editions du Centre National de la Recherche Scientifique, Paris) pp. 279–292.

Deol, M. S. and M. C. Green (1969): Cattanach's translocation as a tool for studying the action of the shaker-1 gene in the mouse. *J. exp. Zool.*, **170**, 301.

Fuller, J. L. and R. E. Wimer (1966): Neural, sensory and motor functions. In: Biology of the Laboratory Mouse, 2nd ed. Green, E. L., ed. (McGraw Hill, New York) pp. 609–628.

Goldin, A., H. A. Noe, B. H. Landing, D. M. Shapiro and B. Goldberg (1948): A neurological syndrome induced by administration of some chlorinated tertiary amines. *J. Pharm. exp. Ther.*, **94**, 249.

Green, M. C. (1966): Mutant Genes and Linkages. In: Biology of the Laboratory Mouse, 2nd ed. Green, E. L., ed. (McGraw Hill, New York) pp. 87–150.

Grüneberg, H. (1952): The Genetics of the Mouse, 2nd ed. (Nijhoff, The Hague).

Kikuchi, K. and D. A. Hilding (1965): The defective organ of Corti in shaker-1 mice. *Acta Oto-laryngol.*, **60**, 287.

Sidman, R. L., M. C. Green and S. H. Appel (1965): Catalog of the Neurological Mutants of the Mouse (Harvard University Press, Cambridge, Mass.).

Ungerstedt, U. (1971): Striatal dopamine release after amphetamine or nerve degeneration revealed by rotational behaviour. *Acta physiol. scand., Suppl.*, **367**, 49.

CHAPTER 11

Personality development in XO-Turner's syndrome

F. J. Bekker

11.1. Introduction

It is nothing new to say that genetic and environmental factors play a role in the development of human behaviour. There is still much obscurity, however, as to the way in which these factors influence behaviour. The time of the nature-nurture controversy can be considered past. The effort to determine in what proportions environmental or genetic factors contribute to a given behaviour pattern has proven to be fruitless. It led to many investigations, but the results could not be coordinated. The same set of data has frequently led to opposite conclusions in the hands of psychologists with different orientations. Most investigations started from the implicit assumption that genetic and environmental factors combine in an additive fashion. Both geneticists and psychologists, however, had repeatedly demonstrated that an interaction model would provide better explanatory possibilities. Anastasi (1958) argued that it is more fruitful to ask *how* genetic and environmental factors interact in behaviour development. The classic approach, based on the question "*how much* of the variance is attributable to genetic and how much to environmental factors?", is comparable to what is called the investigation of quantitative-genetic aspects. The research based on the question "*how*" do genetic and environmental factors contribute to the development of human behaviour?", can be called the investigation of developmental-genetic aspects (van Abeelen, 1972). Although the quantitative-genetic approach in human behaviour genetics is not totally abandoned, many researchers in this field have turned to the developmental-genetic approach.

More and more information now demonstrates the complexity of the interaction between genetic and environmental factors in the development of human behaviour. Nevertheless, it was not so long ago, even in scientific circles, that serious consideration was given to the hypothesis that aggressive, antisocial behaviour might be directly and exclusively genetically determined. The question arose after a high incidence of XYY sex-chromosome configurations was detected in aggressive, male criminals (Jacobs et al., 1965). In 1969 in the U.S.A. a symposium was devoted to the XYY configuration and its implications for personality development. The "Report on the XYY chromosomal abnormality" (1970) gives an account of the conclusions arrived at by this symposium, two of which are: While a greater prevalence rate for XYY males among several groups of institutionalized criminals and mentally disordered offenders is not to be disputed, the widely publicized beliefs concerning aggressivity and antisocial behaviour as typical characteristics of such individuals are premature and subject to error (. . .). The fact that criminal behaviour is a complex and multi-determined phenomenon makes it extremely unlikely that the extra Y chromosome is the only, or even the major, causative factor. Thus even in 1969 it was still necessary to emphasize that it is not possible to attribute human behaviour to genetic factors only.

Unlike animal research in behavioural genetics, this research in man cannot, for obvious reasons, make use of inbred strains, deliberately induced mutations, or well-controlled environmental variables. In human behavioural genetics research one is limited to the study of chance occurrences and the "experiments of nature". Since about 1955, research in behavioural genetics is carried out in cases of aberrant sex-chromosomal patterns such as: chromatin-negative Turner's syndrome (45,XO), chromatin-positive Turner's syndrome (for example 45, XO/46, XX), Klinefelter's syndrome (47, XXY), the already mentioned so called "super males" (47, XYY), and different kinds of hermaphroditism. The research in hermaphroditism and in Turner's syndrome is most elaborated. Mainly on the basis of the information now available about Turner's syndrome, I shall deal with the question of how genetic and environmental factors interact in the development of human behaviour.

11.2. Turner's syndrome

The syndrome was first described in 1938 by the American endocrinologist Henry Turner as a syndrome of sexual infantilism, congenital webbed neck, and cubitus valgus (Turner, 1938). It occurs in phenotypically female patients, and is physically characterized by: (a) hypogonadism, with lack of puberty (amenorrhea) and sterility; (b) stunted growth; (c) not always present abnormalities, like webbed neck and cubitus valgus; and (d) associated congenital anomalies, for example of the kidney or the cardiovascular system.

Cytogenetically, the syndrome is characterized by the absence of one X-chromosome. This variant of the syndrome has a chromatin-negative pattern while the mosaic variants of the syndrome (like 45XO/46XX or 45XO/47XXX) show a chromatin-positive pattern.

It is possible to induce pubertal development by means of cyclic estrogen substitution. However, at least in The Netherlands, the start of the hormonal treatment in most cases comes later than the onset of normal pubertal development, because sex-hormones restrain growth. Nearly all the patients we examined showed a considerable delay in pubertal development in comparison to normally maturing girls.

11.3. Cognitive development in Turner's syndrome

There is still uncertainty with regard to the distribution of intelligence in Turner's syndrome. The reason for this, in our opinion, is that most investigators studied rather small groups of non-institutionalized girls and women. Most of the groups were composed of patients who voluntarily came for advice to specialists or hospitals. Patients hospitalized in institutions for the mentally retarded are seldom represented in the samples. Patients who live intellectually and socially as close to normal as possible certainly can give us more detailed information, but this would mean that less-gifted patients are rather systematically left out. Harms' (1967) study supports this hypothesis. In a survey of 15,938 female neonates in Germany, he found an incidence rate of XO Turner's syndrome of 0.4 per 1000 in the general population. The average incidence rate in three surveys covering 4114 females in institutions for the mentally retarded and in schools for educable subnormals, however, was much higher, 1.2 per 1000. It is very likely, therefore, that many of the least gifted patients are institutionalized and never come to the usual addresses for examination or treatment. The published data on intelligence-level distribution in Turner's syndrome may therefore be biased. The average of 76 IQ's compiled from these publications is 95.6, which is not very far, but still significantly, below the normal average of 100 (Bekker, 1969). In a random sample in which the institutionalized patients are also represented, the IQ distribution may show one of two possible kinds of deviations from the normal curve. Either it will be definitely shifted to the side of the lower IQ's, or it will show a clear bimodality. Several investigators have already found a tendency towards bimodality in their material (Hampson, Hampson and Money, 1955; Lindsten, 1963; Bekker and van Gemund, 1968). But, whichever deviation the IQ distribution shows, it does not mean that a low IQ is obligatory in connection with Turner's syndrome. For the moment we can do no more than conclude that in XO-Turner's syndrome mild and severe retardation are more frequently found than in normals, although cases with average or even superior intelligence do occur also. From the foregoing it should be clear that the exact relationship between the genetic anomaly and the lowering of the IQ is far from having been established.

The same holds for the explanation of a specific deficit in intellectual functioning observed in about 75 % of the patients with Turner's syndrome. This deficit manifests itself in a series of symptoms, not all of which are always present. In everyday life these symptoms include: poor orientation in space and/or in time, inaccurate writing of words and numbers, dyslexia and dyscalculia, difficulties in word-mobilization, and vague or imprecise word usage. Psychological examination reveals the following symptoms: a poorly developed body-image, poor left-right discrimination, difficulties in space-form perception, and poor concentration. This whole syndrome is very similar to what one sees in children with plain dyslexia or dyscalculia. Shaffer (1962) was the first to establish this cognitive deficit in Turner's syndrome; it has since been mentioned in several publications (Money, 1963; Alexander, Ehrhardt and Money, 1966; Bekker and Van Gemund, 1968; Theilgaard, 1972). In our investigations we found the symptoms in 10 out of 14 girls with XO-Turner's syndrome (71 %), in 8 out of 11 girls with mosaic variants of Turner's syndrome (73 %), but also in 5 out of 7 boys with stunted growth and delayed adolescence (71 %) (Bekker, 1969). In her publication of 1972, Theilgaard reports a difference in cognitive functioning between patients with Klinefelter's syndrome and infertile men without sex-chromosomal anomalies. Characteristic symptoms of the Klinefelter patients in this respect are: vague or imprecise word-usage, awkward word-constructions, difficulties in word-mobilization, modifications such as "sort-of", "kind-of", "the like", and the application of demonstrative words without preestablished reference to what these might designate. Furthermore, there are specific difficulties in a figure-composition test such as the subtest Block Design in the WISC. She suspects these symptoms "to be an index of the 47,XXY patient's difficulties in articulating and structuring experience". On the basis of a comparison between the intelligence-test results of a patient with 47,XXY Klinefelter's syndrome and a patient with 45,XO Turner's syndrome, she suggests a difference in cognitive style between these two syndromes as well. The patient with Turner's syndrome had difficulties in the perceptual-analytical and numerical fields in combination with strikingly good verbal capacities, whereas the patient with Klinefelter's syndrome showed poor performance in the latter and normal performances in the former fields. We are not convinced, however, that the difference in cognitive style between Turner's syndrome and Klinefelter's syndrome is so distinct, which may be clear from our description of the cognitive difficulties found in Turner's syndrome. In our opinion, in both syndromes a poorly developed body schema, in combination with a retarded neurological development and/or specific brain damage, causes difficulties in direction sense and in space-form perception. Poor direction sense in particular may cause problems in structuring visual or auditory information. Especially the capacity to arrange properly the elements of a sequence (e.g. words and numbers) seems to be disturbed in these cases (Bakker, 1971). We have been searching for a relation between a physical defect (sex-chromosomal abnormality, lack of sex-hormone production and stunted growth) and the cognitive deficit. The outcome so far is rather disappointing because too many physical symptoms coincide to permit a systematic cross study of the results available at present. The chromosomal aberration may be a causal factor, although the occurrence of the cognitive deficit in boys with stunted growth and delayed adolescence, as well as in dyslectic people, seems to contradict this assumption. The eventual significance of the lack of sex-hormone is denied by the presence of the deficit in both groups just mentioned, while Theilgaard did not find it in the group of men deficient in sex-hormone production. Stunted growth in itself cannot be an explanation because Klinefelter patients tend to be taller than normal.

More generally speaking, in cases of neural maldevelopment combined with stunted growth and failing sex-hormone production, one could think of hypopituitarism. The pituitary gland produces gonadotropic as well as thyrotropic hormones, and thyroid hormone influences neural development. However, patients with Turner's syndrome show no deficiency of thyrotropic, adrenocorticotropic, or gonadotropic hormones, and thus no direct evidence of hypopituitarism (Martin and Wilkins, 1958).

Maybe, close physical examination of dyslectic children with regard to endocrinological or genetical aberrations can clarify the problem. For the moment we can do no more than conclude that individuals with sex-chromosomal aberrations, more than normals, are apt to show a specific cognitive deficit. The relationship between this deficit and the genetic anomaly has not yet been uncovered. The symptoms of the deficit bear a great resemblance to what is seen in people with dyslexia or dyscalculia. It seems likely that in some way the same easily damaged area in the cortex of the brain is affected in patients with sex-chromosomal aberrations, as it is in individuals with dyslexia and/or dyscalculia.

11.4. Personality development

Over the years, a number of investigations have rendered information about various aspects of personality development in girls with Turner's syndrome. The characteristic most frequently mentioned is the retarded personality development. As children and as grown-ups, most of the patients strike us as infantile: they are less independent, show much less self-confidence, are more naive, and are less able to postpone need satisfaction than normals. All people with external physical handicaps have to face more or less serious problems in social adaptation. So the associated physical anomalies in Turner's syndrome, although they are not so obvious in most cases and can be treated by plastic surgery at an early age, can give rise to difficulties in making social contacts. The most conspicuous handicap, however, which cannot be treated and which evokes an infantilizing treatment by others, is the stunted growth (Martin and Wilkins, 1958; van Gemund, 1959; Kind, 1962; Krautschick, 1962). We have demonstrated in a comparative study that this infantilism is not confined to Turner's syndrome, but shows up in every syndrome where short stature is one of the characteristics. The retardation in personality development is not directly related to the genetic anomaly but it varies with the age at which the stunted growth becomes clear and with the extent to which the child is dwarfed. Children who already show a marked degree of stunted growth between 6 and 12 have a greater chance of becoming infantilized than normal children and than children in whom the stunted growth appears after the age of about 12. For children with a height three standard deviations or more below normal (according to the tables of van Wieringen et al., 1968) this chance is greater than it is for children with a less severe dwarfism (Bekker, 1969). With regard to Turner's syndrome, this means that although all the patients have the same genetic anomaly, not all of them show the same degree of infantilism. It also means that infantilism develops as a resultant of the interaction between the child and its social environment. This interaction is well-known to everybody who is familiar with dwarfed children. It is extremely difficult, even for their parents, not to treat these children according to their height instead of their age. In two or three out of 70 families, however, we have seen parents who were able to treat their dwarfed children according to their age; these children had a strikingly more mature personality than the dwarfed children in the other families.

In this connection we should mention an observation by Krautschick (1962), which has received very little attention in the literature on Turner's syndrome. She found a remarkable agreement among the parents of six girls with Turner's syndrome on the fact that none of the girls had shown the typical symptoms of the "trotzperiod" which usually occurs at about the age of three. The essential feature of this period is the child's opposition to the will and the power of the parents and of other people. The child learns to say "no" and practises this new attainment very intensively. According to Krautschick, girls with Turner's syndrome do not show this resisting attitude. In our data we have no reliable information on that developmental period, but from interviews with the children and with the parents we do know that these girls are very resigned during that other period of rebellion, between the ages of about 13 and 18 years. They acquiesce very easily in decisions made by their parents, however upleasant these decisions may be for them. In our opinion, these phenomena are very similar to those described by Money and Mittenthal (1970) as inertia of emotional arousal. They consider this to be a feature of personality shared by many Turner girls and they describe it as "constituted of compliancy, phlegmatism, stolidity, equability, acceptance, resignedness, slowness in asserting initiative and tolerance of personal adversity." The writers suggest a direct effect of the genetic constitution on behaviour. We are not sure, however, that this kind of behaviour is confied to Turner's syndrome. Kind (1962) reports a low general impulse-level and diminished drive-energy in dwarfed children with hypopituitarism. If his observation refers to the same phenomenon as indicated by Money and Mittenthal, this behaviour is probably not solely and directly genetically determined. What both groups of patients have in common is the lack of sex-hormone production, and we suggest, therefore, that in Turner's syndrome this hormonal deficiency might be one of the factors mediating the inertia of emotional arousal.

Even if this hypothesis is valid, we have to remember the fact that many children with stunted growth, whatever the physical background may be, show a behaviour which is very similar to that just mentioned. In fact, we also observed this way of behaving in children with primordial dwarfism in whom, apart from short stature, no other physical anomalies, either hormonal or genetical, have been established. To us this behaviour seems to be related to infantilism and a feeling of helplessness due to short stature. These children tend to rely on the help of others, especially the mother. They do not actively structure and organize their own lifes, but leave this for the most part to people who are, in their eyes, more able to find a way in the struggle for life. In our opinion, this is a learned behaviour pattern which stems from early youth, namely from the time the parents first felt that something was amiss with the child. At that time, motivated by feelings of guilt the parents, the mother in particular, often start to overprotect the child. This can lead to a form of personality absorption which makes the child feel unable to organize things or to make decisions without the help of others.

If this latter behaviour is different from the former – and at present we assume it is – then the "phlegmatism" in Turner's syndrome and in hypopituitarism might consist of two components, one being "inertia of emotional arousal" related to the lack of sex-hormone production, and another which could be named "social dependence" as a sequel of personality absorption. In children with stunted growth and normal sex-hormone production, only the latter component will be present.

The behaviour of a 15-year-old boy of average intelligence with primordial dwarfism may serve as an example. He attended a technical school, but he did not like technics very much; his marks went down and were worse than they could have been. He did not complain of this situation but considered it as rather normal. His parents at last decided to ask the pediatrician for a psychological examination. During this examination the boy did not show any interest in its purposes or results. It was his parents' business, they had to decide what he was going to do. Asked about what kind of job he liked most, he said he had no idea. In fact, he would be very happy to stay at home with his parents and help his mother do the housekeeping. On our remark that this could not last for ever, he answered that after his parents died he could move to his elder married sister. "She always took care of me when I was a little boy."

11.5. Gender role and gender identity

The concepts of gender role and gender identity refer to two closely related psychological phenomena. The individual's awareness of being a boy or a girl, a man or a woman, is called gender identity. This feeling developes on the basis of a number of experiences the child undergoes very early in its life. Hampson and Hampson (1961) have given a systematic survey of the factors involved in this process. We treat the two most important of these: the assignment of sex and the confrontation with the own body. The assignment of sex refers to the fact that usually at birth the baby is assigned to a sex on the basis of the appearance of the external genitals. In this sex the child will be reared and it is here that it encounters the first signpost on its way to psychosexual differentiation. The assigned sex is revealed to the child not only in direct contacts with the parents. All kinds of other more or less implicit environmental factors play their role, like the clothes the child is given, the kind of toys it is allowed or not allowed to play with, the way in which it is treated by peers and adults, etc. As a child grows older, the external genitals are an important bodily feature in determining the childs own feeling of being a boy or a girl. Almost always gender identity is consistent with physiological sex, but in a few cases it is not. In transsexualism, for example, there is no known discrepancy between the sex assigned at birth and the morphology of the external genitals or other measurable somatic criteria of sex. Still these people are deeply convinced that they are actually members of the opposite sex. They often ask for sex reassignment by plastic surgery and hormonal treatment. The etiology of this discrepancy between gender identity, assigned sex, and somatic sex is still obscure. It might be considered as a result of deviations in the learning process, possibly in combination with somatically determined predispositions.

As mentioned before, there is a very close relationship between gender role and gender identity. By gender role is meant all a person says or does in which he reveals himself as boy or girl, man or woman, respectively. The concept thus comprises more than just sexual behaviour in a more restricted sense. The gender role develops on the basis of, and interwoven with, gender identity. Identification with significant others in the environment of the child and the development of cognitive structures play an important role in its establishment. The behavioural habitus which we call gender role thus develops during the early years of life in interaction with the many experiences of the child.

John Money and John and Joan Hampson, in a series of superb studies, have disproved the earlier conjecture that gender identity and gender role are congenitally or genetically determined (Hampson, 1955; Hampson, Hampson and Money, 1955; Money, Hampson and Hampson, 1955a; Money, Hampson and Hampson, 1955b; Hampson, Money and Hampson, 1956; Money, Hampson and Hampson, 1956; Money, Hampson and Hamspon, 1957). From their studies on hermaphroditism it became clear that these people in general develop a gender role which is consistent with the assigned sex, even in the case of contradictory chromosomal or gonadal sex. In girls with Turner's syndrome, despite the sex-chromosomal abnormality, they never observed anomalies either in gender role or in gender identity (Money, 1969). Also in our own study of 15 girls with chromatin-negative and 11 girls with chromatin-positive Turner's syndrome we found no such aberrations (Bekker, 1969).

The meaning of all this is that the gender role is not genetically determined; it is a behaviour pattern which is the result of a social learning process, in interaction with certain somatic cues and dispositions. This learning process does not always lead to complete success. We already mentioned the problem of transsexualism, but other difficulties can occur. Two essential sources of failures of the learning process can be discerned: ambiguity of the external sexual characteristics of the own body and ambivalence of the parents or other people toward the child's sex. The child who is confronted sometime in its early development with the ambiguous shape of its genitals *may* develop feelings of shame and doubt about its sexual identity. For further development of a healthy gender role much will depend on the parents' success in providing a well-balanced support of the child, in combination with the possibilities of plastic surgery.

In every case where the parents feel ambivalent toward the sex of their child, this ambivalence will be induced in the child. We had the opportunity to observe this phenomenon in a 9-year-old hermaphroditic child reared as a girl (46XY/45XO, with a hermaphroditic morphology of the internal and external genitals). She was known to us since her fourth year. Her mother told us that she (the mother) panicked each time the child played with the toy-cars of her brother. The mother tried to prevent this "boyish" behaviour by diverting the child's attention away from the toys of her brother and by overburdening her with all kinds of girls's toys. The mother recognized her own anxiety that this child was not to grow up as a girl. For the child it means that there is too great an emphasis upon a development in a specific direction and a limitation of her normal interest in areas outside the boundaries of her own gender role.

Even when there are no genetic or external somatic aberrations, an ambivalent attitude of the parents toward the child's gender can cause uncertainty in the child about its own gender role. This rather frequently occurs when the child at birth has not the sex the parents had hoped for. Every clinical psychologist is acquainted with the more or less pronounced ambivalence regarding the gender role in daughters who should have been sons and vice versa.

Freud has stated that personality is definitively formed round the age of five. We do not believe that this holds for all aspects of personality. Later in life, continued physical maturation and new social experiences stimulate further personality development and may generate new personality aspects. From the before-mentioned studies by Hampson, Hampson and Money however, it has become clear that particularly the gender role is irreversibly established during the first four years of life. Reassignment of sex after about the fourth year is not possible without great risk of aberrations in personality development. The authors therefore suggest the existence of a critical period early in life, in which gender identity and gender role are definitively and irreversibly established.

This does not mean, however, that under extreme conditions these personality features could not be interfered with later. In one girl we observed a probably transient period of extreme uncertainty as to her gender identity. She is an intelligent (WAIS IQ: 124), chromatin-negative Turner girl who started hormonal substitution treatment shortly after her twelfth birthday. She and her parents were informed about the fact that she suffers from Turner's syndrome, as well as about pubertal failure and the prognosis of sterility. Neither the girl nor her parents knew about the chromosomal background of the syndrome. She had a female gender identity and gender role. When about 22 years old she got a job as a corrector at a publisher's office where medical periodicals were edited. There she eagerly took notice of every article on Turner's syndrome she could lay hands on. She then learned about her chromosomal anomaly, which was a real traumatic experience. Her conclusion was that she was neither a man nor a woman: "I am an it, a nothing." She stopped the estrogen treatment as being of no use to her and as being more dangerous than helpful. During the next four years she had to go through an enormous struggle to rediscover her identity and her role in society. We tried to help her by talking through her problems and her feelings of inferiority and uncertainty. We had the impression, however, of being of very little help to her, because she stuck to the conviction that she was a worthless, incomplete creature. As she was free to come whenever she wanted, she sometimes visited us regularly for weeks or months and then we did not hear from her for a year or more. In the meantime, she visited many medical specialists in search of a new, less hopeless, diagnosis or a fundamental treatment like ovarian transplantation. After such an escapade, with always the same disappointing outcome, she often came back to tell about her experiences and her sorrow. She was impervious to our repeatedly given information that fundamental treatment of the syndrome is as yet impossible. Recently, about four years after her breakdown, she came again for a talk. For the first time in all those years, she seemed a little less down. She talked about herself as a girl again and she was able to face her situation more composedly. Hopefully this improvement is the beginning of a reintegration of her personality.

Among the patients we examined, this girl is exceptional because she is the only one who got such detailed and professional information about the somatic implications of her illness. We learned from her situation that an established gender identity, even in later life, can be deeply disturbed if the individual is bluntly confronted with the fact of having an aberrant sex-chromosome configuration. In itself this knowledge is probably not so traumatic, but it derives its significance from the already existing problems surrounding the stunted growth, the failing puberty, and the sterility. Apart from that, it is important to note that the girl, in spite of her rational doubts concerning her gender identity, looked at her problems from a feminine standpoint. Utterances such as: "I cannot give birth to a child" and "I am not attractive for boys" illustrate this. Moreover, she never suggested that she could as well be a boy instead of a girl, and now after a long period of uncertainty she returned to a female gender identity.

Because of the lack of sex-hormone production, Turner's syndrome offers an opportunity to study the relationship between hormonal and behavioural adolescence. Such a study can provide information about how sex-hormones and environmental factors are involved in the rather spectacular behaviour changes during adolescence. Money (1961) reports that among hypogonadal patients childhood eroticism did not change into adulthood eroticism if sex-hormone failure persisted beyond the age of 16. This observation agrees with the observation of Krautschick (1962) that in girls with Turner's syndrome real sexual interest fails at an age at which it is obvious in normally maturing girls. She mentions in addition that this cannot be attributed to a rejecting attitude towards sexuality in the families the girls came from. In our study of Turner girls, we arrived at the same conclusion (Bekker, 1969). At the time of examination the girls were 12 years or older, and hormonal treatment had not yet begun. The girls, not their parents at that time, were informed about the chromosomal anomaly and the prognosis of pubertal failure and sterility. As long as they are not treated with estrogens, these girls show an almost complete lack of interest in sexual matters. The parents stated that they seldom ask for information about sexuality; they even had the impression that information about sex, when given to the child, was passed over as irrelevant. Very often the erotic or sexual impact of events in their surroundings escapes them, and they do not show a real sexually motivated interest in members of the opposite sex. This does not imply that they should not show the romantic imageries and daydreams normal in prepubertal girls. They really do, but their imageries bear an infantile character.

We are not sure whether these observations are in complete agreement with the conclusions of Money and Mittenthal when they state that: "the development of romantic interests in teenage is independent of the presence of a second X-chromosome and of the influence of ovarian hormones in fetal life. Romantic interests may appear in advance of hormonal puberty, but the odds are against their growth and expansion, through lack of response from boy-friends, until the body looks hormonally mature" (Money and Mittenthal, 1970). We wholly agree with the first part of this conclusion, but we are not convinced that the retarded development of romantic interests is caused by lack of environmental response alone. We have three reasons for this. First of all, we have observed in non-maturing girls, as well as in boys, round the age of 14, the development of a behaviour pattern which on the surface was very much like normal pubertal behaviour. They want to dress in teenage clothes, girls sometimes use make-up or wear padded bras, boys sometimes shave despite the absence of facial hair. They sometimes even have an imaginary boy-friend or girl-friend. Upon closer inquiry it becomes clear that this behaviour is not really sexually motivated; it has an as-if character. It is not meant to bring them into closer relations with the opposite sex, or to make themselves more attractive. Sometimes the children admit that they do all this in order to be better accepted by their age-mates of the same sex. This was clearly expressed by a 16-year-old prepubertal boy when telling about his girl-friend: "She is my girl-friend but she does not know; I did not tell her. I must have a girl-friend to boast about to the other boys." Money (1961) and Theilgaard (1972) state that these childrens' short stature and child-like appearance cause people to treat them as younger than they are. Their pretreatment erotic development differs from that of normally maturing boys and girls because of the inhibitory training in sexual matters and the restricted opportunities for social development. This seems to be correct, but in addition their behaviour lacks internal motivation. The growth and expansion of romantic interests is not only obstructed by the absence of environmental response, but also by the absence of endocrinologically controlled sexual needs.

Our second reason stems from a comparison between non-maturing and late maturing boys and girls. Children who produce sex-hormones, but in whom pubertal development is delayed, do show a heightened interest in sexual matters before there is any visible sign of the onset of puberty. At that time they have to face nearly the same problems of short stature and prepubertal somatic appearance as non-maturing children have to. Nonetheless, late-maturing children are more concerned about the impression they make upon the opposite sex and they are more interested in dating than are non-maturing children.

Thirdly, these observations are supported by the results of a psycho-sexual-development test (Bekker and Koornstra 1967), given to three groups of children, aged 9 to 18 (Bekker, 1972). The children with failing sex-hormone production (none of whom was substituted) were retarded 4 years in psychosexual development, compared with the normal group. The children with delayed adolescence (all prepuberal) were retarded 2 years, as compared to the normals.

We think there is good reason to say that the typical sexual component is missing in the cognitive style of children who do not produce sex-hormones. Without hormonal treatment, sexual stimuli are not appreciated as such, although they can learn to display a quasi-sexually motivated behaviour. The normal psychological development during adolescence seems to be induced by changes in the pattern of somatic needs brought about by sex-hormone production. The ways in which the somatic needs are structured and satisfied are learned by the child from its environment. As a result of the interaction between its own possibilities, physical and psychological, and the possibilities provided by society, the child learns to integrate this new perspective in the whole of its personality. Thus, psychosexual development during adolescence can be seen as a process which is the result of a complex interaction of at least two basic factors: a hormonally determined sexual motivation, on the one hand, and a broad spectrum of environmental influences on the other.

11.6. Concluding remarks

It is very unlikely that a direct causative relationship exists between any form of human behaviour and one or more genetic factors. For example, the lack of one X-chromosome in Turner's syndrome seems to have no direct effect on behaviour development. The failing sex-hormone production as a sequel of gonadal dysgenesis which in its turn is caused by the chromosomal anomaly, is a mediating factor between this anomaly and certain deviations in personality development. External morphology is another important mediating factor, not only in cases of physical handicaps but also in normal psychological development.

The physical functions developing under genetic control constitute a biological substrate of possibilities and limitations which interacts with the social environment. The relationship between biological factors and behaviour may be more or less direct, to the extent that it may be more or less mediated by social learning processes. Personality features such as activity, arousal threshold, and basic emotional colour, which we usually call temperament, may be more directly connected to biological functions, whereas the development of a personality aspect like gender role is to a very high degree mediated by social learning.

Personality development is not a continuous and intrinsically undifferentiated process. Different periods in which different personality aspects come to maturity have always been recognized, and the existence of critical periods in cognitive development has been known for a long time. Research in human behaviour genetics has made it plausible that in personality development, too, critical periods are present. Hampson, Hampson and Money have demonstrated that the first years of life are critical for the establishment of gender identity and gender role. We have the impression that the possibility to integrate sex into the total personality structure during adolescence is also limited to a certain critical period. From the literature as well as from a few observations in our own study, we have concluded that this integration becomes difficult, if not impossible, when puberty comes too late. Substitution therapy should start before the age of 17 or 18 lest the child has become too accustomed to the role of an asexual outsider; otherwise it may not be able to switch to the role of a sexually mature person.

In our opinion, further investigations, especially by geneticists, should clarify the problem of the relationship between genetic anomalies and resulting aberrant physical development. The impact of physical development and physical processes upon behaviour development in many cases is another unsolved problem. An important set of questions concerns the role of the sex-hormones, for example, in cognitive development and in the development of personality features such as inertia of emotional arousal.

If the details of this relationship can be uncovered in the future, the foregoing might lead to suggestions for an earlier start of substitution treatment, with small amounts of sex-hormones, of girls with Turner's syndrome, in order to prevent early aberrations in cognitive or personality development.

In this connection, we should point out that in this chapter no attention was paid to such important environmental variables as climate, diet, and cultural differences. Especially in the training of cognitive functioning and in the development of social behaviour patterns like dating, cultural differences may be quite influential. The way in which an individual realizes his personality in his behaviour is a result of the extremely complex interactions between his own physical and psychological potentialities and the choices offered by cultural norms. It is amazing rather than self-evident that the results of the studies on Turner's syndrome in different parts of the world show up till now so much consistency.

References

Abeelen, J. H. F. van (1972): *Gedragsgenetica Vakbl. Biol.*, **52**, 318.

Anastasi, A., (1958): Heredity, environment and the question "how"? *Psychol. Rev.*, **65**, 197.

Alexander, D., A. A. Ehrhardt and J. Money (1966): Defective figure drawing, geometric and human, in Turner's syndrome. *J. nerv. ment. Dis.*, **142**, 161.

Bakker, D. J. (1971): Temporal order in disturbed reading: developmental and neuropsychological aspects in normal and reading-retarded children. (Thesis, Free University Amsterdam).,

Bekker, F. J. (1969): Dwerggroei en sexueel infantilisme (Stenfert Kroese, Leiden).

Bekker, F. J. (1972): The relation between genetic factors and social influence in psychosexual development during adolescence. In: Determinants of Behavioral Development. Mönks, F. J., W. W. Hartup and J. De Wit, eds. (Academic Press, New York and London).

Bekker, F. J. and J. J. van Gemund, (1968): Mental retardation and cognitive defects in XO-Turner's syndrome. *Maandschr. Kindergeneesk.*, **36**, 148.

Bekker, F. J. and M. J. Koornstra (1967): Test voor de psychosexuele ontwikkeling in de puberteit, Rapport no. O008-67, Psychologisch Instituut Rijksuniversiteit, Leiden.

Gemund, J. J. van (1959): Klinische en psychosociale aspecten van achterstand in groei en ontwikkeling. *Maandschr. Kindergeneesk.*, **27**, 3.

Hampson, J. G. (1955): Hermaphroditic appearance, rearing and eroticism in hyperadrenocorticism. *Bull. Johns Hopkins Hosp.*, **96**, 265.

Hampson, J. L. and J. G. Hampson (1961): The ontogenesis of sexual behavior in man. In: Sex and Internal Secretions. Young, W. C., ed., (Williams and Wilkins, Baltimore).

Hampson, J. L., J. G. Hampson and J. Money (1955): The syndrome of gonadal agenesis (ovarian agenesis) and male chromosomal pattern in girls and women: psychologic studies. *Bull. Johns Hopkins Hosp.*, **97**, 207.

Hampson, J. G., J. Money and J. L. Hampson (1956): Hermaphrodism: recommendations concerning case management. *J. Clin. Endocrinol.*, **16**, 547.

Harms, S. (1967): Anomalien der Geschlechtschromosomenzahl (XXX- und XO-Zustand) bei Hamburger Hilfsschülerinnen. *Pädiat. Pädol.* **3**, 34.

Jacobs, P., M. Brunton and M. Melville (1965): Aggressive behavior, mental subnormality and the XYY male. *Nature*, **208**, 1351.

Kind, H. (1962): Die Persönlichkeit und ihre Störungen bei präpuberal einsetzender Hypophyseninsuffizienz, speziell beim hypophysären Zwerg- und Kleinwuchs. *Arch. Psychiat. Nervenkr.*, **203**, 545.

Krautschick, A. (1962): Psychologische Untersuchungen an sechs Kindern mit Gonadendysgenesie. *Prax. Kinderpsychol. Kinderpsychiat.*, 11, 33.

Lindsten, J. (1963): The nature and origin of X-chromosome aberrations in Turner's syndrome (Almquist and Wiksell, Stockholm).

Martin, M. M. and L. Wilkins (1958): Pituitary dwarfism; diagnosis and treatment, *J. Clin. Endocr.*, **18**, 679.

Money, J. (1961): Sex hormones and other variables in human eroticism. In: Sex and Internal Secretions Young, W. C., ed., (Williams and Wilkins, Baltimore).

Money, J. (1963): Cytogenetic and psycho-sexual incongruities with a note on space-form blindness. *Amer. J. Psychiat.*, **119**, 820.

Money, J. (1969): Sex reassignment as related to hermaphroditism and transsexualism. In: Transsexualism and Sex Reassignment. Green, R. and J. Money, eds. (Johns Hopkins Press, Baltimore).

Money, J., J. G. Hampson and J. L. Hampson (1955a): An examination of some basic sexual concepts: the evidence of human hermaphroditism. *Bull. Johns Hopkins Hosp.*, **97**, 301.

Money, J., J. G. Hampson and J. L. Hampson (1955b): Hermaphroditism: recommendations concerning assignment of sex, change of sex and psychologic management. *Bull. Johns Hopkins Hosp.*, **97**, 284.

Money, J., J. G. Hampson and J. L. Hampson (1956): Sexual incongruities and psychopathology: the evidence of human hermaphroditism. *Bull. Johns Hopkins Hosp.*, **98**, 43.

Money, J., J. G. Hampson and J. L. Hampson (1957): Imprinting and the establishment of gender role. *Arch. Neurol. Psychiat.*, **77**, 333.

Money, J., and S. Mittenthal (1970): Lack of personality pathology in Turner's syndrome: relation to cytogenetics, hormones and physique, *Behav. Genet.* **1**, 43.

Report on the XYY chromosomal abnormality (1970): The National Institute of Mental Health, Chevy Chase, Maryland.

Shaffer, J. W. (1962): A specific cognitive deficit observed in gonadal aplasia (Turner's syndrome), *J. Clin. Psychol.*, **18**, 403.

Theilgaard, A. (1972): Cognitive style and gender role in persons with sex chromosome aberrations. *Danish Med. Bull.*, **19**, 276.

Turner, H. H. (1938): A syndrome of infantilism, congenital webbed neck and cubitus valgus. *Endocrinology*, **23**, 566.

Wieringen, J. C. van, F. Wafelbakker, H. P. Verbrugge and J. H. de Haas (1968): Groeidiagrammen 1965 Nederland, (Wolters-Noordhoff N.V., Groningen).

CHAPTER 12

Emotional responsiveness and interspecific aggressiveness in the rat: Interactions between genetic and experiential determinants

P. Karli, F. Eclancher, M. Vergnes, J. P. Chaurand and P. Schmitt

12.1. Introduction

In the development of behavioural processes, more than in the development of any other biological process, the connections between the genes and their expression in the phenotype are complex and indirect and therefore liable to various kinds of modulation. This is due to the fact that, despite a high degree of specialization which limits their developmental potentialities, nerve cells maintain for a more or less extended period great plasticity as regards their interplay and, particularly, the formation and consolidation of selective routes within the genetically determined neural circuitry.

Even in the lower vertebrates with their rather rigid organization of behaviour (the relation between a given sensory input and a given behavioural output is preorganized to a very large extent), full maturation of the mechanisms that ensure reception and processing of sensory information and of those that ensure motor responses and their feedback-control, requires that all these mechanisms be brought into play and be fully specified by adequate stimulations from the environment. It may be briefly recalled that the complex interactions between genetic and environmental factors have been extensively studied during the maturation of the visual system (cf. Karli, 1974). The full development of the links between sensory input and behavioural output also requires experience, i.e. interactions with the environment, but the genetic determination of these links appears clearly from the ease with which this learning is achieved. As the behaviour takes on a more "intentional" character, it is through the repeated use of preorganized behaviour patterns (i.e. again through experience) that the correspondence between the underlying intention and the result achieved can be progressively verified, confirmed and refined.

In mammals, and not only in the higher species, the picture is somewhat more complicated. Of course, as regards behaviours which are integral parts of some basic biological regulation, the bonds between consummatory behaviour patterns (feeding, drinking, copulation, etc.) and perception of the relevant goal objects are genetically preorganized to a very large extent. Moreover, since the detection of changes in the internal milieu predominantly controls the initiation and arrest of the consummatory behaviour, genetic factors can act by modulating the sensitivity of neural detector-systems. Therefore, the essential aspects of these behaviours are primarily genetically determined, and the life-history merely adds rather minor individual variations.

The situation is different if we consider social behaviour. In many instances, there does not seem to be a change in the internal milieu which could evoke an internal drive state (for instance, "aggressiveness"). The release as well as the nature of a behavioural response depend on the motivational properties, particularly on the affective significance, which have been conferred upon a given stimulus or situation through accumulated previous experiences. It by no means follows that heredity has little to do with the ontogenetic determination of such behavioural responses. For experiential influences are obviously mediated by neural mechanisms which depend on genetic factors, as they involve the action of neurotransmitters the metabolism of which depends on genetically-controlled enzymatic activities. In other words, genetically preorganized systems determine to what extent and in what way a given experience will contribute to the shaping of a particular social behaviour just as, conversely, environmental conditions in the broadest sense will determine to what extent and in what way a given genetic system will express itself in the organism's behavioural phenotype.

Things become even more complicated because of the intimate relations which exist between emotional reactions and social responses in the ontogenesis of behaviour and in the behaviour of the adult organism.

12.2. The concept of "emotional and social responsiveness"

In the analysis of an organism's overt behaviour, it is not too difficult to describe separately the actual social response and the emotional reaction that goes with it, but it is difficult to make the same clear distinction between social responsiveness and emotional responsiveness, with their respective physiological determinants.

There is little doubt that the nature of the affective experience, produced by a given stimulus or situation according to the organism's individual emotional responsiveness, takes an important part in determining both the orientation and the intensity of the social response eventually elicited by this stimulus or situation. An emotional state of "anger" thus often precedes and is responsible for some kind of aggressive response, either directed towards the anger-inducing stimulus or "displaced". Tantrum behaviour in young chimpanzees and in human infants may be considered as a precursor of aggressive behaviour which is gradually replaced by more elaborate modes of aggressive behaviour, with less violent emotional manifestations (Hamburg and van Lawick-Goodall, 1974). On the other hand, social interaction in its turn results more or less directly in a modification of the emotional state, which may be positively or negatively reinforcing, thus leading to consolidation or extinction of the behavioural response. Because of this, the efforts made by the parents to find ways of terminating the tantrums might well provide the infant with an initial model of reward for aggression.

If we turn now to a lower mammal like the rat, one may ask whether affective states also play an important role as causal links in the elicitation and the positive or negative reinforcement of social behaviour; these affective states and their modifications might be just epiphenomena of a more direct and rigid causal relationship between the sensory input and the behavioural output. This does not seem likely in any mammalian species. Aggression can be elicited by a variety of aversive situations as well as by the sudden withdrawal of a variety of rewarding conditions (Hutchinson, 1973). All these circumstances have in common that they give rise to a state of "frustration" and anger. Furthermore, various rewards or the sudden withdrawal of a variety of aversive conditions *following* aggressive behaviour result in positive reinforcement of the latter, and here again these circumstances may act through the mediation of a common affective state. The intimate relations between emotional and social responsiveness are also apparent from the concomitant elicitation of emotional and social responses by brain stimulation and from the concomitant and parallel changes in emotional and social responsiveness following brain lesions (Karli, 1968, 1972).

The fact that emotional states can intervene as important causal links in the elicitation of social behaviour increases the number of genetic and experiential determinants to be taken into consideration. On the one hand, the initial intensity of the elicited emotional response and the extent to which this response may be amplified and prolonged by humoral and neural feedback, are certainly under genetic control. On the other hand, experiential factors play an important role in two closely related ways: the organism learns through experience how to adapt to, and to cope with, aversive emotional states; the anticipation of a particular adaptive social behaviour, by reference to previous experiences, may lessen the emotional response to a given situation.

12.3. Genetic and experiential determination of aggressive behaviour

It follows from the above general considerations that there are many complex interactions between genetic and experiential factors in the determination of any kind of social behaviour. Moreover, since we shall deal mainly with a special kind of aggressive behaviour (namely mouse-killing behaviour in the rat), it should be stressed that any unitary concept of aggressiveness and aggressive behaviour is bound to be misleading. There are different motivational states which may express themselves in different kinds of aggressive behaviour, and behavioural and neurophysiological analyses show that the relative importance of the determinant factors may be quite different from one state to another (Karli et al., 1972).

This means that genes which take a predominant part in the determination of some forms of aggressive behaviour and their underlying "aggressiveness" may well take only a trivial part in the determination of other forms of aggression with their specific underlying "aggressiveness". A concrete example can illustrate this. As Kirsti and Kari Lagerspetz report in the present volume, spontaneous inter-male aggressiveness differs from one inbred strain of mice to the other; these authors have succeeded in developing special lines of mice by selective breeding for aggressiveness and non-aggressiveness. Considering that androgenic hormones play an essential role in the determination of inter-male aggressiveness in mice, it is conceivable that this selective breeding mainly sorts out genes which control the sensitivity of brain structures to androgenic hormones. Since the latter play a less important role in controlling mouse-killing behaviour in the rat, one cannot conclude that such genes contribute much to the determination of the aggressiveness displayed by the experienced killer-rat. In other words, looking for "genes determining aggressiveness" would be misleading, unless these genes are clearly specified in terms of their influence on the probability that an animal of a given species and sex will manifest a given kind of aggressive response when confronted with a given situation.

The ontogeny of aggressive behaviour and its manifestation in the adult organism can be modified by a great variety of environmental factors. Lagerspetz and Lagerspetz (1971) for instance have shown that in mice the differences created by 19 generations of selective breeding for aggressiveness can be completely masked by environmental factors. In the adult mouse-killing rat, interspecific aggression can be suppressed by a number of aversive training procedures (Baenninger, 1970). The ways in which experiential factors modify aggressive behaviour in mice are dealt with by Kirsti and Kari Lagerspetz in the present volume, and we shall come back to this point later with regard to mouse-killing behaviour in the rat. It must be stressed again that, whether considering the effects of genetic factors or experiential factors, aggressiveness is not a unitary concept. Hence, it is not surprising that Tsai and Dexter (1970) found that fighting reinforced by the termination of painful electric shocks developed into an efficient operant response in respect of terminating a specific kind of aversive stimulation, but without any increase in the "overall aggressiveness level" of the trained rats. On the other hand, the changes in emotional reactivity brought about by early social interactions or by early handling do not necessarily affect all kinds of aggressive behaviour in the same way. It seems likely that a low emotional reactivity can reduce the probability that an aversively motivated aggression is elicited, while facilitating the elicitation of an appetitively motivated aggressive behaviour.

Taken together, these considerations lead to the following conclusion: if we are to understand the neurochemical and neurophysiological mechanisms through which genetic and experiential determinants affect the ontogenetic shaping and later manifestation of aggressive behaviour, our experimental studies must deal with clearly specified kinds of aggressive behaviour together with the relevant aspects of the organism's emotional responsiveness, and certainly not with "overall aggressiveness and emotional responsiveness". It is evident that such studies presuppose a precise knowledge of at least some of the mechanisms which underlie a given kind of aggressive behaviour, mechanisms which are liable to modulation by genetic and environmental factors.

For nearly twenty years, we have studied mouse-killing behaviour in the rat. Lesion experiments not only have provided a better understanding of the brain mechanisms underlying this interspecies aggression, but in some instances it has also been possible to study the effects of the environment upon the neural and behavioural reorganization provoked by the lesion (on the basis and within the limits of genetically-determined potentialities). The data thus obtained may throw light upon some aspects of the environmental control over the neural and behavioural patterning occurring under normal circumstances. These lesion experiments, in the adult rat, point to another kind of interaction between experiential factors and genetically-determined neural functions (or dysfunctions): a brain lesion may have different effects depending on whether the rat has had experience with mice prior to the operation. More recently, we have started a study which analyzes the role of various brain structures and mechanisms in the shaping of defined aspects of the rat's emotional and social responsivenesss, and particularly the mediation of environmental influences upon behavioural ontogeny. The evidence already obtained will be reported, together with some of the data obtained in the adult rat. This material will be presented with two aims in mind. Firstly, we intend to show the great variety of mechanisms which may take part in the elicitation of a behavioural response as "simple" as mouse-killing behaviour appears to be at first sight; secondly, our material should make it clear that, although our experimental analysis has so far little to do with behavioural genetics as such, it is a prerequisite for a meaningful study of the genetic determination of the behaviour at the neurochemical level, more precisely, in terms of local enzymatic activities, which are the most directly genetically-determined biological parameters.

12.4. *Mouse-killing behaviour in the rat and its underlying brain mechanisms*

When observing rat behaviour under natural and experimental conditions (stimulation and lesion experiments), two different types of mouse-killing behaviour, with different biological significances can be distinguished: appetitively-motivated attack-behaviour and aversively-motivated defence-behaviour. Although this distinction is rather clear-cut for killing-responses elicited by brain stimulation, it is certainly too schematic with regard to the natural, spontaneous killing-behaviour. In the latter case, as we shall see, the two different motivational states succeed each other in time, but they may also overlap and interact to a variable extent.

12.4.1. Appetitively motivated attack-behaviour

When presented with a mouse, the experienced killer-rat attacks in a rather stereotyped and "cold-blooded" manner. The aggression is well-oriented, the rat seeks out the mouse and usually kills it within a few seconds. There is little emotional display. The latency of the killing response decreases with experience (Karli, 1956), and there is also a progressive decrease in the emotional components of the aggressive response.

In the experienced killer-rat, the appetitive motivation is certainly largely preponderant, as is evident from the following findings:

1. In the experienced killer-rat, the availability of a mouse for killing can act as a reinforcer in instrumental learning situations (Myer and White, 1965; Van Hemel, 1972). The observation (Myer, 1967) that killing experience increases the resistance of this behaviour to the suppressive effects of punishment, lends further support to the idea that killing is self-strengthening.

2. An experienced killer-rat looks for the mouse and kills it, even when its reactivity (in particular the emotional responsiveness to aversive experiences) has been considerably lowered by doses of chlorpromazine or reserpine as high as 15–20 mg/kg (Karli, 1959), or after it has been deprived of olfactory, visual, and auditory afferents (Karli, 1961).

Positive reinforcement of killing may well to some extent derive from the repeated association of the aggressive behaviour with eating part of the mouse (often the brain). In those experienced killer-rats which normally kill after some delay (up to several hours), food deprivation induces immediate killing-responses. The association of eating (with its sensory and humoral feedback) with the mouse-killing behaviour does not seem to be essential for the positive reinforcement of the latter, since killer-rats go on killing mice even if they are regularly prevented from eating a part of them. It should be recalled here that in a rodent like the rat, gnawing is rewarding in itself (Roberts and Carey, 1965), which means that a positive reinforcement can be derived from the execution of a species-typical behaviour with its specific proprioceptive feedback. In the experienced killer-rat, biting the live mouse and gnawing at the skull of the dead animal may be self-reinforcing, all the more so if we assume that biting has been associated repeatedly with the effective termination of an aversive experience, and gnawing of the skull with highly palatable food.

If we now turn to the underlying brain mechanisms, it appears that the appetitive components of killing rest on complex interactions between the lateral hypothalamus, the ventral mesencephalic tegmentum, and the amygdala.

Bilateral lesions placed within the posterior two-thirds of the lateral hypothalamic area cause a long-lasting abolition of the killing response. If the lesioned animal is allowed (through tube-feeding) to resume oriented behavioural activities, the recovery of the killing-response invariably precedes that of feeding behaviour by some days, if not by a few weeks (Karli and Vergnes, 1964). Lesions of the ventro-medial mesencephalic tegmentum transiently abolish mouse-killing and self-stimulation at lateral hypothalamic sites (Chaurand et al., 1973).

Conversely, electrical stimulation of lateral hypothalamic sites (Vergnes and Karli, 1969b; De Sisto and Huston, 1971; Panksepp, 1971; Woodworth, 1971) and of a number of ventral tegmental sites (Chaurand, 1973; Chaurand et al., 1974) elicits the "cold-blooded", well oriented killing behaviour which is spontaneously shown by the experienced killer-rat. The effective stimulation sites are also self-stimulation sites. If the rat is presented with both a self-stimulation lever and a mouse, it leaves the lever at some stage (in spite of the highly rewarding effects of the self-stimulation), kills the mouse, and then resumes self-stimulation.

The centromedial amygdala exerts an important facilitating influence upon the release of the killing behaviour. Bilateral lesions limited to the centro-medial region of the amygdala cause a long lasting abolition of the killing response, whereas amygdaloid lesions sparing this region have little or no effect upon the killing behaviour (Karli and Vergnes, 1965; Karli et al., 1972). Electrical stimulation of the amygdala does not induce any obvious facilitation of mouse-killing behaviour, even at sites and with stimulation parameters that otherwise sustain self-stimulation. On the contrary, every time a stimulation proves to be effective, it results in an immediate inhibition of the mouse-killing behaviour, the killer-rat resuming its aggressive behaviour as soon as the stimulation is discontinued. It must be added, however, that these stimulations which inhibit the killing behaviour always prove to be epileptogenic; hence their inhibitory effects most probably result from an interference of the seizure discharge with the normal functioning of the amygdala (Vergnes and Karli, 1969b).

12.4.1.1. Processes and mechanisms involved in the release of appetitively motivated attack behaviour. If we are to understand the genetic and experiential determination of mouse-killing behaviour at the neurophysiological and neurochemical level, it is not enough to draw up a list of central nervous structures that take part in the control of this interspecific aggression. It is important to analyze the separate, though related, behavioural processes and to try to identify the neurotransmitters involved in the various mechanisms underlying such processes. Although our knowledge in this respect is still fragmentary, it is possible to mention a few facts and speculate about them.

Coming back to the finding that bilateral lesions within the lateral hypothalamic area cause a long-lasting abolition of the mouse-killing behaviour, we should now stress that such lesions provoke a profound behavioural deficit, affecting various kinds of motivated behaviour. This is not surprising in view of the fact that the lateral hypothalamus (and the fibre systems running through it) is involved in a number of processes which generally give rise to appetitive motivation. Thus, activation of the lateral hypothalamic area – through experimental stimulation or as a consequence of some natural humoral fluctuation – may bring simultaneously into play the following processes:

a. A selective facilitation of the reception and of the processing of sensory information arising from relevant goal-objects. Thus, the nutritional state of the organism (hunger or satiety) modulates the bioelectrical response of the olfactory bulbs to an alimentary odour, but not their response to an odour having no alimentary significance (Giachetti et al., 1970; Pager et al., 1972). The nutritional state also modulates in a selective way the discharge of certain tegmental neurons in response to a signal which allows the animal to anticipate the delivery of food (Phillips and Olds, 1969). The preferential response of the male rat to odours emanating from a receptive female (vs. those from a non-receptive one) depends on the level of androgenic hormones in the internal milieu (Stern, 1970). In relation to attack behaviour, lateral hypothalamic stimulation induces a very sensitive field around the muzzle in the cat (MacDonnell and Flynn, 1966) as well as in the rat (Smith, 1972). On the other hand, a unilateral lesion at this site entails an ipsilateral sensory deficit, the nature of which is not yet clearly understood (Marshall et al., 1971; Turner, 1973).

When considering eating, drinking, or sexual behaviour, we know exactly the humoral factors (glycemia, osmolarity, sex hormone levels) the changes of which are detected by highly sensitive neuronal sets which may therefore activate the lateral hypothalamic area. As for aggressive behaviour, it would certainly be of interest to know to what extent and how humoral changes associated with a given emotional state (for instance, the state of "anger") bring about the kind of selective facilitation of information processing that can be provoked by an experimental stimulation of the lateral hypothalamus.

b. Through complex interactions with mesencephalic structures (particularly the reticular activating system), the lateral hypothalamus, whether activated by experimental stimulation or by spontaneous humoral changes, produces: (1) Cortical as well as behavioural arousal. A simple increase in locomotor activity (Campbell, 1964; Hughes, 1965; Mathews and Finger, 1966) is of obvious interest for the appetitive phase of motivated behaviour (for the organism in search of food, water, or a sexual partner). This increase in locomotor activity may also result in more frequent aggressive encounters. (2) A descending facilitation of spinal reflexes, which in its turn facilitates the consummatory phase of any kind of motivated behaviour (Wayner et al., 1966; White et al., 1970) and possibly also the reward effects derived from the execution of a species-typical behaviour with its specific proprioceptive feedback.

c. Activation of the lateral hypothalamic area may also entail a more or less selective facilitation of processes of positive reinforcement, the reward effects being either anticipated or actually derived from an already ongoing behaviour. It should be recalled here that the effective stimulation sites (in respect of the elicitation of "cold-blooded", well-oriented mouse-killing behaviour) are also self-stimulation sites.

The reward effects (as measured by self-stimulation rates) associated with the elicitation of stimulus-bound feeding behaviour at lateral hypothalamic sites are modulated by the nutritional state of the organism (Margules and Olds, 1962; Hoebel, 1968; Hoebel and Thompson, 1969). Similarly, the level of circulating androgenic hormones modulates the reward effects associated with the elicitation of copulatory behaviour at other sites within the lateral hypothalamic area (Caggiula, 1970). Again, as already stated in relation to the facilitated processing of relevant sensory information, one would like to know to what extent and how humoral changes, associated with specific emotional states, affect the reward effects derived from activation of those lateral hypothalamic sites where stimulus-bound aggression can be elicited.

As for the mechanisms underlying the facilitating influence exerted by the centromedial amygdala upon the release of mouse-killing behaviour in the experienced killer-rat, we may speculate briefly about just two of them:

a. Positive reinforcement resulting from lateral hypothalamic stimulation is thought to involve an ascending reward system, with the amygdala as an important target where the reward effects could antagonize the suppressing effects of punishment (Stein, 1969; Margules, 1971). The amygdala may play an important role in the anticipation of any reward to be derived from mouse-killing, through reference to the traces of past experiences. Recent experimental evidence shows that local anaesthesia of the amygdala profoundly suppresses both self-stimulation and elicitation of stimulus-bound feeding behaviour at lateral hypothalamic sites (Kelly, personal communication; Rolls, 1974). These results are in agreement with some of our previous observations. In amygdaloid-lesioned killer-rats, interspecific aggression was not facilitated by lateral hypothalamic stimulations that otherwise clearly facilitate mouse-killing behaviour (Vergnes and Karli, 1969b).

Since lesions of the centromedial amygdala abolish spontaneous mouse-killing behaviour without abolishing spontaneous feeding behaviour, one may suggest that anticipation of a reward to be derived from mouse-killing presupposes some reference to previous experience, whereas such a reference would not be necessary in the case of feeding. It is of interest to note that amygdaloid lesions do interfere with feeding behaviour if the animal is to reject new or poisonous foods, that is in cases where some reference to previous experience must intervene (Rolls and Rolls, 1973).

b. There is a somewhat different (but, most probably, closely related) amygdaloid function which may contribute to the facilitation of mouse-killing behaviour. The amygdala seems to be an important link in the chain of mechanisms through which intra-central and peripheral excitatory feedback amplifies and prolongs the emotional response and its possible consequences for social behaviour. In fact, extensive amygdaloid lesions invariably induce a general leveling-down of the emotional responsiveness as well as a general reduction in the amount of social interactions exhibited (cf. Karli, 1968, 1972).

For further progress, it is necessary to identify the neural circuits and transmitters involved in the processes which give rise to an appetitive motivation and its behavioural expression. Within the scope of this chapter, we may just indicate briefly that great strides can be made in two ways. Firstly, the recording of unit activities and the study of some functional property (e.g. the refractory period) of individual neurons in relevant brain structures (lateral hypothalamus, ventral mesencephalic tegmentum, amygdala) may define separate neuronal systems involved either in non-specific arousal, or in the elaboration of more or less specific reward effects (Rolls, 1973). Secondly, neurochemical studies may provide further insight into the mechanisms that are liable to be controlled rather directly by genetic factors. The major part taken by a dopaminergic fibre system, ascending through the lateral hypothalamus, in the initiation of motivated behaviour was demonstrated in experiments in which this fibre system was lesioned selectively by intracerebral injection of 6-hydroxydopamine. A bilateral complete degeneration of this ascending dopaminergic pathway produces "severe, long lasting adipsia and aphagia, hypoactivity, difficulties to initiate activity and loss of exploratory behaviour and curiosity" (Ungerstedt, 1971). Furthermore, it was shown that catecholamine-containing neurons play an essential role in the initiation and maintenance of self-stimulation behaviour There is also evidence suggesting that two different ascending systems may be involved: the dopaminergic fibre system arising from cell bodies in the ventral mesencephalon which, once again, seems to be an "incentive motivational" system, and the noradrenergic fibres arising from cell bodies in the locus coeruleus, which are considered to make up the basic "reinforcement system" (Crow, 1972 and 1973). If one accepts that an inherently deficient noradrenergic "reward" system is responsible for certain mental disorders, the following finding is of great (although primarily speculative) interest: the activity of dopamine-β-hydroxylase – the enzyme responsible for the final step in noradrenaline biosynthesis – is deficient in the brain of schizophrenic patients (Wise and Stein, 1973).

12.4.2. *Aversive motivations and their behavioural consequences*

Whether induced by natural circumstances or by an experimental brain stimulation, an aversive emotional experience can have two different behavioural effects with regard to mouse-killing behaviour of the rat. It may induce a tendency to retreat or to flee; such a tendency would obviously interfere, in the experienced killer-rat, with an ongoing appetitively motivated killing-response. It may also induce, as we shall see, an "affective" kind of mouse-killing behaviour which can be regarded as a defence or active-avoidance behaviour, irrespective of whether the rat has ever killed mice before.

12.4.2.1. Interference with spontaneous attack-behaviour in the experienced killer-rat. The importance of the temporal relationships of aversive stimulation to rat behaviour is evident from a study devoted to the "suppression of interspecies aggression in the rat by several aversive training procedures" (Baenninger, 1970). If the onset of an aversive stimulation (painful electric shock) is contingent on the performance of an attack-response, the rat readily learns to passively avoid the shocks and he refrains from killing the mice he is presented with. Furthermore, if termination of the aversive stimulation is contingent on performing an appropriate escape response, such a procedure reinforces operant responses which are incompatible with (and thus apt to interfere with) an appetitively motivated attack-behaviour. Under more natural circumstances, we have observed that when two experienced killer-rats are put into the same cage and then presented with a mouse, the latter is usually killed by the dominant rat, while the dominated (and fearful) rat rarely shows more than a few stray aggressive attempts. In wild Norway rats, the "latency" of the killing response is often considerably increased when the experienced killer-rat is simply transferred from his habitual environment to a new one (Karli, 1956). Conversely, there is a significant increase in aggressive behaviour among male mice when introduced into more and more familiar environments (Jones and Nowell, 1973).

A number of brain stimulations can also induce an aversively motivated tendency to avoid or retreat, which may then interfere with the appetitively motivated tendency to approach and to kill. The medial hypothalamus, the dorsomedial thalamus, and the mesencephalic central grey seem to constitute a medial periventricular system, the activation of which has aversive effects. At many sites scattered throughout these periventricular structures, electrical stimulation frequently inhibits ongoing attack behaviour in the experienced killer-rat. The aversive nature of the stimulation appears from the fact that the animal readily learns to switch-off (by pressing a lever) such stimulation administered to any of these brain sites.

The inhibitory effect of the stimulation is somewhat different from one brain structure to the other. Central grey stimulations inhibit ongoing attack and at slightly higher intensities they provoke flight responses. In the absence of a mouse (in the absence of a competing appetitive motivation?), the lower stimulation intensities may already provoke a clear flight response (Chaurand et al., 1972). Dorsomedial thalamic stimulations inhibit the killing-behaviour at once, but they do not induce flight-responses even at higher stimulation intensities (Vergnes and Karli, 1972). Finally, medial hypothalamic stimulations often provoke flight-responses which, of course, interrupt ongoing aggression, but the flight-response is usually not preceded by an actual inhibition of the killing-behaviour (Vergnes and Karli, 1970). Further research is needed to enable us to interpret in clear terms the differential behavioural effects of aversive brain stimulations, all of which the rat can quickly learn to turn off by pressing a lever.

As a counterpart of the inhibitory effects of central grey stimulations, we may now consider briefly the behavioural effects of central grey lesions. Such lesions greatly reduce fear responses: the animals defecate much less in the open-field and when handled; in the latter case vocalization is often completely abolished (Chaurand et al., 1972; Vergnes and Chaurand, 1973). These lesions clearly facilitated the killing-behaviour in animals that, prior to the operation, used to kill after a more or less prolonged delay. Because the same animals showed a moderate but significant hyperphagia (Chaurand et al., 1972) and a lasting facilitation of lateral hypothalamic self-stimulation (Schmitt, 1972), one is tempted to conclude that a decreased responsiveness to fear-inducing aversive stimulations (or situations) brings about a general facilitation of various appetitively motivated behaviours.

12.4.2.2. Elicitation of an "affective" kind of mouse-killing behaviour in the non-killing rat. As indicated above, the killing behaviour spontaneously shown by the experienced killer-rat is "cold-blooded" and well-oriented, and can be immediately provoked by electric stimulation of lateral hypothalamic and ventral tegmental sites. A different kind of mouse-killing can be elicited by stimulation of medial hypothalamic (Vergnes and Karli, 1970) and a few periaqueductal sites (Chaurand et al., 1972); it is an "affective" kind of aggression with marked emotional display (the aggression is poorly oriented: the rat does not look for the mouse, but attacks only when it comes near). This behaviour can be easily elicited in the natural non-killer. The effective stimulation sites are in every case "switch-off" sites. Once the rat has learned to stop the brain stimulation either by fleeing or by pressing a lever ("switch-off" response), it is difficult to induce it to kill a nearby mouse. This "affective" kind of mouse-killing might be considered as defence-behaviour, a kind of active-avoidance behaviour, killing being a way of putting an end to aversive experience.

In the cat, too, a distinction was made by Wasman and Flynn (1962) between a "quiet biting" type and an "affective" type of attack towards a nearby rat as elicited by electrical stimulation of hypothalamic sites. In both cats and rats, the "affective" type of aggression seems to be an aversively motivated one, as is suggested by the finding that a learned escape response (escape from tail shock by jumping onto a stool) could be transferred to hypothalamic stimulations that yielded "affective" attack accompanied by vocalization, but never to those stimulations that yielded "quiet biting" attack (Adams and Flynn, 1966).

The problem now arises of the determinants which induce some rats (and only some of them) to kill the mouse when encountering one for the first time. Even in the absence of prior killing experience, appetitive components may be involved to some extent. Anticipation of a reward, of a positive reinforcement, may play a role in the elicitation of the initial killing-behaviour through reference to previous experiences with situations somewhat similar to the mouse-presentation situation. Observation of the rat's behaviour shows that it is usually quite excited by the presence of the mouse. On the other hand, as we shall see later, most if not all of the brain lesions that may cause mouse-killing behaviour in the natural non-killer, are lesions that entail some kind of increased responsiveness, often an increased emotional responsiveness to aversive experiences. The rat's aggressive behaviour is possibly directed at getting rid of something new, strange, and aversive. If so, the rat's initial mouse-killing behaviour (under natural conditions or after a number of brain lesions) might to a large extent reflect an aversively motivated defence or active-avoidance behaviour.

It is of obvious interest to analyze the elicitation of mouse-killing behaviour by means of brain lesions made in the adult or early in life in terms of changes affecting one and/or the other of two closely related groups of mechanisms: (1) the mechanisms which control the adult rat's emotional responsiveness and, more specifically, its emotional responsiveness to aversive stimulations; (2) the mechanisms which underlie the shaping of behavioural-emotional adaptations during ontogenesis, and through which the individual life-history determines the behaviour displayed by the rat when presented with a mouse for the first time.

Although our present knowledge about the behavioural effects of particular brain lesions and their causal relationships is probably too fragmentary to allow us to arrange the experimental data under clearly distinct headings (changes in the emotional responsiveness, interference with the ontogenetic shaping of social-emotional adaptations), we will nevertheless use such headings in order to indicate the conceptual framework of our experimental approach.

12.4.2.3. Changes in the emotional responsiveness of the rat. As already indicated, most if not all of the brain lesions that may elicit mouse-killing behaviour in the natural non-killer are lesions that entail some kind of hyperreactivity. The determinants and the components of such an increased responsiveness can be quite different from one case to the other. We shall examine the effects of bilateral lesions in the following brain structures: olfactory bulbs, septum, ventromedial hypothalamus, dorsomedial thalamus, raphé nuclei, and mesencephalic central grey.

Olfactory bulbs. Ablation of the olfactory bulbs was the first brain lesion shown to induce mouse-killing behaviour in many non-killer rats (Vergnes and Karli, 1963). The outcome of the lesion in the natural non-killer depends on both genetic and environmental factors: The percentage of animals made into mouse-killers was significantly higher when using animals from a stock bred almost exclusively from mouse-killers than when using animals from another stock, bred from the same original Wistar strain, but without selecting mouse-killers for breeding (Karli et al., 1969). Some of the genetic factors for which selection was made may control the secretion of and/or the sensitivity to androgenic hormones during ontogenesis, since castration at 1 month of age significantly reduced the percentage of adult non-killing rats which could be made mouse-killers by olfactory bulb removal (Didiergeorges and Karli, 1967).

The natural non-killer deprived of olfactory bulbs starts killing only if he is kept in isolation after the operation, regardless of the age of ablation (Vergnes and Karli, 1969a). If the bulbectomized animals are kept in groups, vicarious behavioural adaptations can develop on the basis of social interactions and sensory information other than olfactory, thus preventing the appearance of mouse-killing behaviour (Didiergeorges and Karli, 1966). However, the transient or lasting character of the effects of the social interactions depends upon the age at which the lesioned animals are exposed to them. In adult animals, the inhibition (or the non-development) of the killing-behaviour by exposure to social stimulation is only transient: aggressive behaviour appears again in these animals once they have been isolated for a few weeks after grouping for two months following the operation. If, by contrast, animals are lesioned at an early age (4 or 7 weeks) and then kept together until adult age, the inhibition (or the non-development) of the interspecific aggression is very stable and isolation will never induce killing-behaviour.

Anosmia as such is not sufficient to elicit mouse-killing behaviour. We have based this, at first tentative, conclusion mainly on the finding that lesions likely to interfere with the perception of mouse odours (such as partial deafferentations of the olfactory bulbs or subtotal interruptions of the fiber tracts originating in the olfactory bulbs and anterior olfactory nuclei) were ineffective (Vergnes and Karli, 1965; Didiergeorges et al., 1966). The fact that in most cases a few days elapsed between an ablation of the olfactory bulbs and the first manifestation of the rat's interspecific aggressive behaviour, also seemed to support this possibility. More recently, and more conclusively, anosmia caused by destruction or removal of the nasal mucosa has been shown not to elicit mouse-killing behaviour in the natural non-killer (Alberts and Friedman, 1972; Spector and Hull, 1972). Thus it appears that the role played by the olfactory bulbs goes beyond the mere transmission of olfactory cues. This is not surprising because in a macrosmat like the rat the progressive shaping of relations with the environment is based largely upon sensory information of an olfactory nature. Moreover, the olfactory bulbs are closely interconnected with the amygdala, and functional interactions between these two structures may well be important in the behavioural-emotional adaptations of the rat. Ablation of the olfactory bulbs would alter the animal's emotional responsiveness, and this alteration would involve (and partially result from) the loss of behavioural-emotional adaptations based on olfactory input.

Bulbectomized rats are often hyperreactive and difficult to handle. From the start, we were struck by their savage way of killing mice and by their "disinhibited" emotional display, contrasting with the "cold-blooded" behaviour of the experienced killer-rat (Didiergeorges et al., 1966; Karli et al., 1969). Furthermore, ablation of the olfactory bulbs entails an increased elimination of urinary catecholamines, which contrasts with the low rates of elimination found in "spontaneous" killer-rats (Jund et al., 1971). Bulbectomized rats have heavier adrenals than control groups and they show high levels of the adrenal enzyme tyrosine hydroxylase which is involved in catecholamine biosynthesis (Eichelman et al., 1972). Extensive behavioural studies specifically aiming at analyzing emotionality changes possibly induced by an ablation of the olfactory bulbs have clearly substantiated the various signs of increased emotional responsiveness or "irritability": hyperactivity in the open field (which can be more or less masked by freezing behaviour, depending on the test situation), increased defecation in the open field, and facilitated learning of an active avoidance response (Douglas et al., 1969; Bernstein and Moyer, 1970; Richman et al., 1972; Sieck and Gordon, 1972; Sieck, 1973). Enhanced heart rate changes in a startle test (Phillips and Martin, 1971) can be considered as an autonomic component of the increased responsiveness of the bulbectomized rat.

When carried out early in life (for instance, at the age of 25 days), removal of the olfactory bulbs has persistent behavioural effects. At adult age, most bulbectomized rats not only display mouse-killing behaviour (provided they have been kept in isolation), but they also show increased defecation in the open field and facilitated learning of an active avoidance response in a two-way shuttle-box (unpublished observations).

When the rat's emotional responsiveness is judged from handling, there appears to be no predictable relation between its postoperative irritability and its behaviour towards the mouse. We are now trying to clarify the possible correlations between the elicitation of mouse-killing behaviour and the postoperative alterations of the emotional responsiveness, by measuring the latter in more precise and quantifiable terms. Furthermore, an analysis of some local neurochemical correlates has been undertaken in collaboration with the biochemists of the Centre de Neurochimie. In this respect, we may mention a recent result we shall come back to when dealing with the amygdala. Removal of the olfactory bulbs, inducing interspecific aggressiveness, results in increased activity of choline acetyltransferase in the amygdala. More precisely, the local activity of the enzyme reaches the high level which characterizes the natural killer-rat's amygdala (Ebel et al., 1973).

Septum. In rats accustomed to a mouse in their cage, a septal lesion does not induce mouse-killing behaviour (Karli, 1960), but it does so in rats which had a brief (24 hr) experience with mice (Miczek and Grossman, 1972).

It is well known that septal lesions result in increased emotional responsiveness (cf. Karli, 1968 and 1972). The data obtained by Miczek and Grossman (1972) suggest that there is a close correlation between this increased responsiveness and the appearance of mouse-killing behaviour, and that the latter is mainly an aversively motivated one. First of all, the effect of the septal lesion on the rat's behaviour towards the mouse is temporary and its time course parallels more or less that of the "septal rage" syndrome. None of the rats which were first retested 15 days after surgery (i.e. when the lesion-induced hyperemotionality had sharply declined) killed any mice. On the other hand, the interspecific aggression displayed immediately following the lesion is of an "affective" kind and it is observed together with increased fighting in response to electric shock, whereas preoperatively dominant rats become submissive in a food competition test, i.e. in a situation where aggressive behaviour is mainly appetitively motivated. That a septal lesion may induce interspecific aggression due to increased emotional responsiveness to aversive stimulations is also substantiated by the finding that painful electric shock which does not elicit mouse-killing behaviour in the intact rat (Karli, 1956) proves effective when combined with a septal lesion (Miley and

Baenninger, 1972). Since the mouse-killing behaviour is self-strengthening, it is not surprising that, once it has been induced by a brain lesion, it is in most instances irreversibly established in spite of the progressive decline of the initial hyperemotionality; this was shown to be the case in septal-lesioned rats by Miczek and Grossman (1972).

A septal lesion administered at the age of 8 days has permanent behavioural effects. When adult, the lesioned rats still show increased irritability when handled, and facilitated learning of an active avoidance response in a two-way shuttle-box. Thus one might ask whether an increased incidence of mouse-killing behaviour in adults accompanies this persistent hyper-irritability. Although the histological check on the septal lesions is not yet available, there seems to be a significant increase in the proportion of killer-rats as compared to non-operated and sham-operated control animals.

Since the outcome of olfactory bulb removal depends upon the level of circulating androgens, as reported above, it is worth mentioning that early castration renders septal lesions ineffective in producing hyperemotionality (Phillips and Lieblich. 1972). Sex hormones, which were recently shown to control the postnatal differentiation of synaptic contacts in the preoptic area of rat brain (Raisman and Field, 1973), seem once again to influence basic aspects of neural and behavioural maturation with regard to emotional and social responsiveness.

Ventromedial hypothalamus. Most rats lesioned in the ventromedial hypothalamus are clearly hyperreactive and often difficult to handle; but this increased responsiveness is obviously not sufficient to provoke by itself the elicitation of interspecific aggressive behaviour, since only about 30 % of the lesioned rats start killing mice. On the other hand, the effects of ventromedial hypothalamic lesions and removal of the olfactory bulbs can be additive: the hypothalamic lesion elicits mouse-killing behaviour in a large proportion (60 %) of non-killing rats whose behaviour remained unchanged after prior removal of the olfactory bulbs (Eclancher and Karli, 1971).

We did not find a correlation between the postoperative development of hyperphagia and mouse-killing behaviour. Grossman and coworkers found that the lesioned rats show a decrease in aggressive interactions in a food competition situation, but a clear facilitation of both pain-induced aggression and acquisition and performance of active avoidance responses (Eichelman, 1971; Grossman, personal communication). Taken together, these findings suggest that the ventromedial hypothalamic lesion does not induce the rat to look for a new source of food and to engage in an appetitively motivated mouse-killing behaviour; his interspecific aggressiveness may rather result from enhanced responsiveness to aversive stimulations.

When the hypothalamic lesion is made at the age of 8 days and the behaviour towards mice is first tested 5 months later, the proportion of killer-rats is significantly greater than that observed in non-operated controls, and it is also greater than that found in rats lesioned at an adult age (Eclancher and Schmitt, 1972). The question arises whether the smaller proportion of killer-rats in the latter group is simply due to the fact that they had prolonged experience with mice prior to the operation, an experience the 8 day old pups did not have. This point is now being investigated.

Dorsomedial thalamus. Bilateral destruction of the dorsomedial thalamic nuclei can elicit mouse-killing behaviour in natural non-killers. This behavioural change does not result from the transection of the epithalamic circuit that usually goes with the dorsomedial thalamic lesion, since bilateral lesions limited to the medial habenular nucleus or to the stria medullaris do not induce mouse-killing behaviour (Eclancher and Karli, 1968 and 1969). Dorsomedial thalamic lesions have been shown to enhance fear responses in the cat (Roberts and Carey, 1963) and to entail lowered jumping thresholds for unavoidable shocks and a trend toward increased shock-induced fighting in the rat (Eichelman, 1971).

There is no evidence for summation of the respective effects of ventromedial hypothalamic and dorsomedial thalamic lesions: firstly, combined lesions of both diencephalic structures are not more effective than the hypothalamic lesion alone and, secondly, a thalamic lesion is ineffective in non-killing rats whose behaviour has remained unchanged following hypothalamic lesion (Eclancher and Karli, 1972).

Raphé nuclei. Brain serotonin depletion caused by administration of p-chlorophenylalanine can elicit mouse-killing behaviour, especially in bulbectomized rats which did not become killer-rats following the operation (Karli et al., 1969; Sheard, 1969; Di Chiara et al., 1971). Conversely, the spontaneous interspecific aggressiveness of the killer-rat can be transiently abolished by injection of 5-hydroxytryptophan, a precursor of serotonin (Kulkarni, 1968).

These findings draw the attention to the serotonergic neurons of the raphé nuclei. In a recent experiment, lesions of the dorsal and medial nuclei of the raphé (which provoked a marked decrease in the levels of serotonin and 5-hydroxyindoleacetic acid in the forebrain, but left the levels of norepinephrine unchanged) induced mouse-killing behaviour in 6 out of 18 originally non-killing rats (Vergnes et al., 1973). Similar results were obtained by Grant et al., (1973), but Sheard (1973) did not observe an elicitation of interspecific aggressiveness in raphé-lesioned rats. It must be stressed that in our own experiment the provoked decrease in the serotonin level in the forebrain, however large it may have been, was not always sufficient to induce mouse-killing behaviour in the natural non-killer.

The raphé-lesioned rats are clearly hyperactive (Kostowski et al., 1968; Vergnes et al., 1973; Grant et al., 1973; Sheard, 1973), and the lesions potentiate the facilitatory effect of amphetamine on locomotor activity (Neill et al., 1972). It is improbable, however, that any close causal relationship exists between the lesion-induced hyperactivity and the development of interspecific aggressiveness: Some of the animals which do not become killer-rats after the operation are just as hyperactive as those which start killing mice (Vergnes et al., 1973), the time-course of the increased motor activity does not parallel the gradual development of the mouse-killing response (Grant et al., 1973), and lesions of the posterior part of the periaqueductal grey matter entail the same kind of hyperactivity without ever inducing mouse-killing behaviour (Vergnes and Chaurand, 1973).

As a matter of fact, raphé lesions also provoke enhanced excitability and, in particular, increased responsiveness to painful stimuli. A further behavioural effect consists of a less than normal decrease in activity over time during the open-field test, which might indicate a deficient habituation to a novel situation (Vergnes et al., 1973). It is rather in this direction (increased responsiveness to aversive stimulation, lack of habituation) that one might find causal relationships with the elicitation of mouse-killing behaviour in raphé-lesioned rats.

Mesencephalic central grey. As reported earlier in this chapter, central grey lesions markedly reduce the fear responses of rats and clearly facilitate various appetitively motivated behaviours (feeding, self-stimulation, mouse-killing in the experienced killer-rat). Under these circumstances, it seems unlikely that the mouse-killing behaviour induced by central grey lesions in a small proportion (10% or less) of natural non-killers results from an increased responsiveness to aversive stimulations and situations. This brings back the observation that appetitive components may be present and, at times, may even be preponderant in the motivation underlying the killing behaviour displayed by rats when first presented with a mouse. If this is the case, it follows that a possible facilitation of appetitively motivating processes should be taken into consideration when we induce mouse-killing behaviour in naturally non-killing rats.

12.4.2.4. Interference with ontogenetic shaping of social-emotional adaptations. First of all, two facts should be recalled which emphasize the importance of the life-history in determining whether an adult rat will display mouse-killing behaviour. Firstly, early social contacts with mice reduce greatly the incidence of the killing behaviour in the adult rat (Denenberg et al. 1968; Myer, 1969). Early interactions with mice render these animals less novel and strange. Secondly, the incidence of mouse-killing behaviour can be increased in a group of rats by repeatedly exposing the animals to both food deprivation and competition for food (Heimstra, 1965); through the repeated execution of aggressive behaviour, the rats get some experience of its operant value which they may later generalize in more or less similar situations. Furthermore, hunger clearly facilitates the initiation of mouse-killing behaviour, but this phenomenon depends to a large extent upon whether the rats have a history of food deprivation prior to the killing-tests(Paul et al., 1971). In addition, the effects of a particular factor (in this case, the state of hunger) again seem to depend upon the extent to which rats have been exposed to mice prior to the killing-tests. The hunger-induced facilitation of mouse-killing behaviour seems to be much greater in rats with little or no experience with mice (Paul et al., 1971) than in rats which have been in contact with mice for a long period (Karli, 1956) or than in rats given initial non killing experience with rat pups (Paul et al., 1973).

The amygdala seems to play a key-role in mediating the effects of early social interaction upon the development of behaviour traits, in the progressive shaping of social-emotional adaptations during ontogenesis and, more specifically, in the inhibition (or the non-development) of mouse-killing behaviour. This general conclusion is derived from recent (mostly unpublished) experimental evidence which shows that amygdaloid lesions made at an early age greatly interfere with the inhibition (or the non-development) of the killing-behaviour during ontogenesis.

When the amygdala is bilaterally lesioned in 8-day old rat pups, about 90 % of the lesioned animals kill mice when adult, as compared to 15 % killer-rats in the control group. Moreover, early social contacts with mice no longer prevent the development of mouse-killing behaviour in most amygdaloid-lesioned rats, since 70 % of these animals, raised from weaning with a mouse, turn out to be killer-rats. The amygdala also plays an essential role in the vicarious behavioural adaptations which normally occur in rats following olfactory bulb removal. In about 75 % of the rats, grouped after amygdaloid lesions and olfactory bulb removal at the age of 25 days, postoperative social interactions no longer prevent the development of mousekilling behaviour.

Bilateral destruction of the amygdala in 8-day old rats not only yields a much greater proportion of adult killer-rats, but also other lasting behavioural changes. A less than normal decrease in activity over time during the openfield test is observed in 40-day old lesioned animals and this deficient habituation to a novel situation is still present, though less marked, several months later. On the other hand, these amygdaloid lesioned rats (in particular those kept in isolation) show a clear trend toward facilitation of learning of active avoidance responses, which contrasts with the impaired acquisition of such responses normally found when amygdaloid lesions are applied in adult rats.

There is little doubt that the amygdala is essential in the control of both emotional responsiveness and interspecific aggressiveness during development and in the adult animal. The amygdala seems to be a major site of interaction between positively reinforcing "reward" effects and suppressant effects, the latter being due to fear, to punishment, or simply to the absence of an anticipated reward. Since reward and punishment, behaviour facilitation and behaviour suppression, are more and more considered in terms of neurotransmitters released by fibre systems, a major aim of our present and future research is to study neurophysiological and neurochemical correlates of their interactions in the amygdala.

Choline acetyltransferase activity is higher in the amygdala of the killer-rat than in the amygdala of the non-killer; when the latter starts killing after the removal of the olfactory bulbs, the activity of the enzyme goes up to a level characteristic for the amygdala of the natural killer-rat (Ebel et al., 1973). These data can be correlated with some behavioural effects resulting from cholinergic activation of the amygdala. Local application of a cholinesterase inhibitor was shown to increase reactivity and aggressiveness in the rat and to induce in some instances mouse-killing behaviour (Igic et al., 1970). In addition, cholinergic stimulation of the amygdala was found to interfere with the suppressing effect of punishment, since it produced severe deficits in passive avoidance and CER learning (Goddard, 1969). If one accepts that cholinergic activation of amygdaloid neurons is involved in the natural development and release of mouse-killing behaviour, it is important to know more about the possible modulating influences exerted upon this cholinergic amygdaloid system by the olfactory input and by the ascending noradrenergic, dopaminergic, and serotonergic fibre systems.

It must be pointed out that it is difficult to establish clear causal relationships between behavioural and neurochemical data. An example of these difficulties can be seen from results obtained by Salama and Goldberg (1973). In an earlier study, these authors have reported high forebrain norepinephrine levels and turnover rates in killer-rats, and such neurochemical data have been considered to be characteristic of aggressive animals. However, their recent study has shown that the killing episode itself increases forebrain norepinephrine level and turnover, and that these neurochemical effects do not last for more than 24 hours. Without a killing episode, there is no significant difference between killer and non-killer rats with regard to forebrain norepinephrine. Since such difficulties occur often, it is clear that this kind of problem should be studied, whenever possible, by using an interdisciplinary approach. It is only through cooperation between workers in the fields of ethology, experimental psychology, neurophysiology, neurochemistry, and genetics, that we will succeed in uncovering the successive links in the chain of events leading from the gene to its eventual expression in the behavioural phenotype.

References

Adams, D. and J. P. Flynn (1966): Transfer of an escape response from tail shock to brain-stimulated attack behaviour. *J. exp. Anal. Behav.* **9**, 401.

Alberts, J. R. and M. I. Friedman (1972): Olfactory bulb removal but not anosmia increases emotionality and mouse killing. *Nature*, **238**, 454.

Baenninger, R. (1970): Suppression of interspecies aggression in the rat by several aversive training procedures. *J. comp. physiol. Psychol.*, **70**, 382.

Bernstein, H. and K. E. Moyer (1970): Aggressive behaviour in the rat: effects of isolation, and olfactory bulb lesions. *Brain Res.* **20**, 75.

Caggiula, A. R. (1970): Analysis of the copulation-reward properties of posterior hypothalamic stimulation in male rats. *J. comp. physiol. Psychol.*, **70**, 399.

Campbell, B. A. (1964): Theory and research on the effects of water deprivation on random activity in the rat. In: Thirst, 1st International Symposium on Thirst in the Regulation of Body Water, Wayner, M. J., ed. (Macmillan, New York) pp. 317–334.

Chaurand, J. P. (1973): Etude de la participation fonctionnelle de certaines structures mésencéphaliques au déterminisme du comportement d'agression interspécifique Rat-Souris, (Thèse de Doctorat-és-Sciences, Université Louis Pasteur de Strasbourg).

Chaurand, J. P., P. Schmitt and P. Karli (1973): Effets de lésions du tegmentum ventral du mésencéphale sur le comportement d'agression Rat-Souris. *Physiol. Behav.*, **10**, 507.

Chaurand, J. P., M. Vergnes and P. Karli (1972): Substance grise centrale du mésencéphale et comportement d'agression interspécifique du Rat. *Physiol. Behav.*, **9**, 475.

Chaurand, J. P., M. Vergnes and P. Karli (1974): Déclenchement de conduites agressives par stimulation électrique du tegmentum ventral du mésencéphale chez le rat. *Physiol. Behav.*, **12**, 771

Crow, T. J. (1972): Catecholamine-containing neurones and electrical self-stimulation: 1. A review of some data. *Psychol. Med.*, **2**, 414.

Crow, T. J. (1973): Catecholamine-containing neurones and electrical self-stimulation: 2. A theoretical interpretation and some psychiatric implications. *Psychol. Med.*, **3**, 66.

Denenberg, V. H., R. E. Paschke and M. X. Zarrow (1968): Killing of mice by rats prevented by early interaction between the two species. *Psychon. Sci.*, **11**, 39.

De Sisto, M. J. and J. P. Huston (1971): Aggression and reward from stimulating common sites in the posterior lateral hypothalamus of rats. *Commun. behav. Biol.*, (Part A) **6** 295.

Di Chiara, G., R. Camba and P. F. Spano (1971): Evidence for inhibition by brain serotonin of mouse-killing behaviour in rats. *Nature*, **233**, 272.

Didiergeorges, F. and P. Karli (1966): Stimulations "sociales" et inhibition de l'agressivité interspécifique chez le Rat privé de ses afférences olfactives. *C. R. Soc. Biol., Paris*, **160**, 2445.

Didiergeorges, F. and P. Karli (1967): Hormones stéroïdes et maturation d'un comportement d'agression interspécifique du Rat. *C. R. Soc. Biol., Paris*, **161**, 179.

Didiergeorges, F., M. Vergnes and P. Karli (1966): Privation des afférences olfactives et agressivité interspécifique du Rat. *C. R. Soc. Biol., Paris*, **160**, 866.

Douglas, R. J., R. L. Isaacson and R. L. Moss (1969): Olfactory lesions, emotionality and activity. *Physiol. Behav.* **4**, 379.

Ebel, A., G. Mack, V. Stefanovic and P. Mandel (1973): Activity of choline-acetyltransferase and acetylcholinesterase in the amygdala of spontaneous mouse-killer rats and in rats after olfactory lobe removal. *Brain Res.*, **57**, 248.

Eclancher, F. and P. Karli (1968): Lésion du noyau dorso-médian du thalamus et comportement d'agression interspécifique Rat-Souris. *C. R. Soc. Biol., Paris*, **162**, 2273.

Eclancher, F. and P. Karli (1969): Comportement d'agression interspécifique Rat-Souris: effets de lesions du noyau dorso-médian du thalamus et des structures épithalamiques. *J. Physiol., Paris*, **61**, 283.

Eclancher, F. and P. Karli (1971): Comportement d'agression interspécifique et comportement alimentaire du Rat: effets de lésions des noyaux ventro-médians de l'hypothalamus. *Brain Res.*, **26**, (Suppl. 2) 71.

Eclancher, F. and P. Karli (1972): Lésions combinées de l'hypothalamus ventro-médian et du thalamus dorso-médian: effets sur le comportement social interspécifique, la réactivité émotionelle et le comportement alimentaire du Rat. *C. R. Soc. Biol., Paris*, **166**, 439.

Eclancher, F. and P. Schmitt, (1972): Effets de lésions précoces de l'amygdale et de l'hypothalamus médian sur le développement du comportement d'agression interspécifique du rat. *J. Physiol., Paris*, **65**, 231 A.

Eichelman, B. S. (1971): Effect of subcortical lesions on shock-induced aggression in the rat. *J. comp. physiol. Psychol.*, **74**, 331.

Eichelman, B., N. B. Thoa, N. M. Bugbee and K. Y. Ng (1972): Brain amine and adrenal enzyme levels in aggressive, bulbectomized rats. *Physiol. Behav.*, **9**, 483.

Giachetti, I., P. Mac Leod and J. Le Magnen (1970): Influence des états de faim et de satiété sur les réponses du bulbe olfactif chez le Rat. *J. Physiol.*, *Paris*, **62**, (suppl. 2), 280.

Goddard, G. V. (1969): Analysis of avoidance conditioning following cholinergic stimulation of amygdala in rats. *J. comp. physiol. Psychol.*, **68**, Monograph Supplement no° 2, 1.

Grant, L. D., D. V. Coscina, S. P. Grossman and D. X. Freedman (1973): Muricide after serotonin depleting lesions of midbrain raphé nuclei. *Pharmacol. Biochem. Behav.*, **1**, 77.

Hamburg, D. A. and J. van -Lawick-Goodall (1974): Factors facilitating development of aggressive behavior in chimpanzees and humans. In: Determinants and Origins of Aggressive Behavior. Hartup, W. and J. de Wit, eds., (Mouton, The Hague) in press.

Heimstra, N. W. (1965): A further investigation of the development of mouse-killing in rats. *Psychon. Sci.*, **2**, 179.

Hoebel, B. G. (1968): Inhibition and disinhibition of self-stimulation and feeding: hypothalamic control and postingestional factors. *J. comp. physiol. Psychol.* **66**, 89.

Hoebel, B. G. and R. D. Thompson (1969): Aversion to lateral hypothalamic stimulation caused by intragastric feeding or obesity. *J. comp. physiol. Psychol.*, **68**, 536.

Hughes, R. N. (1965): Food deprivation and locomotor exploration in the white rat. *Anim. Behav.*, **13**, 30.

Hutchinson, R. R. (1973): The environmental causes of aggression. In: Nebraska Symposium on Motivation (University of Nebraska Press, Lincoln) pp. 155–181.

Igic, R., P. Stern and E. Basagic (1970): Changes in emotional behavior after application of cholinesterase inhibitor in the septal and amygdala region. *Neuropharmacol.*, **9**, 73.

Jones, R. B. and N. W. Nowell (1973): The effect of familiar visual and olfactory cues on the aggressive behaviour of mice. *Physiol. Behav.*, **10**, 221.

Jund, A., B. Canguilhem and P. Karli, Catécholamines urinaires et comportement d'agression interspécifique du Rat. *C. R. Soc. Biol.*, *Paris*, **165**, 1998.

Karli, P. (1956): The Norway Rat's killing-response to the white Mouse. An experimental analysis. *Behaviour*, **10**, 81.

Karli, P. (1959): Action de substances dites tranquillisantes sur l'agressivité interspécifique Rat-Souris. *C. R. Soc. Biol.*, *Paris*, **153**, 467.

Karli, P. (1960): Effets de lésions expérimentales du septum sur l'agressivité interspécifique Rat-Souris. *C. R. Soc. Biol.*, *Paris*, **154**, 1079.

Karli, P. (1961): Rôle des afférences sensorielles dans le déclenchement du comportement d'agression interspécifique Rat-Souris. *C. R. Soc. Biol.*, *Paris*, **155**, 644.

Karli, P. (1968): Système limbique et processus de motivation. *J. Physiol.*, *Paris*, **60**, (suppl. 1), 3.

Karli, P. (1972): Rôle du système limbique dans le déterminisme physiologique de la réactivité émotionnelle et sociale. In: Actualités Pharmacologiques, 25ème Série (Masson, Paris) pp. 61–90.

Karli, P. (1974): Rôle des stimulations dans le développement du cerveau. In: Prenatal and postnatal development of the human brain. In: Modern Problems in Paediatrics, vol. 13, pp. 203–228 (Karger, Basel).

Karli, P. and M. Vergnes (1964): Dissociation expérimentale du comportement d'agression interspécifique Rat-Souris et du comportement alimentaire. *C. R. Soc. Biol., Paris*, **158**, 650.

Karli, P. and M. Vergnes (1965): Rôle des différentes composantes du complexe nucléaire amygdalien dans la facilitation de l'agressivité interspécifique du Rat. *C. R. Soc. Biol., Paris*, **159**, 754.

Karli, P., M. Vergnes and F. Didiergeorges, (1969): Rat-Mouse interspecific aggressive behaviour and its manipulation by brain ablation and by brain stimulation. In: Aggressive Behaviour, Garattini, S. and E. B. Sigg, eds., (Excerpta Medica, Amsterdam) pp. 47–55.

Karli, P., M. Vergnes, F. Eclancher, P. Schmitt and J. P. Chaurand (1972): Role of the amygdala in the control of mouse-killing behavior in the rat. In: The Neurobiology of the Amygdala, Eleftheriou, B. E., ed. (Plenum Press, New-York) pp. 553–580.

Kostowski, W., E. Giacalone, S. Garattini and L. Valzelli (1968): Studies on behavioural and biochemical changes in rats after lesion of midbrain raphé. *Eur. J. Pharmacol.*, **4**, 371.

Kulkarni, A. S. (1968): Muricidal block of 5-hydroxytryptophan and various drugs. *Lif. Sci.*, **7**, 125.

Lagerspetz, K. M. J. and K. Y. H. Lagerspetz (1971): Changes in the aggressiveness of mice resulting from selective breeding, learning, and social isolation. *Scand. J. Psychol.* **12**, 241.

Mac Donnell, M. F. and J. P. Flynn (1966): Control of sensory fields by stimulation of hypothalamus. *Science*, **152**, 1406.

Margules, D. L. (1971): Localization of anti-punishment actions of norepinephrine and atropine in amygdala and entopeduncular nucleus of rats. *Brain Res.*, **35**, 177.

Margules, D. L. and J. Olds (1962): Identical 'feeding' and 'rewarding' systems in the lateral hypothalamus of rats. *Science*, **135**, 374.

Marshall, J. F., B. H. Turner and P. Teitelbaum (1971): Sensory neglect produced by lateral hypothalamic damage. *Science*, **174**, 523.

Mathews, S. R., Jr. and F. W. Finger (1966): Direct observation of the rat's activity during food deprivation, *Physiol. Behav.*, **1**, 85.

Miczek, K. A. and S. P. Grossman (1972): Effects of septal lesions on inter-and intraspecies aggression in rats. *J. comp. physiol. Psychol.*, **79**, 37.

Miley, W. M. and R. Baenninger (1972): Inhibition and facilitation of interspecies aggression in septal lesioned rats. *Physiol. Behav.*, **9**, 379.

Myer, J. S. (1967): Prior killing experience and the suppressive effects of punishment on the killing of mice by rats. *Anim. Behav.*, **15**, 59.

Myer, J. S. (1969): Early experience and the development of mouse-killing by rats. *J. comp. physiol. Psychol.*, **67**, 46.

Myer, J. S. and R. T. White (1965): Aggressive motivation in the rat. *Anim. Behav.*, **13**, 430.

Neill, D. B., L. D. Grant and S. P. Grossman (1972): Selective potentiation of locomotor effects of amphetamine by midbrain raphé lesions: *Physiol. Behav.*, **9**, 655.

Pager, J., I. Giachetti, A. Holley and J. Le Magnen (1972): A selective control of olfactory bulb electrical activity in relation to food deprivation and satiety in rats. *Physiol. Behav.* **9**, 573.

Panksepp, J. (1971): Aggression elicited by electrical stimulation of the hypothalamus in albino rats. *Physiol. Behav.*, **6**, 321.

Paul, L., W. M. Miley and R. Baenninger (1971): Mouse killing by rats: roles of hunger and thirst in its initiation and maintenance. *J. comp. physiol. Psychol.*, **76**, 242.

Paul, L., W. M. Miley and N. Mazzagatti (1973): Social facilitation and inhibition of hunger-induced killing by rats. *J. comp. physiol. Psychol.*, **84**, 162.

Phillips, A. G. and I. Lieblich (1972), Developmental and hormonal aspects of hyperemotionality produced by septal lesions in male rats. *Physiol. Behav.*, **9**, 237.

Phillips, D. S. and G. K. Martin (1971): Effects of olfactory bulb ablation upon heart rate. *Physiol. Behav.*, **7**, 535.

Phillips, I. and J. Olds (1969): Unit activity: motivation-dependent responses from midbrain neurons. *Science* **165**, 1269.

Raisman, G. and P M. Field (1973): Sexual dimorphism in the neuropil of the preoptic-area of the rat and its dependence on neonatal androgen. *Brain Res.*, **54**, 1.

Richman, C. L., R. Gulkin and K. Knoblock (1972): Effects of bulbectomization, strain, and gentling on emotionality and exploratory behavior in rats. *Physiol. Behav.*, **8**. 447.

Roberts, W. W. and R. J. Carey (1963). Effect of dorsomedial thalamic lesions on fear in cats. *J. comp. physiol. Psychol.*, **56**, 950.

Roberts, W. W. and R. J. Carey (1965): Rewarding effect of performance of gnawing aroused by hypothalamic stimulation in the rat. *J. comp. physiol Psychol.*, **59**, 317.

Rolls, B. J. and E. T. Rolls (1973): Effects of lesions in the basolateral amygdala on fluid intake in the rat. *J. comp. physiol. Psychol.*, **83**, 240.

Rolls, E. T. (1973): Refractory periods of neurons directly excited in stimulus-bound eating and drinking in the rat. *J. comp. physiol. Psychol.*, **82**, 15.

Rolls, E. T. (1974): The neural basis of brain-stimulation reward. *Progr. Neurobiol.*, in press.

Salama, A. I. and M. E. Goldberg (1973): Temporary increase in forebrain norepinephrine turnover in mouse-killing rats. *Eur. J. Pharmacol.*, **21**, 372.

Schmitt, P. (1972): Effets de lésions de la substance grise centrale du mésencéphale sur les réponses de 'switch-off' et d'autostimulation au niveau de l'hypothalamus. *J. Physiol. Paris*, **65**, 501 A.

Sheard, M. H. (1969): The effect of p-chlorcphenylalanine on behavior in rats: relation to brain serotonin and 5-hydroxyindoleacetic acid, *Brain Res.*, **15**, 524.

Sheard, M. H. (1973): Brain serotonin depletion by p-chlorophenylalanine or lesions of raphé neurons in rats. *Physiol. Behav.*, **10**, 809.

Sieck, M. H. (1973): Selective olfactory system lesions in rats and changes in appetitive and aversive behavior. *Physiol. Behav.*, **10**, 731.

Sieck, M. H. and B. L. Gordon (1972): Selective olfactory bulb lesions: reactivity changes and avoidance learning in rats. *Physiol. Behav.* **9**, 545.

Smith, D. A. (1972): Increased perioral responsiveness: a possible explanation for the switching of behavior observed during lateral hypothalamic stimulation. *Physiol. Behav.*, **8**, 617.

Spector, S. A. and E. M. Hull (1972): Anosmia and mouse killing by rats: a nonolfactory role for the olfactory bulbs. *J. comp. physiol. Psychol.*, **80**, 354.

Stein, L. (1969): Chemistry of purposive behavior. In: Reinforcement and behavior. Tapp, J. T., ed. (Academic Press, New-York) pp. 328–355.

Stern, J. J. (1970): Responses of male rats to sex odors. *Physiol. Behav.*, **5**, 519.
Tsai, L. S. and G. E. Dexter (1970): Reinforcing 'displaced aggression' and dominance hierarchy in white rats. *J. gen. Psychol.*, **83**, 97.
Turner, B. H. (1973): Sensorimotor syndrome produced by lesions of the amygdala and lateral hypothalamus. *J. comp. physiol. Psychol.*, **82**, 37.
Ungerstedt, U. (1971): Adipsia and aphagia after 6-hydroxydopamine induced degeneration of the nigro-striatal dopamine system. *Acta physiol. scand.*, suppl. 367, 95.
Van Hemel, P. F. (1972): Aggression as a reinforcer: operant behavior in the mouse-killing rat. *J. exp. Analysis Behav.* **17**, 237.
Vergnes, M. and J. P. Chaurand (1973): Effets comportementaux de lésions de la partie postérieure de la substance grise périaqueducale. *C. R. Soc. Biol., Paris*, **167**, 351.
Vergnes, M. and P. Karli (1963): Déclenchement du comportement d'agression interspécifique Rat-Souris par ablation bilatérale des bulbes olfactifs. Action de l'hydroxyzine sur cette agressivité provoquée. *C. R. Soc. Biol., Paris*, **157**, 1061.
Vergnes, M. and P. Karli (1965): Etude des voies nerveuses d'une influence inhibitrice s'exerçant sur l'agressivité interspécifique du Rat. *C. R. Soc. Biol., Paris*, **159**, 972.
Vergnes, M. and P. Karli (1969a): Effets de l'ablation des bulbes olfactifs et de l'isolement sur le développement de l'agressivité interspécifique du Rat. *C. R. Soc. Biol., Paris* **163**, 2704.
Vergnes, M. and P. Karli (1969b): Effets de la stimulation de l'hypothalamus latéral, de l'amygdale et de l'hippocampe sur le comportement d'agression interspécifique Rat-Souris. *Physiol. Behav.*, **4**, 889.
Vergnes, M. and P. Karli (1970): Déclenchement d'un comportement d'agression par stimulation électrique de l'hypothalamus médian chez le Rat. *Physiol. Behav.*, **5**, 1427.
Vergnes, M. and P. Karli (1972): Stimulation électrique du thalamus dorsomédian et comportement d'agression interspécifique du Rat. *Physiol. Behav.*, **9**, 889.
Vergnes, M., G. Mack and E. Kempf (1973): Lésions du raphé et réaction d'agression interspécifique Rat-Souris. Effets comportementaux et biochimiques. *Brain Res.*, **57**, 67.
Wasman, M. and J. P. Flynn (1962): Directed attack elicited from hypothalamus. *Arch. Neurol.*, **6**, 220.
Wayner, M. J., S. Kahan and W. Stoller (1966): Lateral hypothalamic mediation of spinal reflex facilitation during salt arousal of drinking. *Physiol. Behav.*, **1**, 341.
White, S. D., M. J. Wayner and A. Cott (1970): Effects of intensity, water deprivation, prior water ingestion and palatability on drinking evoked by lateral hypothalamic electric stimulation. *Physiol. Behav.*, **5**, 611.
Wise, C. D. and L. Stein (1973): Dopamine-β-hydroxylase deficits in the brains of schizophrenic patients. *Science*, **181**, 344.
Woodworth, C. H. (1971): Attack elicited in rats by electrical stimulation of the lateral hypothalamus. *Physiol. Behav.*, **6**, 345.

CHAPTER 13

Genetic determination of aggressive behaviour

K. M. J. Lagerspetz and K. Y. H. Lagerspetz

The role of genetic factors in the interindividual variation of aggressiveness has interested both scientists and laymen for a long time. The question as to what extent aggressiveness can be suppressed or generated by environmental conditions and as to what extent it is genetically determined is important for the planning of education and legislation, for example.

Not very much is so far known about the hereditary determination of aggressiveness, and most studies providing more detailed knowledge on the subject have been performed with small laboratory mammals as experimental subjects. In most investigations mice have been used, as they are well-suited for aggression research (Scott and Fredericson, 1951), but some investigators have used other rodents, chickens, and dogs.

Mainly two methods have been employed in these studies. Firstly, comparisons of aggressiveness between different often highly inbred strains of the same species have been made and, secondly, special strains have been developed through selective breeding for aggressiveness and non-aggressiveness. The first method has been much more common, because less work is needed in comparisons of strains already developed for other purposes. However, for reasons to be presented in Section 3, the second method gives clearer information about aggressiveness and its connections with other characteristics of the animals.

In the first section we will discuss questions regarding the measurement of aggressiveness, because the outcome of many studies and some discrepancies between the results depend on the variables chosen for measurement. Next, we shall present studies on strain differences, then the selective breeding experiments, and finally a discussion on the investigations into the interaction of genetic and environmental factors in the development of aggressiveness. The physiological aspects involved will be briefly summarized in a separate section.

13.1. Measurement of aggression

In all experiments dealing with fighting, aggression, dominance, ferocity, or savageness, observations are made of the animals' attacking behaviour towards some external object.

Mostly the observations are made during a short period of 2–15 minutes, and under standard conditions (e.g. Scott, 1940; Lagerspetz, 1964; Craig and Baruth, 1965; Vale et al., 1972), but sometimes the behaviour of the animals is observed in their home cages, usually for longer periods (e.g. Vale et al., 1971; Ewbank and Bryant, 1972); sometimes frequencies of occurrence of specific behaviour patterns during the observation period are counted and recorded (e.g. Lagerspetz, 1964; Ginsburg, 1967), whereas in other investigations a general evaluation of the intensity of fighting is made for pairs of animals or for each animal separately in the form of scoring on a scale (Hall and Klein, 1942; Lagerspetz, 1964); or as an evaluation of the dominance of the rated animal in relation to the opponent (e.g. Graig et al., 1965).

In early studies on the fighting behaviour of rats and mice, the animals' reactions towards the experimenter have usually been observed (Yerkes, 1913; Utsurikawa, 1917; Stone, 1932). In this connection, the tendency to attack has been called "savageness", the occurrence of escape or avoidance reactions is called "wildness", and an absence of attack or avoidance behaviour and/or the presence of approach behaviour is called "tameness" (cf. Scott and Fredericson, 1951). In some connections, "wildness" refers to the behaviour usually called "motor activity", for instance in Dawson's (1932) selective breeding experiments on the inheritance of "wildness" and "tameness" in mice.

In later studies on fighting behaviour, the reactions of the animal towards another individual of the same species (and often of the same sex) have been observed. The animal's attacking behaviour has sometimes been called "dominance" and sometimes "aggressiveness" (or "aggression"), depending on the methods of evaluation of the behaviour. For instance, in the experiments of Craig, Ortman and Guhl (1965) on the "dominance ability" in chickens, the behaviour of a bird in paired contests was rated either as "winning" or as "losing". Winning was further divided into winning by fighting, by pecking, or by threatening. Losing was rated in a comparable fashion. In addition, some contests were rated with "no decision" and others with "no contest". Similar ratings of dominance have been used with rats and mice (e.g. Levine et al., 1965; Porter, 1972).

In experiments where the aim is to investigate "aggressiveness" rather than "dominance", frequencies and latencies of attacks or biting are the usual measures. As compared with evaluations of dominance relationships, this manner of recording yields scores which in statistical calculations can be regarded as interval or ratio scale-ratings. A disadvantage of this procedure is that the behaviour of each animal in a paired encounter is partly determined by the behaviour of the opponent. The assessment of the individual reaction pattern of each animal is thus disturbed by the reactions of the opponent and vice versa. This drawback can be avoided by using especially chosen "standard" partners which are known to never attack their opponents, thus providing a uniform stimulation for the elicitation of aggression in the test animals (Lagerspetz, 1969). Non-aggressive animals caged together can be profitably used for this purpose.

Mettälä (1965) has performed a factorial analysis of the intercorrelations between nine types of behaviour patterns of mice, recorded in fighting situations (in some the behaviour of the opponent was simulated by the movement of a bottle brush). Two factors were recognized: an "aggression factor" including biting, biting to draw blood, and tail-rattling, and a "latency factor" including nosing and running. The results showed that frequencies of biting and tail-rattling can be used to differentiate between degrees of aggression. The analysis also indicated that the amount of investigatory behaviour could not be used as a predictor of initiation or intensity of fighting.

In some investigations rating scales have been used to estimate the degree of aggressiveness of each animal in a contest (Hall and Klein, 1942; Lagerspetz, 1961, 1964, pp. 42–48). These may have advantages over quantitative recordings of separate behaviour patterns in cases where standard partners are not used and where the behaviour of both participants has to be evaluated.

The term "dominance" has also been used in a sense somewhat different from the one referred to above. In situations where the animals compete for food, water, etc., "dominance" sometimes denotes some measure for the possession of the object competed for – the length of time an animal controls the food or the access to the drinking-water, the amount eaten by an animal, etc. (e.g. Uyeno, 1960). Of course, this meaning of "dominance" partly overlaps the one defined above, but in some experiments the control over food does not primarily result from attacks by the "dominant" individual but through some other kind of behaviour, for instance, sitting or laying on the food container. An example of this use of the term "dominance" is found in an experiment by Heimstra (1961).

13.2. Strain differences

Differences in several measures of aggressiveness have been observed in different, often inbred, strains. One of us has reviewed the literature on strain comparisons for aggressiveness up to 1963 (Lagerspetz, 1964, pp. 14–16). The present survey is in part based on that review.

Yerkes (1913) compared "wild" and "tame" rats and their F_1 and F_2 hybrids with regard to "wildness", "savageness", and "timidity". The existence of these traits and their intensity were inferred from the behaviour of the animals when placed on a table. The parameters for "savageness" were: "(1) biting, (2) exposing or gnashing the teeth, (3) jumping at hand or forceps, (4) squealing." Yerkes claims that the results prove conclusively that savageness, wildness, and timidity are heritable behaviour complexes.

Utsurikawa (1917) also used savageness as one of the observed variables in comparing the behaviour of inbred and randomly bred strains of the albino rat. Savageness was tested by grading the behaviour of the animals on a 5-point scale while scratching the floor of their cage with a copper wire. "Some individuals would, at this, dash forward and bite viciously and persistently the wire, whereas others merely noticed the disturbance and were otherwise indifferent to it" (ibid., p. 113). The results showed, surprisingly, that the females surpassed the males in degree of savageness, that inbred rats were more savage than the randomly bred, but that, all in all, individual differences exceeded strain and sex differences. Stone (1932) also found differences in wildness and savageness in rats of different strains. Scott (1940, 1942) was the first to study strain differences in rodents by observing their aggressive behaviour towards other males of the same species. Three pure-bred mouse strains, C57BL/10 (black), C3H (agouti), and BALB/Sc (albino), showed large differences in their reactions towards the partner, e.g. in the latencies to initiate fights, the C57BL/10 animals being the most aggressive and the albinos the least aggressive. Ginsburg and Allee (1942) compared the effects of conditioning on social dominance in the same inbred strains of mice and found similar differences in dominance as reported by Scott.

Hall and Klein (1942) compared the aggressiveness of two strains selectively bred by Hall (1938) for high and low tendency to defecate in strange situations ("emotionality" and "non-emotionality"). Several aspects of the behaviour directed to another male rat were observed and graded by means of a 7-point rating scale. A striking difference in aggressiveness was found between the two strains, the non-emotionals being more aggressive than the emotionals. For instance, the non-emotionals initiated 326 agressive encounters but the emotionals only 68. "In fact it appears as though the two strains have been selectively bred, without our knowledge, for peacefulness and pugnacity as well as for timidity and fearlessness" (Hall and Klein, 1942, p. 377). The inverse relationship between aggressiveness and emotionality, as indicated by defecation in strange situations, has been confirmed in a study by one of the present authors (Lagerspetz, 1964, pp. 70–80), where she found significant differences in defecation between her selected aggressive and non-aggressive mouse strains, the aggressive mice being significantly less emotional than the non-aggressives. She also found negative correlations between aggressiveness and defecation in the open field (rho = -0.30; not significant), and between aggressiveness and defecation in the aggression test situations (rho = -0.36; $P < 0.01$).

When investigating the effects of castration and testosterone propionate administration on the aggressive behaviour of C57 and albino mice, Beeman (1947) also observed differences in the aggressive behaviour of these strains. A significant difference was found with respect to the ratio of tail rattles and attacks, the albino mice rattling their tails more often per attack than the black mice.

Fredericson et al. (1955) showed that two mouse strains, BALB/c and C57BL/10, differed in the relationship between sexual and aggressive behaviour. This study will be discussed later in the present chapter.

When investigating the effects of early social experiences on aggressive behaviour, King (1957) could not confirm the strain differences in latencies to initiate fights originally observed by Scott (1940, 1942). Bauer (1956) recorded the frequencies of 21 behavioural responses when mice of strains C57BL/10 and BALB/c were exposed to opponents of like or unlike strains. The opponents were dangled by the tail in front of the subjects. The two strains differed significantly for frequency of the following responses: strong attack, tail-rattling, defense posture, acquiescent posture, grooming of the dangling mouse, and withdrawal into a corner. Levine et al. (1965) investigated the effects of isolation and defeat on aggressiveness displayed in interstrain contests of ST/J and CBA mice. The CBA males achieved fighting superiority under all conditions, but their higher aggressiveness was more pronounced after a period of social experiences or after a longer fighting-time. Social experiences clearly reduced the aggressiveness of ST/J males but not that of the CBA males. Karczmar and Scudder (1969) compared several strains of mice, also other species than *Mus musculus*, and found the albino mouse strain CF-1 to be the most aggressive one. Southwick (1970) analyzed 14 inbred mouse strains for aggressiveness measured by number of aggressive chases, attacks and fights per hour when four adult males were grouped following a period of isolation. Some strains, such as C57BR, AKR, CFW, and BALB/c, were much more aggressive than others. Southwick reports the order of aggressiveness of the strains and the statistical differences.

TABLE 13.1

Differences in aggressiveness between mouse strains.

Authors	Strains
Scott 1940, 1942	C57BL/10 > C3H > BALB/Sc
Ginsburg & Allee 1942	C57BL/10 > C3H > BALB/c
Beeman 1947	C57BL/10 > BALB/c
Fredericson 1952	C57BL/10 > BALB/c
Bauer 1956	C57BL/10 > BALB/c
King 1957	BALB/c = CBA > C57BL/10
Bourgault et al. 1963	SC-1 > C57BL/10
Levine et al. 1965	CBA > ST
Ginsburg 1967	HS > DBA/2 > DBA/1 > A/J = C57BL/6 = C57BL/10 > C3H > C-albino
Karczmar & Scudder 1969	CF-1 > C57BL/10 > MO > Onychomus > Microtus > Peromyscus maniculatus bairdii
Southwick 1970	C57BR = AKR = CFW = BALB/c > DBA/1 = DBA/2 = C57BL > RF = C57L > C3H = SWR > CBA > AHe > A/J
Vale et al. 1971	BALB/cJ > A/J = C57BL/6J = DBA/2J = C3H/He
Tellegen & Horn 1972	RF/J = BALB/cJ > SJL/J
Porter 1972	C57BL/10 > BALB/c

(for statistical evaluations of the differences, see the original reports)

Table 13.1 shows the results of the comparisons between mouse strains with regard to aggressive behaviour. The many inconsistencies apparent between the results depend partly upon the strains each strain is compared with. For instance, the three strains found to be most aggressive in Southwick's experiment have not been used in the other studies. The inconsistencies are also partly due to differences in the methods used for measuring aggressiveness. In some investigations, groups of animals are observed either in their living-cages with partners they are caged with (Vale et al.) or placed together with groups of strangers (Southwick), whereas in other studies paired encounters are observed either with specially chosen non-aggressive partners (Tellegen and Horn) or by rating both animals belonging to an experimental group (Ginsburg, Scott). All in all, the picture that emerges from the comparisons in Table 1 is far from clear.

Ginsburg (1967) found that handling with forceps at an early age affected differentially the aggressiveness of inbred strains, which is an example of interaction between genotype and experience. A special section below will deal with such interactions. Vale et al. (1971) demonstrated that the BALB/cJ strain differs significantly in aggressiveness from three other inbred mouse strains under various conditions. Scott, et al. (1971) found strain differences for the effects of amphetamine sulphate on fighting behaviour. At a dose of 10 mg/kg, fighting was inhibited in C57 mice but not in BALB animals, except when body temperatures, elevated through fighting, approached lethal levels. Tellegen and Horn (1972) compared three inbred strains of mice with regard to the effectiveness of aggression reinforcement. Male mice were rewarded by giving them the opportunity to fight in the goal box of a T-maze. In the animals of the two most aggressive strains, BALB/cJ and RF/J, the effectiveness of aggression reinforcement approached that of reinforcement by eating, whereas actual fighting did not have a markedly reinforcing effect on the least aggressive SJL animals. This result is in accordance with studies carried out by one of us, using an obstruction apparatus (Lagerspetz, 1964). It was found that animals of a selectively bred aggressive strain (TA, see below) were rewarded by opportunities to fight, provided they were motivated through an immediately preceding fight.

Collins (1970) studied aggressive behaviour in 60 male mice from the 15th generation of selection for high and low brain weight and from an unselected control line. Mice with low brain weights were highly aggressive, as compared with the controls, which were moderately aggressive, and with the clearly non-aggressive high brain weight animals. These results contrast with those obtained by Lagerspetz et al. (1968) who showed that mice selectively bred for aggressiveness (TA) had slightly heavier forebrains than those bred for non-aggressiveness (TNA). This contrast may be due to different methods of measurement of aggression. The TA and TNA animals were reared in isolation and tested individually in encounters with partners which were non-aggressive and had been caged together, whereas in Collins' experiment the aggressiveness of the animals was judged from the scars of inflicted wounds.

Not many other species than rats and mice have been used in strain comparisons for aggressiveness. However, Scott (1958, p. 80) points out that certain breeds of dogs, for instance almost all terrier breeds, have been selected for their fighting ability. In Scott's experiments, the aggressiveness was assessed by observing playful fighting of puppies, for instance over a bone, or their tendency to playfully paw or bite a human handler who patted them in a standard way. Fox-terriers and basenjis showed a marked aggressiveness, whereas beagles and cocker-spaniels were more peaceful. A breed of basenjis and one of cocker-spaniels were crossed. The first generation was intermediate in aggressiveness directed to a human handler. In the backcross, the offspring were more like the cockers or more like the basenjis, depending on the way the cross was made. Scott claims that only a small number of genetic factors is involved in the control of aggressiveness in these dogs, with a minimum of two genes.

13.3. Selective breeding for aggressiveness

In addition to strain comparisons, another method to demonstrate that the interindividual variation of a behavioural trait has a hereditary basis is selective breeding. In each generation, those animals which possess to a high degree the characteristic investigated are selected for mating. The selection is often made in two directions, giving a "high" and a "low" line. Of course, selection must always occur within these lines. It is not always necessary to select in more than one direction if an unselected control group is maintained (Broadhurst 1960, p. 13).

The aim of the selection is to obtain lines which, if possible differ from each other only with regard to the investigated trait. The lines will also differ for other characteristics dependent upon the same pleiotropic genes and for those traits inherited through genes linked with those determining the trait for which selection is made. Thus, selectively bred non-inbred lines offer an opportunity to investigate correlations between the investigated trait and other characteristics (Lagerspetz, 1964, p.16).

In selection experiments, inbreeding is undesirable because it will result in sub-lines differing for heritable characteristics for which the animals were not selected. As Broadhurst (1960, p. 19) points out, in selective breeding experiments the purpose is to obtain lines which are homozygous for the genes determining the characteristic for which selection is made, but not for other genes.

The aim of most of the selective breeding experiments performed in the area of behavioural genetics has so far been to demonstrate the existence of a hereditary effect upon the investigated trait. Some authors, however, have presented genetical analyses based on crosses between selected lines.

An important problem is whether it is possible, on the basis of selection, to assess the number of relevant genes, to establish their dominance interaction, and to estimate quantitatively the contribution of the genetic variation to the phenotypic variation (Lagerspetz and Lagerspetz, 1974).

Most, if not all, of the behavioural characteristics used in selective breeding experiments are not unit characteristics but quantitative variables. They are probably polygenically controlled and the minimum number of genes involved has in some cases been estimated to be 2 or 3. Accordingly, it is difficult to establish the dominance relations of the genes, and some such attempts have been criticized.

The estimates of heritability in the narrow sense, i.e. the reliability of the phenotype as a guide to the breeding value, and of heritability in the broad sense, i.e. the degree of genetic determination, vary widely even for the same behavioural characteristic with the method of measuring the characteristic and with the model on which the calculation of heritability is based. For instance, heritabilities between 0.16 and 0.28 were found by Guhl et al. (1960) and by Craig et al. (1965) for aggressiveness and social dominance in chickens.

Many of the heritability values found are surprisingly low and, in addition, their meaningfulness is doubtful (Hirsch, 1967, pp. 421–429). In conclusion, selective breeding experiments, beyond establishing the fact that hereditary factors determine the interindividual variation of behaviour, have contributed rather little to the understanding of the nature and size of the genetic effects on behaviour. The same can be said about comparisons of behaviour between inbred strains, which in addition have the disadvantage that the animals within a strain are alike for all other characteristics, too, so that the investigated trait can only be studied in that particular genotype.

There is another aspect, however, for which the selective breeding method and, to a lesser degree, also strain comparisons can be valuable. When effects of environmental variation on behaviour are studied, it is of importance to use test animals with a controlled genetic architecture. If we know that newborn animals have a genetic predisposition for high or low aggressiveness, we can, by varying the rearing and other conditions, obtain information about the extent to which different environmental factors affect its expression.

Domesticated animals have been subjected to selection for behavioural characteristics. Usually the selection has been for peacefulness, but in some dog breeds the animals have been intentionally chosen for matings on the basis of their fighting ability (Scott, 1958). In the following, we shall review selection experiments carried out with respect to aggressiveness.

Coburn (1922) made crosses comprising three generations of mice in order to investigate whether the Mendelian laws could be applied to the segregation of the traits wildness, savageness, and tameness, as classified by Yerkes (see above). This was not an investigation of aggressiveness in the usual meaning of the concept. The characteristics seemed to have a hereditary basis. When investigating the influence of a dominant versus a submissive mother on dominant and submissive behaviour of the offspring, Uyeno (1960) performed a breeding experiment with one generation of selection. Because here the emphasis was on environmental factors, the study will be discussed later.

Since 1959, selection for aggressiveness and non-aggressiveness has been performed by one of us at the Institute of Psychology, University of Turku, and since 1970 at the Institute of Psychology at Åbo Akademi, the Swedish University of Turku. Starting from a Swiss albino mouse stock, with a relatively normal distribution for aggressiveness, the most aggressive and least aggressive individuals were selected for breeding. As females did not show much interindividual differences in aggressive behaviour, they were selected on the basis of their brothers' aggression scores. The males were selected on the basis of the mean score of seven tests made on different days. The difference in aggressiveness between the lines was already significant in the second generation of selective breeding (S_2) (Lagerspetz, 1961, 1964; Lagerspetz and Lagerspetz, 1971).

Craig, Ortman and Guhl (1965) carried out a bidirectional selection of mature male chickens for social dominance scores, comprising five generations of two different breeds, White Leghorn and Rhode Island Red. Large line differences were obtained within the breeds. The aggressiveness was observed in paired contests in which a bird was rated either to be a "winner" or a "loser" in case fights occurred.

In order to show that the variability of behaviour within a particular inbred mouse strain, C57 BL/10, is due to nongenetic causes, Ginsburg (1967) performed an interesting selective breeding experiment for some behavioural traits, among which aggressiveness. Aggressiveness was measured as the latency to fighting and success in fighting among males that had been isolated from weaning to 75 to 80 days of age. After seven generations of selection for low and high aggressiveness, there were no significant differences between the progeny of the selected animals and those of the original generation preceding the selection, thus giving proof that the variability in the original strain was nongenetic (Ginsburg 1967, p. 138).

In her early generations of selective breeding for aggressiveness and non-aggressiveness in albino mice, Lagerspetz (1964) has studied correlations between aggression scores and some other behavioural measures. Significant positive correlations were found between aggressiveness and motor activity measured in a revolving drum ($+0.76$), between aggressiveness and ambulation in an open field ($+0.46$), and between aggressiveness and maze learning ability ($+0.41$). A significant negative correlation was found between aggressiveness scores and defecation in the aggression test situations (-0.36). Defecation in an open field has been used as an index for emotional responsiveness (Broadhurst, 1960). The number of fecal boluses deposited in open-field test situations was significantly ($P < 0.01$) higher in the males of the non-aggressive (TNA) strain. This difference in tendency for emotional defecation was even more clear-cut during encounters with another male in the aggression tests. The 22 TNA males from generation S_3 deposited an average of 12.3 boluses, while the 29 S_3 aggressive (TA) males defecated only an average of 5.2 boluses in these tests. The difference was highly significant ($P < 0.001$). The correlations and the strain differences were interpreted as indicating a tendency for a higher sympathetic tone and a higher level of arousal in the aggressive animals (Lagerspetz, 1964, pp. 115–116).

If a positive correlation would exist between sexuality and aggressiveness, the aggressive line (TA) should be sexually more active than the non-aggressive line (TNA). This possibility was investigated in two separate experiments (Lagerspetz and Hautojärvi, 1967; Lagerspetz and Lagerspetz, 1971a). In the first study, no difference in sexual behaviour was found between the lines, whereas in the second experiment such a difference did emerge, the non-aggressive line having the higher mean sexuality scores. However, the frequencies and latencies of mounting and copulation did not differ. The idea that aggressiveness and sexuality may be inversely related, contrary to a rather popular belief, even among scientists, is also supported by the mutually inhibitory effects of the learning of aggressive and the learning of sexual behaviour (see below).

13.4. Variation in the physiological mechanisms of aggressive behaviour

Apparently, at least all higher vertebrates possess the physiological mechanisms necessary for aggressive behaviour. Much knowledge about these mechanisms has been summarized in two recent collections of review articles (Eleftheriou and Scott, 1971; Moyer, 1971).

Interesting from the standpoint of the genetic determination of aggressive behaviour are those studies which pertain to variations in the physiological mechanisms of aggression. The observed interindividual phenotypic variation may in some cases manifest itself in the interstrain variation to be detected by physiological and biochemical comparisons between strains, often inbred ones, which differ with respect to their aggressiveness, or in the interstrain variation found between strains selectively bred for aggressiveness and non-aggressiveness. On the other hand, many studies have dealt with the physiological and biochemical changes found in animals, the aggressive behaviour of which has been altered by environmental influences such as isolation, grouping, fighting, defeat, drug or hormone administration, handling, castration, olfactory deafferentation, and viral encephalitis infection.

Obviously, studies on interstrain variation are of immediate concern to the problem of the genetic determination of aggressive behaviour. The physiological and biochemical correlates found for the environmentally modified aggressive behaviour may also be relevant, at least as far as they mask or mimic interstrain differences.

Relatively much attention has been paid to strain differences in the concentrations of neurotransmitter substances in the brain. The chemical characterization of the neurones involved in the regulation of aggressive behaviour is of obvious physiological as well as pharmacological and clinical importance; the interactions of transmitter levels and turnover rates with nervous activity furnish a basis for easily discernible variation in the neurotransmitter chemistry of the brain.

Some of the comparisons of transmitter levels in the brain have been made between inbred mouse strains, but in these studies no data on the aggressiveness of the animals have been given. Since the aggressiveness estimates obtained by different authors with the same strains sometimes differ from each other (see Table 1 and Section 2 above), it is difficult, on the basis of these studies, to draw conclusions about the relations between hereditary aggressiveness and brain chemistry. The C57BL/10 mice which have often, but not always, been found to be more aggressive than BALB/c mice, and which also differ from the latter in other behavioural respects, contained less 5-HT (serotonin, 5-hydroxytryptamine) in their brain stem (Maas, 1962) and more NA (noradrenaline) in their hippocampus and pyriform cortex (Sudak and Maas, 1964). According to Al-Ani et al. (1970), C57BL mice which are more active and less reactive to mild stress than A2G mice, also showed a higher dopamine (DA) level in their brain stem than A2G mice, while the NA and 5-HT levels were similar. The gamma-aminobutyric acid (GABA) production was higher in C57BL mice and the acetylcholinesterase (AChE) activity higher in A2G mice. Unfortunately, there are no estimates available on the aggressiveness of the A2G mice. Karczmar and Scudder (1969) extended their earlier neurochemical and behavioural comparisons between three mouse strains and three other small rodent species to cover also the aggressive behaviour of these animals. They suggest that aggression in groups, exploration, good learning ability, mobility, and flexibility all may form a behavioural complex concomitant with relatively low brain levels of catecholamines and acetylcholine (ACh). As Karczmar and Scudder (1969) point out, the localization of the differences in the distribution of the neurotransmitter substances, as well as measurements of their turnover rates, are needed for further conclusions about the neurochemistry of the mechanisms of aggressive and other behaviour.

In addition to the neurochemical variation, variations in the functions of the adrenals and the gonads would seem to be of interest for the study of aggressive behaviour. Vale et al. (1971), studying male mice of five inbred strains, found a high positive correlation between aggressiveness and adrenal weight, while the testis and seminal vesicle weights were not related to aggressiveness but to social grooming. The relation between aggressiveness and adrenal weight probably reflects the fact that fighting-stress activates adrenocortical functions.

The advantages of using animals selectively bred for aggressiveness and non-aggressiveness for biochemcial and physiological comparisons are obvious. Not only are these animals selected exclusively on the basis of these behavioural features and are thus expected to show random variation for traits which are not linked with the relevant genes, but their aggressiveness is also to some degree predictable, so that previous disturbing aggressiveness tests are not always needed.

Using male animals from the more recent generations of selective breeding ($S_{13} - S_{15}$), Lagerspetz et al. (1968) studied line differences in 5-HT and NA concentration in the forebrain and brain stem. The animals of the aggressive strain (TA) showed a higher concentration of NA in the brain stem, and a lower concentration of 5-HT in the forebrain than the animals of the non-aggressive strain (TNA). The animals had been living in isolation since weaning. No significant interstrain differences were found in the activity of AChE in cortical tissue and the hypothalamus of mice of the aggressive and non-aggressive line (Lagerspetz, unpublished observations).

Individual caging of mice is known to render them more aggressive (see Section 5, below). This method has been used to make laboratory mice aggressive, e.g. for use in the pharmacological screening of psychoactive drugs (Yen et al., 1959). Because isolation also makes mice more susceptible to the toxic effects of amphetamine, aggressiveness has been associated with increased amphetamine toxicity (Consolo et al., 1965; Welch and Welch, 1966). However, we failed to find a significant difference in the toxicity of d-amphetamine, when males from generations $S_{16} - S_{19}$ of the TA and TNA lines were compared, although the animals differed highly significantly in their aggressiveness scores (Lagerspetz and Lagerspetz, 1971b). Thus, high amphetamine toxicity is not invariably connected with high aggressiveness in mice. It seems more probable that the increased amphetamine toxicity is associated with changes in other variables than aggressiveness produced by isolation; some of the neurochemical changes produced by isolation do not seem to be related to aggressiveness at all.

Some endocrinological parameters were also studied in the TA and TNA animals (Lagerspetz et al., 1968). No line difference was found in adrenal weight, although the TA animals had slightly more adrenaline in their adrenal glands. For different generations, the mean weights of the testes were 26 to 32 per cent larger in the aggressive animals than in those of the non-aggressive line. No differences were found in sexual behaviour or in the weights of the seminal vesicles. Moreover, since histological examination of the testes did not reveal any differences in the number and size of the interstitial cells, the levels of androgen production are probably equal in the lines studied. On the other hand, the amount of tubular tissue was larger in the testes of the TA mice, which may indicate a difference in production or release of follicle stimulating hormone in the pituitary. This might, in turn, depend on neurochemical differences in the hypothalamic-pituitary system.

It is interesting to compare the neurochemical results obtained through the comparisons of the mouse lines selected for aggressiveness with the results of similar comparisons between animals of which the aggressiveness has been altered by specific social, surgical, or chemical manipulation. Eleftheriou and Church (1968) subjected isolated mice to repeated defeats by means of short exposures to trained fighters and analyzed the changes in the 5-HT and NA concentrations in the amygdala, hypothalamus and frontal cortex after different numbers of encounters. After 2 days, with two fights of 5 minutes each day, the subjects showed decreased 5-HT levels, whereas the NA concentrations had increased. Because the very aggressive trained fighters stimulate the aggressiveness of isolated animals during the first encounters, it may be assumed that the aggressiveness of the subjects was still at a high level, although no behavioural data were given. After 4 days with two fights a day (and also after 8 and 16 days), the situation was reversed. Now the 5-HT concentrations were increased in the amygdala and in the hypothalamus above control level, while the NA levels were much reduced in these parts of the brain. In this phase of the experiment, the aggressiveness of the subjects had certainly already reached a low level (see Lagerspetz, 1964, pp. 58–69). These neurochemical changes are possibly associated with changes in aggressiveness caused by the repeated encounters with trained fighters and resemble the differences found between genetically aggressive and non-aggressive animals (Lagerspetz et al., 1968), i.e. high 5-HT levels in the entire forebrain or amygdala and low levels of NA in the brain stem or hypothalamus are associated with non-aggressiveness, either genetically determined or learned.

In addition, Karli, Vergnes and Didiergeorges (1969) found that the removal of the olfactory bulbs tends to convert non-mouse-killing rats to mouse-killers with a simultaneous decrease of the 5-HT content of the amygdala. Also, injections of parachlorophenylalanine, which inhibit the endogenous synthesis of 5-HT and thus cause a decrease in the 5-HT level in the brain, were effective in inducing mouse-killing in non-killing animals.

An interesting observation on the induction of aggressive behaviour by neurochemical manipulation was made by Lycke et al., (1969). Mice infected with Herpes simplex virus will develop encephalitis during which the synthesis of both dopamine (DA) and 5-HT in the brain is increased. If parachlorophenylalanine is given to these animals, their 5-HT synthesis is decreased but the synthesis of DA remains unaffected. These injected animals behave aggressively and become even more aggressive if L-DOPA, the precursor of dopamine, is given to them.

In conclusion, the neurochemical findings in the selected mice and mice affected by defeats or by surgical or chemical manipulation all point to a common neurophysiological basis for aggressiveness, whether genetically or environmentally controlled. Apparently, the aggressive behaviour is dependent on a balance between the activities of serotoninergic and, on the other hand, adrenergic and/or dopaminergic tracts in the brain. Many problems await their solution in this field.

13.5. Interaction of genotype and environment in the development of aggressiveness

Analyses of environmental effects can be divided into four categories according to the time period during which the factors affect the organism: 1) fetal and neonatal influences, 2) experiences in infancy, 3) experiences immediately preceding the test trials, and 4) stimulation in the test situation.

Investigations into the interaction between prenatal environmental effects and genetic factors require transplantations of embryos from one strain to the other. Another method which, however, does not exclude effects of sex-linkage, is the reciprocal crossing of selected lines and the subsequent comparison of both types of F_1 progeny (Broadhurst, 1961). No such experiments have yet been performed in the field of aggression research.

As for neonatal influences, Vale et al. (1972) studied the interaction of genetic and hormonal factors in the development of aggressiveness. Aggressiveness has been found to be strongly affected by neonatal androgen in mice; castration on the first day of life leaves them much less aggressive than the controls, even following testosterone supplementation in adulthood. Androgen, administered neonatally, has earlier been shown to increase aggressive behaviour in adult female mice. Vale et al. showed that the aggressiveness of the females was increased to a degree commensurate to their genetic constitution. They compared the effects of testosterone propionate injections on day 3 of life in females which were taken from strains known to differ in male aggressive behaviour, viz. C57BL/6, BALB/c, and A. The tests consisted of observing the aggressive reactions of the females directed toward a male attempting to mate and their reactions in a dangling test. The effects of the treatment were seen only in BALB/c and C57BL/6 females, which became nearly as aggressive as the males. The males of these two strains were more aggressive than those of A strain. The effect upon the A females was negligible. The authors interpret the results as being due to "activation of genes ordinarily repressed in females during the period of maximal 'organizational' effects of sex hormone ... The same genes, whose products serve to organize certain neural structures in a masculine direction, would normally be active in males because of the presence of testicular androgens" (ibid., p. 330).

Cross-fostering experiments cannot be regarded as an approach to neonatal influences because they are extended until the age of weaning. Fredericson (1952) moved soon after birth 43 mice of the aggressive C57BL/10J strain to mothers of a submissive strain, BALB/cScJ, and 60 offspring of the latter strain to foster mothers of the aggressive strain. The objective was to investigate possible maternal effects. The results showed that the high aggressiveness of the offspring of the former strain persisted when they were fostered to mothers of the submissive strain, and that the offspring of the submissive strain still behaved submissively when fostered to aggressive mothers. No tests of the aggressiveness of the foster mothers were carried out, however. It is a known fact that female mice of all strains do not usually display much aggression.

Uyeno (1960) performed a breeding experiment with one generation of rats selected for dominance and submissiveness. The offspring of six matings of dominant and six matings of submissive parents were divided such that each mother nursed four pups, none of which was her own. The following groups were obtained: (a) offspring of dominant parents nursed by dominant mothers, (b) offspring of dominant parents nursed by submissive mothers, (c) offspring of submissive parents nursed by dominant mothers, and (d) offspring of submissive parents nursed by submissive mothers. The dominance of the animals was tested in food competition situations. The results showed that the offspring of dominant parents were significantly more dominant than the offspring of submissive parents, when fostered to either kind of mother. One interesting feature was seen in the offspring from dominant parents: the more dominant were those that had been fostered to the submissive mothers and not those fostered to the dominant ones. This might be explained by the finding that defeat and failure in competition tend to reduce dominance and aggression (see below). If there has been competition over food in the living cages, the dominant mothers probably won from the pups they were nursing, whereas the pups nursed by submissive mothers might have had experiences of winning.

When considering the results of Fredericson and those of Uyeno, the question arises whether the aggressive and the submissive mothers did in fact treat the pups differently in respects relevant to the development of aggressiveness. Attention was paid to this question in a cross-fostering experiment by Lagerspetz and Wuorinen (1965) with the TA and TNA mouse lines. The mothers of the experimental groups were eight female mice of the aggressive and eight of non-aggressive line, taken from S_7. These were mated with males of the respective lines. The pups were exchanged on the day of birth so that every mother in the experimental group nursed foster pups. In the two control groups, the offspring were nursed by their own mothers. The nursing behaviour of all mothers was observed. Carrying and retrieving the young occurred significantly more frequently among TA mothers, whether they were nursing their own offspring or TNA pups. Trampling on the pups was more common in TNA mothers. It thus seems that the pups received different treatments. The genotypic constitution of the offspring influenced the mothers' behaviour: licking and carrying occurred more often with their own offspring than with fostered pups. Both in the control and the experimental groups, the offspring of the aggressive line was significantly more aggressive than the offspring of the non-aggressive line. This shows that the line difference is not attributable to postnatal maternal influences. However, in both lines there appeared a slight but significant tendency for animals nursed by their own mothers to be more aggressive than those nursed by mothers of the other line. This might be a result of the greater amount of handling they receive from their mothers.

Denenberg et al. (1964, 1966) and Lagerspetz and Heino (1969) made cross-fostering studies in which mouse pups were reared by rat mothers and were housed with rats after weaning. In both investigations the rat-reared mice were less aggressive when placed together in pairs. Lagerspetz and Hieno (1969) also observed the behaviour of the mice toward rats and found a decrease in aggressiveness even in these interspecific contacts. The authors assume that these mice have learned from their foster species the habit of not fighting, because the rat strain used in the study was very peaceful. Investigations using foster animals with different levels of aggression would be useful for answering the question of early learning of inter- and intraspecific aggression.

Handling by humans at an early age influences later aggressiveness of rats and mice. Levine (1959) found that the aggressiveness of C57BL/10 mice increased as a result of handling in infancy. Aggressiveness was measured as latencies to initiate fights. In other investigations it has been found that emotionally non-reactive individuals are more aggressive than emotionally reactive ones (Hall and Klein, 1942). Levine assumes that the non-handled subjects have a lower threshold of emotional susceptibility, i.e. they are more fearful. According to him, handling causes the animals to be emotionally more stable and thus more aggressive. Since the experiments of Levine, doubts have arisen about how early handling will affect later aggressiveness. In some experiments with mice, a decrease in aggressiveness has been found after handling of young animals, but in other cases results have been obtained that are in accordance with those of Levine.

Ginsburg's (1967) study mentioned above is extensive but the results are difficult to interpret. Aggressiveness was measured as the number of fights initiated by the subjects, which were handled and non-handled animals from strains C57BL/10, C3H, and C-albino (ibid., p. 148). This was the order of aggressiveness of the strains when non-handled, but handling decreased the number of fights initiated by the C57BL/10 mice, having no effect on the fighting of the other more peaceful strains. Aggressiveness was also measured using latencies to attack and a few other behavioural features (ibid., p. 150). This time six strains were used: HS, DBA/2, DBA/1, A/J, C57BL/10, and C57BL/6, of which the first three were, when non-handled, the most aggressive ones. Here handling reduced aggressiveness in the three most aggressive strains, while it augmented it in some of the others, for instance C57BL/10 (cf. Levine, 1959). The only consistency to be found is that the strains that happen to be more aggressive lose their aggressiveness when handled, whereas the least aggressive strains tend to show more fighting as a result of handling.

Porter's (1972) experiments seem to fit into this scheme, because he found that BALB/c mice – which in his experiments originally were less aggressive C57BL/10 animals – became more aggressive when handled. Porter investigated the differential effects of infantile handling at the age of 2–16 days on dominance behaviour in these two strains of mice. This was not a study of aggressive behaviour proper, but of dominance relationships in interstrain contacts. "Dominance was operationally defined as occurring when one animal repeatedly attacked the other with impunity, while the subordinate animal adopted a submissive posture or fled from its attacker." (ibid., p. 417). In encounters between non-handled animals, the C57BL/10 mice were dominant over the BALB/c mice. The BALB/c strain was, however, more susceptible to behavioural modification through handling. Their dominance increased as compared with non-handled animals, while such manipulations had little effect on the dominance behaviour of the C57BL/10 mice. If anything, handled mice of this strain appeared to be somewhat subordinate to non-handled strain-mates.

Ginsburg (1967, p. 150) also found that handling is more effective when executed at certain age periods. This period was somewhat different for different strains. These results, however, will become more interesting as soon as the question of increases versus decreases in aggressiveness as a result of handling is settled. It is conceivable that handling may exert different effects on genetically aggressive and non-aggressive animals. When fearful, the former will react with aggression and the latter will show the typical "freezing" behaviour, described for instance by Levine (1959).

The effects of isolation and fighting experiences on aggressiveness have been investigated by many. Since raising of mice and rats in complete isolation prior to weaning is technically difficult, all studies of such genotype-environment interactions have until now used adult animals.

Aggressiveness of mice and rats can be altered by repeated encounters with highly aggressive or non-aggressive partners (Allee and Ginsburg, 1941; Ginsburg and Allee, 1942; Scott, 1946, 1958). If the animals are defeated on successive occasions by an aggressive partner, they become non-aggressive. On the other hand, the animal's aggressiveness may be elevated by giving it several opportunities to win from a less aggressive partner; see also Ginsburg and Allee (1942). Low-ranking albinos could be trained to be dominant and high-ranking blacks could be conditioned toward low dominance.

Lagerspetz (1964) used this method to modify the aggressiveness of her TA and TNA mice. After daily contests of 2 minutes, the aggressiveness of the test animals was drastically changed within a few days in the two expected directions. These effects of defeats or victories were still present after one week of isolation, but the aggressiveness returned to the original level in two weeks.

Using the selected mice of generation S_{15}, Lagerspetz and Lagerspetz (1971a) found that the levels of aggressiveness and sexual activity increased in young previously isolated mice after a few encounters with non-aggressive males or with receptive females. This result again demonstrates the role of learning in the development of emotional behaviour.

Social deprivation is known to enhance aggressiveness, while grouping reduces it (Allee, 1942; Scott and Fredericson, 1951; Bourgault et al., 1963; Banerjee, 1972). Neither the psychological nor the physiological mechanisms controlling this phenomenon are fully understood yet (see Lagerspetz, 1971).

Lagerspetz and Lagerspetz (1971a) kept 88 male TA and TNA mice of generation S_{19} in groups of three or more from weaning up to the age of 8 months. At this age, neither the TA nor the TNA animals showed any signs of aggression toward a strange male. After these tests, the animals were placed in individual cages. After one week of isolation, one-third of the TA animals showed signs of aggressiveness (biting, tail-rattling) in the aggression test, while no TNA animals showed this type of behaviour. After 8 weeks of isolation, about 60 % of the TA males displayed aggressive behaviour. The authors interpret the effects of grouping as due to the unlearning of aggressive responses, reinforced by punishment received from cage mates. Isolation seems to cause a return to the inter-individually variable aggressiveness level, which is determined by genetic variation and by early experience. Apparently, recent social experience can to a great extent mask the hereditary differences in behaviour.

Another factor that has been shown to lower intraspecific aggressiveness in male mice is previous sexual activity (Lagerspetz and Hautojärvi, 1967). On the other hand, previous fighting experience was found by the same authors to reduce very significantly the sexual behaviour of sexually inexperienced males. This decrease in sexual responses was only partially due to aggressive behaviour directed towards the female. In another study, Hautojärvi and Lagerspetz (1968) observed that inhibition of aggression caused by successive defeats did not inhibit sexual behaviour, not even in inexperienced animals. These effects were all found in socially naive animals which had been reared in isolation since weaning. After sexual and aggressive activities had been carried out over a longer time-span, they no longer influenced each other. It is concluded that the results do not substantiate the idea that aggressive and sexual behaviour are controlled by a common general state of arousal which is channelled by environmental stimuli. In the learning, but not in the inhibition, of these two types of activity there seems to be a certain amount of mutual negative interference.

The genetic constitution plays a part in determining which conditions function as rewards or punishment for the animal. The opportunity to fight acts as a reward for genetically aggressive animals but not for non-aggressive ones, as was shown by Tellegen and Horn (1972) and supported by the results of Lagerspetz (1964). This is also reflected in the findings of Ginsburg and Allee (1942) and Lagerspetz (1964): it is easier to condition aggressive animals to become non-aggressive through defeats, than to condition non-aggressive mice to become aggressive by using opportunity to fight as reinforcement. Thus, genetic differences can interact with the learning of emotional responses not only through differential learning capacities but also through differences in motivational variables.

Fredericson et al. (1955) studied the effects of the presence of a female in estrus on the fighting behaviour in two inbred mouse strains. The data indicated that when the female was put into the pen, the C57BL/10 mice interrupted their fight in favour of attempts to copulate. The BALB/cSc mice were not affected by the presence of the female and continued to fight. This experiment can be regarded as a kind of preference test to decide the relative predominance of aggressive and sexual tendencies. The authors conclude that the balance of sexuality and aggressiveness is influenced by the hereditary background of the organism.

13.6. Conclusion

Future studies on interactions between genetic and environmental factors will probably promote the understanding of the development of behaviour. These interactions take place on different functional levels: neuro-ontogeny, perceptual processes, motivation, learning, and social phenomena. For instance, Miller's (1969) results on the learning of autonomic responses may have important implications for aggression research.

Similar behavioural phenocopies can be produced by different treatments. As described earlier (Lagerspetz, 1969; Lagerspetz and Lagerspetz, 1971a), non-aggressive mice could be produced in four different ways: by selective breeding for non-aggressiveness, by having aggressive animals defeated, by grouping aggressive animals for long time-periods, and by giving naive animals (even descendants of the aggressive TA line) sexual experiences. The latter three methods imply that even in small laboratory mammals like mice social experiences may cause complex and psychologically interesting effects and interactions. They also show that a genetic disposition for aggressiveness can be modified by different social factors. In man, the development of behaviour rests to a larger extent on learning than in other animals. It seems probable that in humans environmental influences account to a high degree for the variation in aggressiveness.

13.7. Summary

Literature on selective breeding and on strain comparisons concerning aggressive behaviour and its physiology and biochemistry is reviewed. Special attention is paid to studies where the effects of environmental factors on animals with a controlled genetic background are studied, because this is seen as a means to estimate the modifiability of the traits in question. For instance, isolation and social learning have been observed to modify significantly even a genetically determined variation of aggressiveness level: isolation increases the aggressiveness of animals selectively bred for aggression, and punishment through defeats abolishes temporarily aggressive behaviour in animals selectively bred for aggressiveness. Sexual contacts have also been found to decrease the aggression level of these animals.

References

Al-Ani, A. T., G. Tunnicliff, J. T. Rick and G. A. Kerkut (1970): GABA production, acetylcholinesterase activity and biogenic amine levels in brain for mouse strains differing in spontaneous activity and reactivity. *Life Sci.*, **9**, 21.

Allee, W. C. (1942): Group organization among vertebrates. *Science*, **95**, 289.

Banerjee, U. (1971): An inquiry into genesis of aggression in mice induced by isolation. *Behaviour*, **40**, 86.

Bauer, F. J. (1956): Genetic and experiential factors affecting social relations in male mice. *J. comp. physiol. Psychol.*, **49**, 359.

Beeman, E. A. (1947): The effect of male hormone on aggressive behaviour of mice. *Physiol. Zool.*, **20**, 373.

Bourgault, P., A. Karczmar and C. L. Scudder, (1973): Contrasting behavioural, pharmacological, neurophysiological, and biochemical profiles of C57BL/6 and SC-I strains of mice. *Life Sci.*, **8**, 533.

Broadhurst, P. L. (1960): Experiments in psychogenetics. Applications of biometrical genetics to inheritance of behaviour. In: Experiments in Personality. Eysenck, H. J., ed. (Routledge and Kegan Paul, London.) pp. 3–102.

Broadhurst, P. L. (1961): Analysis of maternal effects in the inheritance of behaviour. *Anim. Behav.*, **9**, 129.

Coburn, C. A. (1922): Heredity of wildness and savageness in mice. *Behav. Monogr.*, **4**, 1.

Collins, R. A. (1970): Aggression in mice selectively bred for brain weight. *Behav. Genet.*, **1**, 169.

Consolo, S., S. Garattini and L. Valzelli (1965): Amphetamine toxicity in aggressive mice. *J. Pharm. Pharmacol.*, **17**, 53.

Craig, J. V. and R. A. Baruth (1965): Inbreeding and social dominance ability in chickens. *Anim. Behav.*, **13**, 109.

Craig, J. V., L. L. Ortman and A. M. Guhl (1965): Genetic selection for social dominance ability in chickens. *J. Anim. Behav.*, **13**, 114.

Dawson, W. M. (1932): Inheritance of wildness and tameness in mice. *Genetics*, **17**, 296.

Denenberg, V. H., G. A. Hudgens and M. X. Zarrow (1964): Mice reared with rats: modification of behaviour by early experience with another species. *Science*, **143**, 380.

Denenberg, V. H., G. A. Hudgens and M. X. Zarrow (1966): Mice reared with rats: effects of mother on adult behavior patterns. *Psychol. Rep.*, **18**, 451.

Eleftheriou, B. E. and R. L. Church (1968): Brain levels of serotonin and norephinehrine in mice after exposure to aggression and defeat. *Physiol. Behav.*, **3**, 977.

Eleftheriou, B. E. and J. P. Scott, eds. (1971): The Physiology of Aggression and Defeat. (Plenum Press, New York.)

Ewbank, R. and M. J. Bryant (1972): Aggressive behaviour amongst groups of domesticated pigs kept at various stocking rates. *Anim. Behav.*, **20**, 21.

Fredericson, E., A. W. Story, N. L. Gurney and K. Butterworth (1955): The relationship between heredity, sex, and aggression in two inbred mouse strains. *J. genet. Psychol.*, **87**, 121

Ginsburg, B. E. (1967): Genetic parameters in behavioral research. In: Behavior-Genetic Analysis. J. Hirsch, ed. (McGraw-Hill, New York) pp. 135–154.

Ginsburg, B. and W. C. Allee (1942): Some effects of conditioning on social dominance and subordination in inbred strains of mice. *Physiol. Zool.*, **15**, 485.

Guhl, A. M., J. V. Craig and C. D. Mueller (1960): Selective breeding for aggressiveness in chickens. *Poultry Sci.*, **39**, 970.

Hall, C. S. (1938): The inheritance of emotionality. *Sigma Xi Quart.*, **26**, 17.

Hall, C. S. and S. J. Klein (1942): Individual differences in aggressiveness in rats. *J. comp. Psychol.*, **33**, 371.

Hautojärvi, S. and K. M. J. Lagerspetz (1968): The effects of social induced aggressiveness or non-aggressiveness on the sexual behaviour of inexperienced male mice. *Scand. J. Psychol.*, **9**, 45.

Heimstra, N. W. (1961): Effects of chlorpromazine on dominance and fighting behaviour in rats. *Behaviour*, **18**, 313.

Hirsch, J. (1967): Behavior-genetic analysis. In: Behavior-Genetic Analysis. J. Hirsch, ed. (McGraw-Hill, New York) pp. 416–435.

Karczmar, A. G. and C. L. Scudder (1969): Aggression and neurochemical changes in different strains and genera of mice. In: Aggressive Behaviour. Proceedings of the International Symposium on the Biology of Aggressive Behaviour, Milan, 1968. S. Garattini, and E. B. Sigg, eds. (Excerpta Medica, Amsterdam.)

Karli, P., M. Vergnes and F. Didiergeorges (1969): Rat-mouse interspecific aggressive behaviour and its manipulation by brain ablationand by brain stimulation. In: Aggressive Behaviour, Proceedings of the International Symposium on the Biology of Aggressive Behaviour, Milan 1968, S. Garattini, and E. B. Siggs, eds. (Excerpta Medica, Amsterdam).

King, J. A. (1957): Relationship between early social experience and adult aggressive behavior in inbred mice. *J. genet. Psychol.*, **90**, 151.

Lagerspetz, K. M. J. (1961): Genetic and social causes of aggressive behaviour in mice. *Scand. J. Psychol.*, **2**, 167.

Lagerspetz, K. M. J. (1964): Studies on the aggressive behaviour of mice. *Ann. Acad. Sci. Fenn., Ser. B*, **131**, 3. 1.

Lagerspetz, K. M. J. (1969): Aggression and aggressiveness in laboratory mice. In: Agressive Behaviour. Proceedings of the International Symposium onthe Biology of Aggressive Behaviour. Milan. 1968, S. Garattini, and E. B. Sigg, eds. (Excerpta Medica, Amsterdam).

Lagerspetz, K. M. J. and S. Hautojärvi (1967): The effect of prior aggressive or sexual arousal on subsequent aggressive or sexual reactions in male mice. *Scand. J. Psychol.*, **8**, 1.

Lagerspetz, K. M. J. and T. Heino (1970): Changes in social reactions resulting from early experience with another species. *Psychol. Rep.*, **27**, 255.

Lagerspetz, K. M. J. and K. Y. H. Lagerspetz (1971a): Changes in the aggressiveness of mice resulting from selective breeding, learning, and social isolation. *Scand. J. Psychol.*, **12**, 241.

Lagerspetz, K. M. J. and K. Y. H. Lagerspetz (1971b): Amphetamine toxicity in genetically aggressive and non-aggressive mice. *J. Pharm. Pharmacol.*, **23**, 542.

Lagerspetz, K. M. J. and R. Portin (1968): Simulation of cues eliciting aggressive responses in mice at two age levels. *J. genet. Psychol.*, **113**, 53.

Lagerspetz, K. M. J. and K. Wuorinen (1965): A cross-fostering experiment with mice selectively bred for aggressiveness and non-aggressiveness. Reports from the Institute of Psychology, University of Turku, **17**, 1–6.

Lagerspetz, K. Y. H. and K. M. J. Lagerspetz (1974): The search for mechanisms of behaviour through selective breeding experiments. In: Actual Problems of Animal Behaviour Genetics. Fyodorov, V. K., ed. (Physiological Section of the Academy of Sciences of the U.S.S.R., Leningrad) (in press).

Lagerspetz, K. Y. H., R. Tirri and K. M. J. Lagerspetz (1968): Neurochemical and endocrinological studies of mice selectively bred for aggressiveness. *Scand. J. Psychol.*, **9**, 157.

Levine, L., C. A. Diakow and G. E. Barsel (1965): Interstrain fighting in male mice. *Anim. Behav.*, **13**, 52.

Levine, S. (1959): Emotionality and aggressive behaviour in the mouse as a function of infantile experience. *J. genet. Psychol.*, **94**, 77.

Lycke, E., K. Modigh and B.-E. Roos (1969): Aggression in mice associated with changes in the monoamine-metabolism of the brain. *Experientia*, **15**, 951.

Maas, J. W. (1962): Neurochemical differences between two strains of mice. *Science*, **137**, 621.

Mettälä, R. (1965): A factorial study of the behaviour of mice in simulation experiments

eliciting aggressive responses. Reports from the Institute of Psychology, University of Turku, **15**, 1–9.
Miller, N. E. (1969): Autonomic learning: Clinical and physiological implications. Proc. XIX Internat. Congr. Psychol., London. (British Psychological Society, London), p. 117.
Moyer, K. E. (1971): The Physiology of Hostility, (Markham, Chicago).
Porter, R. H. (1972): Infantile handling differentially affects interstrain dominance interactions in mice. *Behav. Biol.*, **7**, 415.
Scott, J. P. (1940): Hereditary differences in social behavior (fighting of males) between, two inbred strains of mice. *Anat. Rec.*, **78**, 103 (abstr.).
Scott, J. P. (1942): Genetic differences in the social behavior of inbred strains of mice. *J. Hered.*, **33**, 11.
Scott, J. P. (1958): Aggression. (University of Chicago Press, Chicago.)
Scott, J. P. and E. Fredericson (1951): The causes of fighting in mice and rats. *Physiol. Zool.*, **24**, 273.
Scott, J. P., C. Lee and J. E. Ho (1971): Effects of fighting, genotype, and amphetamine sulphate on body temperature of mice. *J. comp. physiol. Psychol.* **76**, 349.
Southwick, C. H. (1970): Genetic and environmental variables influencing animal aggression. In: Animal Aggression. Selected readings. Southwick, C. H. ed. (Van Nostrand, New York.) pp. 213–229.
Stone, C. P. (1932): Wildness and savageness in rats of different strains. In: Studies in the Dynamics of Behaviour. Lashley, K. S. ed. (Univ. of Chicago Press, Chicago.) pp. 3–55.
Sudak, H. S. and J. W. Maas (1964): Central nervous system serotonin and norepinephrine localization in emotional and non-emotional strains in mice. *Nature*, **203**, 1254.
Tellegen, A. and J. M. Horn (1973): Primary aggressive motivation in three inbred strains of mice. *J. comp. physiol. Psychol.*, **78**, 297.
Utsurikawa, N. (1917): Temperamental differences between outbred and inbred strains of the albino rat. *J. Anim. Behav.* **7**, 111.
Uyeno, E. T. (1960): Hereditary and environmental aspects of dominant behaviour in the albino rat. *J. comp. physiol. Psychol.*, **53**, 138.
Vale, J. R., D. Ray and A. Vale (1972): The interaction of genotype and exogenous neonatal androgen: Agonistic behaviour in female mice. *Behav. Biol.*, **7**, 321.
Vale, J. R., A. Vale and J. P. Harley (1971): Interaction of genotype and population number with regard to aggressive behaviour: social grooming, and adrenal and gonadal weight in male mice. *Commun. Behav. Biol.*, **6**, 209.
Welch, B. L. and A. S. Welch (1966): Graded effect of social stimulation upon d-amphetamine toxicity, aggressiveness and heart and adrenal weight. *J. Pharmacol. exp. Ther.*, **151**, 331.
Yen, C. Y., R. L. Stagner and N. Millman (1959): Ataractic suppression of isolation-induced aggressive behaviour. *Arch. Internat. Pharmacodyn.*, **123**, 179.
Yerkes, R. M. (1913): The heredity of savageness and wildness in rats. *J. Anim. Behav.*, **7**, 11.

CHAPTER 14

Genotype and the cholinergic control of exploratory behaviour in mice

J. H. F. van Abeelen

14.1. Introduction

The house mouse (*Mus musculus* L.) has been used extensively as material for investigations in the field of behaviour genetics (see Lindzey and Thiessen, 1970). This is not surprising, because there are several reasons which make this species well-suited for this line of research. Firstly, much knowledge has been accumulated about the Mendelian genetics and cytogenetics of mice, and many selected and inbred laboratory strains (Staats, 1972) and mutant types (Green, 1966) have become available. Secondly, these animals are fairly easy to keep, they reproduce rapidly, and they are fit for physiological and pharmacological experiments. Thirdly, from a behavioural point of view, mice show a rich and varied repertoire of behaviour patterns, which has been analyzed in great detail (Eibl-Eibesfeldt, 1950 and 1958).

One of the most conspicuous features of the behaviour of mice is their investigativeness, also called curiosity or neophilia. Unlike rats which tend to be more neophobic (Barnett, 1963), mice are attracted by novel stimuli and they spend long periods of exploration when exposed to a novel environment. This tendency is expressed in acts and postures such as sniffing in the air and at objects, rearing, leaning against objects and walls, and probably also in ambulation. Rearing responses (vertical activity) and locomotion (horizontal activity) seem to be influenced by a common internal factor, viz. the level of arousal (Lát and Gollová-Hémon, 1969). With the passage of time, the animal becomes more and more familiar with its surroundings and exploratory activities occur less and less frequently. In all likelihood, the gradual decline of these nonreinforced behavioural responses reflects a special kind of learning process which is termed habituation.

The genetic analysis of any behavioural phenotype can start from two different angles; one can distinguish between the quantitative genetic aspects and the phenogenetic aspects of behaviour. The first approach is primarily concerned with the usually polygenic determination of behavioural variation as it is observed in different breeds of animals or human families. Phenotypes like ambulation and defecation rates, used as indices of emotionality in rats and mice (Broadhurst, 1969; Bruell, 1969), the activity scores of mice in running-wheels and open fields (Bruell, 1964; DeFries et al., 1970; Newell, 1970), the learning of conditioned emotional responses in mice (Henderson, 1968), and parameters of intelligence in man (Fulker, 1973) have been subjected to biometrical analyses and in some cases it has been possible to arrive at conclusions about their evolutionary histories. The other approach deals with the mechanisms through which the genotype controls the different kinds of behaviour. Behaviour depends on structures and functions of the nervous system, hormone actions, and on the properties of sense organs and muscles; the objective of each phenogenetic behavioural study is to trace these intermediate pathways from genes to traits. In the present investigation a combination of both approaches was attempted, focusing on exploration in mice.

Behavioural phenotypes can be manipulated in several ways. Indirect genetic methods are: crossing of inbred strains, followed by an analysis of different hybrid types, and selection for a behavioural characteristic, starting with a genetically heterogeneous population. In both cases, concomitant biochemical and physiological alterations may be studied and may yield information on the causation of the behaviour. Traits can also be manipulated more directly by interfering with the functions of the physical substrata of behaviour. In this respect, drugs can be particularly useful for assaying physiological and psychological differences between animals of different genotype. Inbred strains often do not respond to a given drug in the same way, as has been demonstrated, for example, by Bovet et al. (1966) for the effects of nicotine, and by Fuller (1970) for the effects of chlorpromazine and chlordiazepoxide, upon avoidance conditioning in mice. In principle it must be possible, by using drugs of which the mode and the site of action are known, to attenuate or correct inherited behavioural deviations (see also Green and Meier, 1965).

As far as the control of exploratory behaviour is concerned, Carlton (1963, 1968) has suggested that a cholinergic mechanism in the brain of rodents acts to inhibit nonreinforced orientational responses. The finding that anticholinergic drugs tend to prolong and intensify exploratory activity in mice (Oliverio, 1968; Wakeley and O'Sullivan, 1969), may be regarded as evidence that these drugs disrupt the animals' ability to habituate to novelty. This postulate of response disinhibition, however, has not remained unchallenged; Warburton and Brown (1971) have obtained results which are more consistent with the hypothesis that anticholinergics impair attention mechanisms, involving ascending cholinergic pathways. Moreover, since each of the above experiments was carried out with only one strain of mice or rats, the question of the generality of anticholinergic – induced decreases in exploration arises. Strain differences have been established for exploratory behaviour in mice (van Abeelen, 1970) and it is conceivable that these differences are mediated by a cholinergic system, the function of which is, in turn, genetically controlled. It seems furthermore worthwhile to examine also the effects of anticholinesterase drugs upon exploration in different genotypes.

The effectiveness of administering drugs in the study of central synaptic functions depends on their ability to pass the blood–brain barrier (see Herz et al., 1965). There are several tertiary ammonium compounds which act as anticholinergics or anticholinesterases and which penetrate easily into the brain. Their quaternary analogues are believed not to pass the blood–brain barrier or to do so only to a very limited extent, but they mimic any additional peripheral effects the tertiary compounds may have and can therefore serve as the appropriate controls. In order to locate the central sites of action more precisely, injections with drugs can be given intracerebrally. The hippocampus seems to be a likely candidate; it has been suggested in the literature that this brain region is involved in the control of nonreinforced behaviour (for critical reviews, see Douglas, 1967; Kimble, 1968; Routtenberg, 1971).

The present study employs the psychopharmacogenetic approach to investigate the cholinergic regulation of exploratory behaviour in two inbred mouse strains and in two lines of mice, derived from a cross between the inbreds, and selected for rearing frequency.

14.2. Methods

14.2.1. Breeding conditions

The inbred and selected mice were housed in plastic breeding cages with a metal cover and a bedding of peat dust or, in later years, wood-shavings. They had free access to laboratory food pellets and water. Their cages were cleaned every week. Due to some nursing problems, the neonate mice from most generations of the selected lines had to be fostered to lactating mothers from a random-bred stock. Weaning took place at 5 weeks. At the age of about 8 weeks, the animals were dipped in an ectoparasite-killing solution. After this, they were kept usually two or three in a cage in the air-conditioned mouse room where a normal light–dark cycle prevailed. The environment probably was not constant over time; there may have been seasonal influences. In order to avoid possible effects of the female oestrus cycle on behaviour, only male mice were used for observation.

14.2.2. Observation techniques

Three exploratory acts and locomotor activity of mice introduced into a problem-free novel situation were recorded. For this purpose, two large observation cages were used, which, except for the transparent front panes, were painted in pale green. One of these measured $133 \times 53 \times 58$ cm and, apart from its bedding of peat dust, was empty. Three light beams, directed upon three photo-electric cells, divided the floor into four equal areas. Interruption of any beam actuated a counter so that ambulation (horizontal activity) could be measured. The size of the other observation cage was $108 \times 49 \times 49$ cm; in later stages of the study, its hardboard floor was divided by white painted lines into 21 rectangles. Line crossings were scored with a hand tally counter. Attached to its back wall, an empty metal food hopper was present for exploratory object-sniffing. Single animals were placed in the centre of the cage and observed directly and continuously in one session of 15 min or, in some cases, 20 min. The frequencies of rearing, leaning against the wall, and object-sniffing (vertical activities) were registered in a standard fashion on checklists; these brief behavioural components can be recorded easily and reliably. By rearing (Figs. 14.1 and 14.2) is meant standing upright on the hind-legs, often using the tail as a support, while visually exploring the environment (scanning, reconnoitering). In this

Fig. 14.1. Rearing.

Fig. 14.2. Rearing.

Fig. 14.3. Leaning against the (glass) wall.

Fig. 14.4. Sniffing at the object.

posture, the forepaws are not in contact with any surface and sniffing movements of the nasal skin may occur. In leaning (Fig. 14.3), the animal stems itself on its hind-legs and places one forepaw or both against the wall. Leaning is not always combined with sniffing at the wall. When sniffing at the object (Fig. 14.4), the animal has approached the empty hopper; the front paws may be placed against it and the nose is held close to it or is actually touching it and movements of the nasal skin take place, the mouse itself still being on the floor. Occasionally, freezing can be seen. If in the course of an observation period an animal froze for a total duration of k min, that session was prolonged with k min.

14.2.3. Determination of enzyme activity

The brains of inbred, hybrid, and selected mice from a number of generations were assayed for acetylcholinesterase activity by means of the sensitive photometric method of Ellman et al. (1961). The whole brain, except the bulbi olfactorii, was taken from the skull, weighed, and homogenized in buffer for 3.5 min with an Ultra-Turrax homogenizer set at an indicated speed of 5.0–5.5. Of each brain two measurements were taken and averaged. At first the temperature in the cuvettes was not strictly controlled, but in later determinations it was kept constant at 35.4 °C. Naturally, measuring whole brain homogenates can only provide a crude first estimate of the enzyme activity levels in different genotypes.

14.2.4. Drug administration

Animals were injected with different dose-levels of the transmitter-blocking agent scopolamine hydrobromide or the enzyme inhibitor physostigmine sulphate, or with their quaternary congeners scopolamine methylbromide and neostigmine bromide (Fig. 14.5) which pass the blood–brain barrier only with great difficulty. The drugs were freshly dissolved in isotonic saline. Peripheral injections with a volume of 1 ml/kg were given either intravenously, in the base of the tail, or intraperitoneally 30 min prior to the behavioural observation. The i.v. injections were administered under light ether anesthesia and their success was checked by adding a pharmacologically neutral dye, Evans blue, to the solvents. For intracerebral injections a stereotaxic instrument was used, consisting of a head holder, with ear bars and snout clamp, and a manipulator which could be adjusted by a system of three

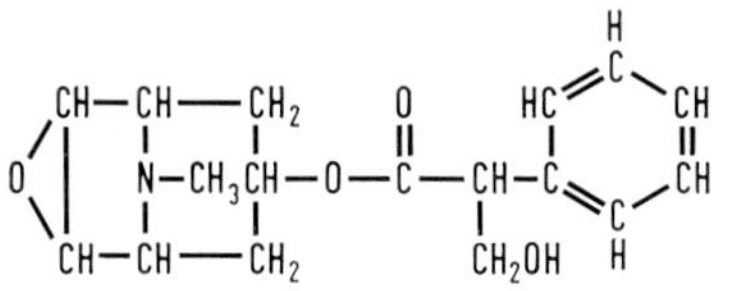

scopolamine (hyoscine)

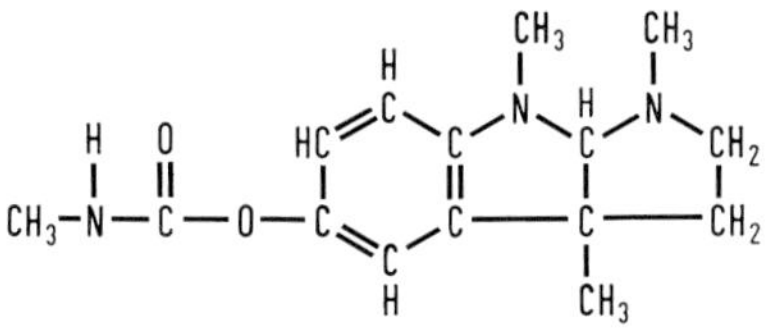

physostigmine (eserine)

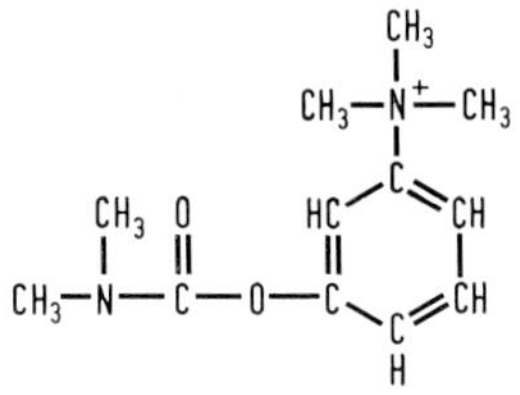

neostigmine

Fig. 14.5. Chemical structure of the anticholinergic drug scopolamine and the anticholinesterases physostigmine and neostigmine.

microscrews and to which a microsyringe was attached. The needle was introduced, under light ether narcosis, by piercing the skin, the skull, and the cerebral cortex. Fifteen min before observation, the animals received a total volume of 0.8 μl of fluid into the left dorsal hippocampus. Mice injected unsuccessfully (31 %), as judged by the distribution of Evans blue staining in their brains immediately after completion of the behavioural tests, were excluded from the data. Only methylscopolamine and neostigmine were used for intrahippocampal injections.

14.2.5. Statistical evaluation

For the statistical testing of drug effects, strain and line differences, and correlations, nonparametric procedures were employed exclusively (see Siegel, 1956; Sachs, 1969).

14.3. Results and discussion

14.3.1. Inbred strains and hybrids

14.3.1.1. Behavioural differences. The origin of the present investigation was a strain comparison made in 1964 between the highly inbred mouse strains C57BL/6J and DBA/2J and their F_1 hybrid. During these tests a novel object was introduced into the observation cage; this was done by lowering a wooden cube by means of a pulley quietly into the cage 15 min after the beginning of the session. Table 14.1 shows the results of this experiment. Strain C57BL/6 is characterized by its high locomotor activity count and its high scores for rearing, sniffing, and climbing, as compared to the low-scoring DBA/2 mice, whereas the DBA/2 strain demonstrated higher grooming frequencies. The strains did not differ with regard to leaning against the wall and the novel object. The values of the F_1 indicated dominance for some behavioural components (rearing, sniffing, grooming, and locomotor activity) and heterosis for others (leaning and climbing).

In a subsequent experiment the attention was concentrated on rearing behaviour, a response which is clearly oriented in an upward direction, and on locomotor activity, which takes place in a horizontal plane. In order to determine whether the strain differences for these behaviours are reproducible characteristics which are amenable to genetic manipulation, seven consecutive generations of C57BL/6JNmg and DBA/2JNmg mice were studied. The strains were obtained from The Jackson Laboratory,

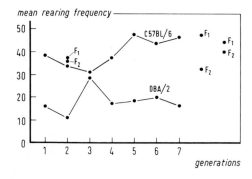

Fig. 14.6. Mean rearing frequencies exhibited by male mice from two inbred strains and their F_1 and F_2 hybrids. The difference between the parental strains, as evaluated by means of the Mann-Whitney U test, was nonsignificant only in the third generation.

TABLE 14.1

Mean frequencies of behavioural components and mean locomotor activity counts observed in male mice from two inbred strains and their F_1 hybrid, and the results of the statistical tests.

Behaviour	Mean			P (Mann-Whitney U test)		
	DBA/2	C57BL/6	$F_1(B6 \times D2)$	DBA/2 v. C57BL/6	DBA/2 v. $F_1(B6 \times D2)$	C57BL/6 v. $F_1(B6 \times D2)$
Rearing	7.8	27.6	32.0	< 0.01	< 0.01	n.s.
Leaning against wall	26.1	27.4	46.5	n.s.	< 0.01	< 0.01
Sniffing at object	13.9	25.6	26.6 *	< 0.01	< 0.01	n.s.
Climbing object	2.7	8.7	13.0	< 0.01	< 0.01	n.s.
Leaning (novel object)	2.2	3.3	4.8	n.s.	< 0.01	n.s.
Sniffing (novel object)	2.6	7.2	7.8 *	< 0.01	< 0.01	n.s.
Climbing (novel object)	0.3	2.2	4.1	< 0.05	< 0.01	< 0.05
Grooming	4.0	2.1	5.0	< 0.01	n.s.	< 0.01
Locomotor activity	43.7	142.4	135.1	< 0.01	< 0.01	n.s.

N = 20 in each group, except for cases marked by * (N = 10); for the activity counts, N = 12. Duration of observation: 20 min per animal, except for components directed towards the novel object (5 min). Observations carried out in 1964 at The Jackson Laboratory, Bar Harbor, Maine, U.S.A.

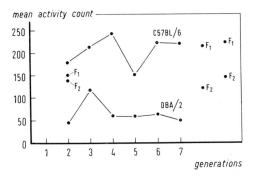

Fig. 14.7. Mean locomotor activities exhibited by male mice from two inbred strains and their F_1 and F_2 hybrids.

Bar Harbor, Maine, U.S.A., in 1965 and since then propagated by sib-mating. At three points in this experiment, F_1 and F_2 generations were produced by crossing the inbreds. As is evident from Fig. 14.6, the parental strains differ consistently with respect to rearing behaviour, except for the third generations where the difference is nonsignificant (this lack of significance is largely due to a chance deviation in the low strain). The frequency of rearing by C57BL/6 mice is generally more than twice that of DBA/2 mice. The performances of the hybrids suggest dominance of alleles favouring

TABLE 14.2

Significance levels of correlations between rearing frequency and locomotor activity observed for 15 min in inbred and crossbred male mice*.

Generation	DBA/2		C57BL/6	
	N	P	N	P
2	10	n.s.	10	n.s.
3	10	< 0.01	10	n.s.
4	10	< 0.01	10	n.s.
5	10	< 0.05	10	n.s.
6	10	< 0.05	10	n.s.
7	10	< 0.01	10	n.s.
	F_1		F_2	
B6 × D2	10	n.s.	27	< 0.01
D2 × B6	10	n.s.	20	< 0.05
B6 × D2	20	n.s.	25	< 0.01

* Correlations were calculated according to Spearman's ranking method; all significant correlations were positive.

high scores. The data for locomotor activity are given in Fig. 14.7. In all generations observed, a significant strain difference is found between the high-scoring C57BL/6 animals and the low-scoring DBA/2 animals. Here, too, dominance of alleles contributing to high activity is indicated. Correlations between rearing frequency and locomotor activity count were calculated (see Table 14.2). Of the (nonsegregating) parentals and F_1, only the low-scoring DBA/2 strain shows in most generations a positive correlation, which must have been brought about by environmental factors to which only this particular genotype is vulnerable. It seems improbable that the positive correlations found between rearing and locomotion in the three segregating, relatively high-scoring, F_2 populations are of purely environmental origin. Underlying genetic factors might operate in the control of some common internal system which is responsible for the regulation of both vertical and horizontal exploratory activities (see also Lát and Gollová-Hémon, 1969).

14.3.1.2. Acetylcholinesterase-activities. An attempt was made to relate these genetically-determined differences in exploratory behaviour to cholinergic synaptic functioning. The data concerning this aspect of brain chemistry obtained from two generations of the inbred strains and the third group of crossbred mice, are entered in Table 14.3. The parental strains clearly differ in enzyme activity of whole brain homogenates, the DBA/2 mice exhibiting the highest values. Brain weights were also measured: C57BL/6 mice turned out to have heavier brains (407 ± 10 mg) than DBA/2 animals (373 ± 9 mg); this difference is significant ($P < 0.002$; Mann-Whitney test).

TABLE 14.3

Means, standard deviations, and significance levels of acetylcholinesterase activities (μmoles/min) in whole brain homogenates of male mice from two inbred strains and their hybrids.

Group	N	Mean	SD	P (Mann-Whitney)
C57BL/6 (gen. 13)	16	6.06	0.51	< 0.02
DBA/2 (gen. 13)	13	6.38	0.28	
C57BL/6 (gen. 15)	5	5.54	0.35	< 0.03
DBA/2 (gen. 15)	5	6.07	0.37	
F_1 (B6 × D2)	16	5.60	0.52	n.s.
F_2 (B6 × D2)	24	5.51	0.45	

These differences in enzyme functioning are purely quantitative in character; Smissaert (personal communication), using the extensive methods described by Smissaert and Hitman (1971), failed to find any structural differences between the acetylcholinesterases of DBA/2 and C57BL/6 animals. Similar differences in acetylcholinesterase (AChE) activity have been reported by Ebel et al. (1973) for homogenized temporal cortices of DBA and C57 mice (substrains not specified).

Looking at the behavioural scores of the inbreds, it would seem that the amount of exploration and the activity level of AChE in the brain are inversely related. By itself, such a negative association may be produced by chance alone. However, although it is somewhat precarious to compare the values of the parentals, on the one hand, with those of the hybrids, on the other (on account of the long time-intervals between the measurements given in Table 14.3), the hybrid types again seem to be more similar to the C57BL/6 mice than to the DBA/2 mice, suggesting dominance for low AChE activity, as was observed for high exploration score. Within the groups considered, no significant correlations were found with Spearman's test between the enzymological parameter and the behavioural characteristic, but at the present stage a possible connection between the two variables cannot be excluded.

An obvious drawback of using tissue homogenates is that one cannot distinguish between the bound pool of AChE which is not directly involved in synaptic transmission and the functional free portion of the enzyme.

14.3.1.3. Responses to drugs. The next step in the analysis of the mechanism controlling exploratory behaviour was a series of three experiments in which mice from the two inbred strains were subjected to treatments with anticholinergic or anticholinesterase drugs and compared with mice receiving saline alone. In the first experiment, four doses of scopolamine and four doses of physostigmine were given i.v. The left part of Table 14.4 shows that scopolamine affected the frequencies of rearing, leaning, and object-sniffing in C57BL/6 mice; the influences of this drug upon the scores of the DBA/2 animals approached significance. From Fig. 14.8 it appears that the anticholinergic exerts opposite effects in the two genotypes. Rearing performance dropped sharply in the C57BL/6 strain and it tended to increase in the DBA/2 strain, so that the original strain difference was reversed. (The low rearing score of 20.5 in the DBA/2 mice at dose-level 5 mg/kg is probably a chance deviation; in an earlier pilot study a mean score of 31.8 was found for them at that dose, which would fit in better with the present curve.) Although under

TABLE 14.4

P-values of frequencies of three exploratory acts observed for 15 min in male mice from two inbred strains treated with different doses of two drugs.*

Behaviour	Scopolamine		Physostigmine	
	DBA/2	C57BL/6	DBA/2	C57BL/6
Rearing	0.14	0.001	0.03	0.03
Leaning	0.08	0.004	0.006	0.06
Sniffing	0.08	0.005	0.004	0.01

* Variations within strains were evaluated using the Kruskal-Wallis one-way analysis of variance by ranks. Since no corrections were made for ties, the level of confidence was maintained at $P = 0.06$.

saline the strains did not differ for leaning, which confirms previous findings, the drug reduced this behaviour in strain C57BL/6 and enhanced it in DBA/2 leading to a systematic difference between them. A similar picture was obtained for sniffing: the original strain difference was completely reversed under the influence of scopolamine. The right part of Table 14.4 shows that physostigmine caused significant variation over dosages (0, 18.75, 37.5, 75, and 150 μg/kg) for all three exploratory acts in both strains. These effects

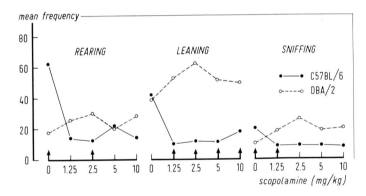

Fig. 14.8. The effects of peripheral injections with scopolamine upon three exploratory acts in inbred mouse strains DBA/2 (open circles) and C57BL/6 (closed circles). For each saline control group, $N = 12$; for each dose level, $N = 6$. Significant differences between strains under different drug conditions, found by means of the Mann-Whitney U test, are indicated by arrows on the abscissae. The strain differences for rearing (10 mg/kg) and sniffing (2.5 and 5 mg/kg) just missed significance ($P < 0.07$ or 0.08).

could be attributed mainly to the largest dose which markedly reduced the frequencies in both genotypes.

From these observations it can be concluded that there exists a genotype-dependent cholinergic mechanism which regulates exploratory behaviour in mice. It is a question whether this system acts to inhibit or to facilitate exploration. If we assume its action to be inhibitory, the low scores in DBA/2 mice could be explained by its more efficient synaptic transmission of impulses, due to a functionally well-balanced acetylcholine/acetylcholinesterase ratio. In that case, the high scores in the C57BL/6 strain might reflect some imbalance in this ratio. However, the data seem to contradict this assumption, because scopolamine and physostigmine both would be expected to disrupt the balanced ACh/AChE ratio in the DBA/2 animals, thereby augmenting exploration in this strain (response disinhibition; Carlton, 1968), whereas in the C57BL/6 animals either scopolamine or physostigmine would induce even higher scores than present in the saline controls and a suitable dose of the other of the two drugs would then reduce the scores by correcting the imbalance. The results obtained are more in favour of a facilitative action of the cholinergic mechanism concerned (cf. Warburton and Brown, 1971): in the C57BL/6 mice, a functionally well-balanced ACh/AChE ratio will promote efficient synaptic transmission and, as a consequence, produce high exploration scores, which will be depressed if anticholinergics or anticholinesterases upset that equilibrium in either of two directions. For the normally low-scoring DBA/2 animals a disequilibrium in the ACh/AChE ratio, with a relatively too large amount of transmitter, is hypothesized. The enzyme inhibitor physostigmine apparently worsens the disequilibrium and thus aggravates the lack of exploration in this strain. Under the influence of suitable doses of scopolamine, however, the imbalance seems to be corrected, resulting in higher scores. Possibly, the functional disturbance in DBA/2 mice is one of enzyme inhibition by excess substrate (see Kato et al., 1972), an inhibition that can be relieved by scopolamine. In addition, their higher AChE activity level, as compared to C57BL/6 mice, might be interpreted as a partial compensation for a surplus of transmitter.

So far for the mode of action of the cholinergic mechanism. In the second experiment, it was tried to obtain information on its site of action. By administering an equimolar dose of the quaternary analogues of the drugs used earlier, methylscopolamine and neostigmine, it is possible to decide whether the mechanism is located centrally or peripherally. With identical experi-

TABLE 14.5

Mean frequencies per 15 min and significance levels of three exploratory acts in male mice from two inbred strains after peripheral administration of saline alone, scopolamine, or an equimolar dose of methylscopolamine.

Behaviour	Strain	Treatment			P (treatment)
		saline	methylscopolamine (5.2 mg/kg)	scopolamine (5.0 mg/kg)	
Rearing	DBA/2	12.8	9.9	*44.3*	< 0.002
	C57BL/6	73.8	72.5	*25.0*	< 0.002
	P (strains)	< 0.002	< 0.002	< 0.02	
Leaning	DBA/2	26.1	19.6	31.3	n.s.
	C57BL/6	33.8	21.9	*14.2*	< 0.02
	P (strains)	n.s.	n.s.	< 0.02	
Sniffing	DBA/2	17.3	11.8	18.6	n.s.
	C57BL/6	27.1	26.2	23.6	n.s.
	P (strains)	< 0.002	< 0.02	n.s.	

N = 12 in each group, except for methylscopolamine-C57BL/6 (N = 11). Values in italics differ significantly (Mann-Whitney-U test) from those for saline.

TABLE 14.6

Mean frequencies per 15 min and significance levels of three exploratory acts in male mice from two inbred strains after peripheral administration of saline alone, physostigmine, or an equimolar dose of neostigmine.

Behaviour	Strain	Treatment			P (treatment)
		saline	neostigmine (58.4 µg/kg)	physostigmine (125.0 µg/kg)	
Rearing	DBA/2	23.0	23.6	2.1	< 0.002
	C57BL/6	67.7	65.5	5.0	< 0.002
	P (strains)	< 0.002	< 0.002	n.s.	
Leaning	DBA/2	38.3	37.6	*5.1*	< 0.002
	C57BL/6	52.5	42.1	*6.3*	< 0.002
	P (strains)	< 0.02	n.s.	n.s.	
Sniffing	DBA/2	4.9	4.1	0.6	< 0.002
	C57BL/6	15.3	13.7	0.7	< 0.002
	P (strains)	< 0.002	< 0.002	n.s.	

N = 15 in each group, except for physostigmine DBA/2 (N = 14). Values in italics differ significantly (Mann-Whitney-U test) from those for saline.

mental procedures, clear evidence was found for a central location. From Table 14.5 it can be seen that scopolamine tends to reduce the frequencies of rearing, leaning, and – only slightly – sniffing in strain C57BL/6, and that it tends to increase them in strain DBA/2 (significantly only for rearing), while methylscopolamine is ineffective, as follows from the comparison between the values for saline and methylscopolamine. Under scopolamine a reversal of the strain difference for rearing is conspicuous. A strain difference for leaning emerged after the treatment with this drug. For all three exploratory acts, physostigmine diminished drastically the scores in both genotypes (Table 14.6), but neostigmine had no such effects (compare the neostigmine values with those for saline). Apparently, the cholinergic system controlling exploration has to be searched for in the brain. The present results also lend support to the conclusions drawn from the first experiment.

To locate the central site of the system more precisely, a third experiment was performed involving, to start with, injections with two doses of methylscopolamine and two doses of neostigmine given into the hippocampus. Fig. 14.9 shows that the effects of intrahippocampal methylscopolamine are most marked at the dose-level of 7.8 μg. The anticholinergic depressed the scores for the three acts in the C57BL/6 mice and, by contrast, augmented

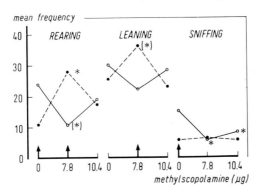

Fig. 14.9. The effects of intrahippocampal administration of methylscopolamine upon the frequencies per 15 min of three exploratory acts in inbred mouse strains DBA/2 (closed circles) and C57BL/6 (open circles). The DBA/2 groups comprise 13 (saline), 11 (7.8 μg), and 13 (10.4 μg) animals; the C57BL/6 groups comprise 13, 12, and 12 animals, respectively. Significant differences ($P \leq 0.05$; Mann-Whitney U test) between strains under different drug conditions are indicated by arrows on the abscissae. Drug groups differing significantly ($P < 0.02$) from their saline controls are marked by an asterisk; indications of such differences ($P < 0.07$) are marked by an asterisk between brackets. From: Psychopharmacologia (Berl.) Vol. 24, pp. 470–475 (1972), by permission.

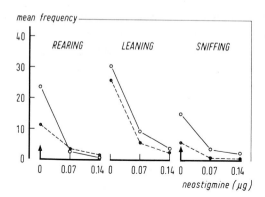

Fig. 14.10. The effects of intrahippocampal administration of neostigmine upon the frequencies per 15 min of three exploratory acts in inbred mouse strains DBA/2 (closed circles) and C57BL/6 (open circles). The DBA/2 groups comprise 13 (saline), 12 (0.07 µg), and 14 (0.14 µg) animals; the C57BL/6 groups comprise 13, 14, and 14 animals, respectively. Significant strain differences ($P < 0.05$; Mann-Whitney U test) are indicated by arrows on the abscissae. All drug groups differ significantly ($P < 0.02$) from their saline controls. From: Psychopharmacologia (Berl.) Vol. 24, pp. 470–475 (1972), by permission.

rearing and leaning in the DBA/2 animals. This particular dose seems to be a critical one. A complete reversal of the strain difference for rearing was found again. Fig. 14.10 demonstrates the reductions in the frequencies of the three behaviours caused by intrahippocampal administration of neostigmine in both strains. The above findings suggest that cholinergic pathways in the hippocampus play a role in the control of exploratory tendencies. Their actual location still needs further clarification because the dorsal and ventral hippocampus have been reported to be functionally distinct (Nadel, 1968; see also Jarrard, 1973). The observed genetically-determined differences in exploration may be mediated by this facilitative nervous mechanism which seems to operate efficiently in strain C57BL/6 and less so in strain DBA/2.

14.3.2. Selected lines

14.3.2.1. Behavioural differences. The method of genotype manipulation by selection was employed to study the genetic determinants of exploratory behaviour more closely. This section deals with the results obtained in mice selected bi-directionally for rearing frequency. Using the F_2 generation produced by the first cross between the above inbred strains as the foundation population, an attempt was made to transfer polygenes responsible for high rearing scores in C57BL/6 mice to a DBA/2 background. For this purpose, extreme scorers were repeatedly back-crossed with DBA/2 females in five consecutive generations, thus initiating two sublines: the S-high and the S-low. This early stage of the selection aimed at developing two lines of mice with similar genetic backgrounds, but differing for alleles affecting rearing behaviour. During this backcrossing process, a number of these alleles may not be maintained in the lines due to genetic drift. After the fourth generations S_4, selection was continued with concomitant inbreeding by sib-mating; since then, the lines have reached S_{19} and the remaining relevant alleles must be close to complete fixation. Full details of this procedure can be found elsewhere (van Abeelen, 1970). All animals have become homozygous for maltese dilution (d/d) and brown (b/b). In the low line, segregation for d and b did not occur anymore after S_2. The high line ceased segregating for d after S_4, but the B allele (black) persisted up to S_{11}. No clear-cut effect of the latter locus upon rearing frequency was detected in the F_2 and the high line, however.

As is evident from Fig. 14.11, a clear separation between a high-scoring and a low-scoring line arose during the early generations of selection combined with backcrossing, the S–high staying initially at a level comparable with the C57BL/6 strain (see also Fig. 14.6), and the S-low declining toward a DBA/2 level. This line difference collapsed at S_5. The lines remain well-apart in all subsequent generations of selection combined with inbreeding; the S-high shows a tendency to increase gradually. These results demonstrate

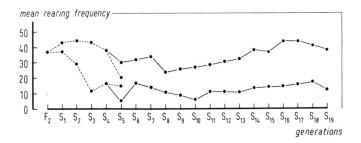

Fig. 14.11. Mean rearing frequencies exhibited by male mice from an F_2 generation (C57BL/6 × DBA/2) and from two lines selected for high and low rearing scores respectively, produced first by selection with backcrossing (dashed lines), and, in later generations, by selection with inbreeding (solid lines). Except for S_1, S_2, and the backcrossed S_5, the lines differ significantly in all generations (Mann-Whitney U test).

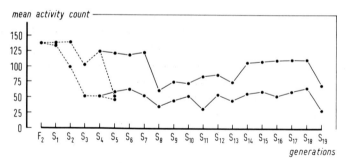

Fig. 14.12. Mean locomotor activities exhibited by male mice from an F_2 generation (C57BL/6×DBA/2) and from two lines selected for high and low rearing scores respectively, produced first by selection with backcrossing (dashed lines), and, in later generations, by selection with inbreeding (solid lines). Except for S_1, S_2, the backcrossed S_5, S_9 and S_{10} the lines differ significantly in all generations (Mann-Whitney U test).

TABLE 14.7

Significance levels of correlations between rearing frequency and locomotor activity observed for 15 min in male mice selected in consecutive generations for high and low rearing scores.*

Generation	S-high		S-low	
	N	P	N	P
1 (backcrossed)	10	< 0.01	10	< 0.025
2	10	n.s.	10	< 0.05
3	10	< 0.01	10	n.s.
4	7	n.s.	10	n.s.
5	10	< 0.025	10	n.s.
6 (sib-mated)	20	n.s.	9	n.s.
7	20	< 0.01	10	< 0.001
8	20	< 0.001	10	n.s.
9	20	< 0.001	10	< 0.025
10	20	< 0.025	10	< 0.001
11	20	< 0.025	9	n.s.
12	20	< 0.001	10	< 0.01
13	20	< 0.005	17	< 0.005
14	20	< 0.001	17	< 0.001
15	20	< 0.001	20	< 0.025
16	20	< 0.001	20	< 0.005
17	20	< 0.001	20	< 0.01
18	20	n.s.	20	< 0.005
19	20	n.s.	20	< 0.025

* Correlations were calculated according to Spearman's ranking method; all were positive.

that this particular behavioural phenotype can be altered by genetic manipulations. As Fig. 14.12 shows, the above changes in rearing frequency are accompanied by similar changes in locomotor activity. However, although the lines differed significantly for locomotion in practically all generations, the S-high did not by a long way attain the level characteristic of strain C57BL/6 (see Fig. 14.7). Positive correlations between rearing and locomotion were found in many generations of S-high as well as S-low (Table 14.7). Especially in the earlier stages of the selection process, loci can be expected to be still segregating. If traits vary dependently, as seems to be the case here, the differences observed for the two behavioural phenotypes may be the result of variation in the same genetic factors. Our correlations can be taken as additional evidence that the alleles contributing to differences in vertical activity exert their effects through a system which also controls horizontal activity.

14.3.2.2. Acetylcholinesterase-activities. Lines selected for a behavioural phenotype are potentially of value for the analysis of phenogenetic mechanisms underlying it. On the basis of the previous findings in the inbred strains and hybrids, one would predict an inverse relationship between exploration rate and AChE activity in the selection lines. The data concerning the cerebral enzyme activities in four generations of the lines can be found in Table 14.8. Long time-intervals elapsed between these line comparisons and, due to the initial technical problems mentioned earlier, the measurements made in the S_{16} and S_{17} undoubtedly are the more reliable

TABLE 14.8

Means, standard deviations, and significance levels of acetylcholinesterase activities (μmoles/min) in whole brain homogenates of male mice selected for high and low rearing frequency.

Group	N	Mean	SD	P (Mann-Whitney)
S_{11}-high	8	4.70	0.20	n.s.
S_{11}-low	8	4.77	0.19	
S_{13}-high	9	5.53	0.19	< 0.08
S_{13}-low	10	5.72	0.23	
S_{16}-high	5	6.43	0.36	< 0.06
S_{16}-low	8	6.79	0.29	
S_{17}-high	19	6.43	0.21	< 0.002
S_{17}-low	19	6.72	0.18	

ones. There is a general tendency for the lines to differ in respect of enzyme activity in whole brain homogenates, the S-low showing the highest values. The selected lines differ also for wet brain weight: S_{16}-high, 376 ± 11 mg; S_{16}-low, 394 ± 7 mg ($P < 0.01$; Mann-Whitney test), and S_{17}-high, 375 ± 10 mg; S_{17}-low, 399 ± 8 mg ($P < 0.01$). Thus it appears that the breeding procedure has led to a difference in brain weight which is independent of the difference produced for rearing performance. In fact, the above negative association between exploratory score and enzyme activity vanishes completely if the AChE calculated per mg of brain tissue is considered. Spearman's test did not reveal any significant correlations between AChE activity and behaviour – rearing and locomotion – within the groups observed. Whole brain determinations naturally cannot yield much information on any functional imbalances in the ACh/AChE ratio which, moreover, may be more subtle in the selected lines than in the original strains from which they were derived. Future assays of ACh contents and AChE activity levels in separate brain regions should clarify this.

14.3.2.3. Responses to drug treatment. Meanwhile, S-high and S-low mice were subjected to experiments involving i.p. injections with scopolamine. In a small-scale preliminary experiment with the S_{14} using 5 mg/kg, the drug tended to exert different effects upon exploration in the two lines: reductions were observed in the high line for rearing ($P < 0.01$) and leaning frequency ($P < 0.01$), but not for sniffing, and an increase was found in the low line for rearing ($P < 0.05$), while leaning and sniffing remained unaffected here. Since scopolamine produced such-like opposite effects in the ancestral C57BL/6 and DBA/2 strains, it was decided to extend and replicate the tests, using generations S_{16} and S_{17}.

Apart from saline injections, three different doses of the anticholinergic were administered to the S_{16} mice. Table 14.9 shows that treatment affected two out of three vertically-oriented behaviours in each line; the influences on locomotion almost reached significance. The results are presented graphically in Fig. 14.13. In this set of data, the saline controls did not show a significant line difference for rearing, but they did for leaning, sniffing, and locomotion. According to expectation, scopolamine depressed the scores for rearing, leaning, and (nonsignificantly) locomotion in the S-high mice; no such effect was obtained for sniffing, however. The drug certainly does not cause a response disinhibition here. In the S-low animals reductions were either absent (leaning) or less pronounced (rearing, locomotion). At the dosage of 1 mg/kg, the sniffing rate of S-low animals exceeded the corresponding

TABLE 14.9

P-values of frequencies of three exploratory acts and locomotor activity counts observed for 15 min in male mice from two selected lines, generation S_{16}, treated with different doses of scopolamine.*

Behaviour	S_{16}-high	S_{16}-low
Rearing	0.001	0.01
Leaning	0.001	n.s.
Sniffing	n.s.	0.01
Locomotion	0.08	0.10

* Variations within lines were evaluated using the Kruskal-Wallis one-way analysis of variance by ranks.

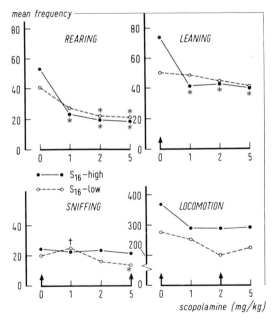

Fig. 14.13. The effects of peripheral injections with scopolamine upon three exploratory acts and locomotor activity in mice selected for high (closed circles) and low (open circles) rearing frequency; generation 16. In the high line, N = 20 (saline), 17 (1 mg/kg), 18 (2 mg/kg), and 17 (5 mg/kg) animals; in the low line, N = 19, 14, 15, and 14 animals, respectively. Significant differences between lines under different drug conditions, found by means of the Mann-Whitney U test, are indicated by arrows on the abscissae. Drug groups differing significantly from their saline controls are marked by an asterisk. The mean sniffing score found at 1 mg/kg in the low line, as indicated by a cross, differs from the values obtained for saline ($P < 0.07$), 2 mg/kg ($P < 0.05$), and 5 mg/kg ($P < 0.01$).

values for saline and the higher doses, which might be indicative of the suitability of that particular dosage for correcting a subtle ACh/AChE disequilibrium caused by a slight surplus of transmitter. Higher doses are apt to overcorrect this. The above findings do not contradict the notion that the cholinergic mechanism acts to facilitate exploratory tendencies.

In the observations carried out with the S_{17}, mice from the high line and the low line received either saline or one of two doses of scopolamine which were lower than those used in the previous experiment. From Table 14.10 it appears that, except for sniffing in the S-high and leaning in the S-low, marked variation over dosages was produced by the drug. The detailed results are shown in Fig. 14.14. S-high mice clearly are more mobile animals than S-low mice: significant line differences emerged between the undrugged groups for rearing, leaning, and locomotion. The rearing score dropped sharply in the scopolamine–injected S-high mice so that the original line difference was practically reversed (at 0.75 mg/kg it just missed significance). The expected increase in the scores of the S-low did not materialize. The frequencies of leaning in drugged S-high and S-low animals became very similar.

TABLE 14.10

P-values of frequencies of three exploratory acts and locomotor activity counts observed for 15 min in male mice from two selected lines, generation S_{17}, treated with different doses of scopolamine.*

Behaviour	S_{17}-high	S_{17}-low
Rearing	0.001	0.06
Leaning	0.01	n.s.
Sniffing	n.s.	0.05
Locomotion	0.02	0.05

* Variations within lines were evaluated using the Kruskal-Wallis one-way analysis of variance by ranks.

Scopolamine given in a dose of 0.50 mg/kg tended to augment sniffing in the low line, suggesting that in the control animals the ACh/AChE ratio is not at its optimum value. As for locomotor activity, the drug exerted opposite effects in the two lines, resulting in a disappearance of the line difference. The general impression is that the differential effects of the drug upon exploration can be more easily distinguished in the selected lines if fairly low doses are employed.

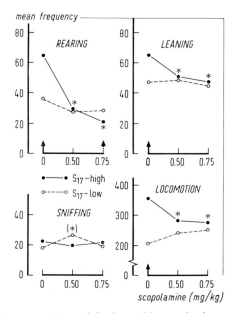

Fig. 14.14. The effects of peripheral injections with scopolamine upon three exploratory acts and locomotor activity in mice selected for high (closed circles) and low (open circles) rearing frequency; generation 17. In the high line, N = 20 for each group; in the low line, N = 19 (saline), 19 (0.50 mg/kg), and 18 (0.75 mg/kg). Significant differences between lines under different drug conditions, found by means of the Mann-Whitney U test, are indicated by arrows on the abscissae. $P = 0.07$ for the line difference for rearing at the dose of 0.75 mg/kg. Drug groups differing significantly from their saline controls are marked by an asterisk; an indication of such a difference ($P = 0.06$) is marked by an asterisk between brackets.

14.4. Summary and prospects

The overall picture provided by the data is compatible with the idea of a genotype-dependent cholinergic mechanism, probably located in the hippocampus, which controls the responses of mice to novelty. A functionally optimal ACh/AChE ratio will guarantee efficient synaptic transmission and thereby facilitate exploration. Any genetically-determined imbalances, such as probably present to a certain degree in DBA/2 and S-low mice, may be restored by administration of an anticholinergic drug. Phylogenetically, a connection between cholinergic pathways in the rhinencephalon and the exploratory sniffing displayed by mice in a novel environment seems plausible. However, in order to seek further support for the hypothesis, detailed determinations of the rates of acetylcholine synthesis and degradation – for instance by measuring choline acetyltransferase and acetylcholinesterase activities in a number of different brain regions cf. Ebel et al. (1973) – are needed. The possibility that alternative noncholinergic systems play a role as well must be kept in mind. Finally, it is intended to cross the two selected lines and to analyze the behaviour of their hybrids so that perhaps the number of loci contributing to the difference in rearing frequency can be estimated.

Acknowledgement

The author wishes to acknowledge the enthusiastic participation of many doctoral students in various phases of the work. Thanks are due to Mr. A. J. M. Buis for his expert biotechnical assistance. The collaboration of Dr. H. F. P. Joosten and Mr. C. A. van Bennekom on the determination of enzyme activity is gratefully acknowledged.

References

Abeelen, J. H. F. van (1970): Genetics of rearing behavior in mice. *Behav. Genet.*, **1**, 71.
Barnett, S. A. (1963): A Study in Behaviour. Principles of Ethology and Behavioural Physiology, Displayed Mainly in the Rat. (Methuen, London).
Bovet, D., F. Bovet-Nitti and A. Oliverio (1966): Effects of nicotine on avoidance conditioning of inbred strains of mice. *Psychopharmacologia (Berl.)*, **10**, 1.
Broadhurst, P. L. (1969): Psychogenetics of emotionality in the rat. *Ann. N.Y. Acad. Sci.*, **159**, 806.
Bruell, J. H. (1964): Inheritance of behavioral and physiological characters of mice and the problem of heterosis. *Amer. Zoologist*, **4**, 125.
Bruell, J. H. (1969): Genetics and adaptive significance of emotional defecation in mice. *Ann. N.Y. Acad. Sci.*, **159**, 825.
Carlton, P. L. (1963): Cholinergic mechanisms in the control of behavior by the brain. *Psychol. Rev.*, **70**, 19.
Carlton, P. L. (1968): Brain acetylcholine and habituation. In: Anticholinergic Drugs and Brain Functions in Animals and Man, Progress in Brain Research, vol. 28. Bradley, P. B. and M. Fink, eds. (Elsevier, Amsterdam) pp. 48–60.
DeFries, J. C., J. R. Wilson and G. E. McClearn, (1970): Open-field behavior in mice: Selection response and situational generality. *Behav. Genet.*, **1**, 195.
Douglas, R. J. (1967): The hippocampus and behavior. *Psychol. Bull.*, **67**, 416.

Ebel, A., J. C. Hermetet and P. Mandel, (1973): Comparative study of acetylcholinesterase and choline acetyltransferase enzyme activity in brain of DBA and C57 mice. *Nature New Biol.*, **242**, 56.

Eibl-Eibesfeldt, I. (1950): Beiträge zur Biologie der Haus- und Aehrenmaus nebst einigen Beobachtungen an anderen Nagern. *Z. Tierpsychol.*, **7**, 558.

Eibl-Eibesfeldt, I. (1958): Das Verhalten der Nagetiere. Handbuch Zool. Bd. 8, Lief. 12 (De Gruyter, Berlin) pp. 1–88.

Ellman, G. L., K. D. Courtney, V. Andres Jr. and R. M. Featherstone, (1961): A new and rapid colorimetric determination of acetylcholinesterase activity. *Biochem. Pharmacol.*, **7**, 88.

Fulker, D. W. (1973): A biometrical genetic approach to intelligence and schizophrenia. *Soc. Biol.* **20**, 266.

Fuller, J. L. (1970): Strain differences in the effects of chlorpromazine and chlordiazepoxide upon active and passive avoidance in mice. *Psychopharmacologia (Berl.)*, **16**, 261.

Green, E. L. and H. Meier (1965): Use of laboratory animals for the analysis of genetic influences upon drug toxicity. *Ann. N.Y. Acad. Sci.*, **123**, 295.

Green, M. C. (1966): Mutant genes and linkages. In: Biology of the Laboratory Mouse, 2nd ed. Green, E. L., ed. (McGraw-Hill, Blakiston Division, New York) pp. 87–150.

Henderson, N. D. (1968): Genetic analysis of acquisition and retention of a conditioned fear in mice. *J. comp. physiol. Psychol.*, **65**, 325.

Herz, A., H. Teschemacher, A. Hofstetter and H. Kurz (1965): The importance of lipid-solubility for the central action of cholinolytic drugs. *Int. J. Neuropharmacol.*, **4**, 207.

Jarrard, L. E. (1973): The hippocampus and motivation. *Psychol. Bull.*, **79**, 1.

Kato, G., E. Tan and J. Yung, (1972): Allosteric properties of acetylcholinesterase. *Nature New Biol.*, **236**, 185.

Kimble, D. P. (1968): Hippocampus and internal inhibition. *Psychol. Bull.*, **70**, 285.

Lát, J. and E. Gollová-Hémon, (1969): Permanent effects of nutritional and endocrinological intervention in early ontogeny on the level of nonspecific excitability and on lability (emotionality). *Ann. N.Y. Acad. Sci.*, **159**, 710.

Lindzey, G. and D. D. Thiessen, eds. (1970): Contributions to Behavior-Genetic Analysis: The Mouse as a Prototype. (Appleton-Century-Crofts, New York).

Nadel, L. (1968): Dorsal and ventral hippocampal lesions and behavior. *Physiol. Behav.*, **3**, 891.

Newell, T. G. (1970): Three biometrical genetic analyses of activity in the mouse. *J. comp. physiol. Psychol.*, **70**, 37.

Oliverio, A. (1968): Some additional data on the effects of scopolamine on habituation and conditioning. *Psychopharmacologia (Berl.)*, **13**, 356.

Routtenberg, A. (1971): Stimulus processing and response execution: A neurobehavioral theory. *Physiol. Behav.*, **6**, 589.

Sachs, L. (1969): Statistische Auswertungsmethoden. (Springer, Berlin).

Siegel, S. (1956): Nonparametric Statistics for the Behavioral Sciences. (McGraw-Hill, New York).

Smissaert, H. R. and K. T. Hitman (1971): Acetylcholinesterase from Tryon-maze-bright and Tryon-maze-dull rat strains is identical. *Physiol. Behav.*, **6**, 729.

Staats, J. (1972): Standardized nomenclature for inbred strains of mice: Fifth listing. *Cancer Res.*, **32**, 1609.

Wakeley, H. G. and D. O'Sullivan (1969): Drug effects on mouse exploratory behavior. *Psychon. Sci.*, **16**, 27.

Warburton, D. M. and K. Brown (1971): Attenuation of stimulus sensitivity induced by scopolamine. *Nature*, **230**, 126.

CHAPTER 15

Genetic factors in the control of drug effects on the behaviour of mice*

A. Oliverio

The existence of individual differences in the reactivity to many psychotropic agents, evident in humans and in laboratory animals, is related to (1) differences in rate of absorption, distribution, metabolism, or elimination of drugs or (2) differences in the sensitivity of target systems to compounds acting at the C.N.S. level.

In man, one of the best known examples illustrating the influence of genetic factors upon drug responses concerns the effects of primaquine. During the war it was observed that in coloured soldiers, but not whites, severe hemolysis occurred after the ingestion of this antimalaric agent. The cause of this reaction to primaquine was found to be connected with a deficiency of glucose-6-phosphate dehydrogenase, an enzyme present in the erythrocytes. Since then a number of genetically determined differences in drug sensitivity due to abnormal enzymatic pathways were recognized to account for slow or rapid inactivation of isoniazid, anticoagulants, barbiturates, or antidepressant agents (Kalow, 1966; Koppanyi and Avery, 1966). In addition, a series of twin studies on drug elimination from the plasma have indicated that intratwin differences in metabolism are consistently greater in fraternal than in identical twins. It has thus been demonstrated that the mechanisms responsible for individual differences in rates of elimination of drugs such as phenylbutazone, antipyrine, coumarin, and ethanol are under genetic control rather than environmental (Vesell, 1972).

The blood concentration of a given drug has been shown to be responsible

* The experiments reported in this study were partially supported by a grant of the Italian and French National Research Councils, and by a grant of the European Training Program in Brain and Behaviour Research.

for a specific effect at its site of action. To maintain a particular dose level in animals of different species, a distinctive dose schedule is required. A number of data indicate that, after the same dose per kilogram, blood concentrations of drugs such as sulfonamides may differ ten to twentyfold (Spinks, 1965). For example, after a dose of 100 mg/kg of sulfadimidine, the plasma concentration of the drug attains a level of about 2 mg per ml in mice, chicks, or rabbits and 15–16 mg per ml in cats and men. The problem is whether drug concentration determines the variation over species or whether this variability is independent of drug levels: the plasma concentration of a drug is in fact a product of different mechanisms such as absorption, storage, metabolism, and excretion. The large differences between the enzyme patterns of species may account for differences in the speed of metabolism and the number of metabolic products: humans are able to form conjugated glucuronides, to acetylate aromatic amines, and to acetylate S-arylcysteins to mercapturic acid; by contrast, glucuronic transferase is almost absent in cats, while dogs do not acetylate arylamines such as para-aminobenzoic acid or sulfonamides (Williams, 1962). The persistent depression exhibited by cats and dogs following administration of barbiturates is due to the retention of barbital for a longer period of time in the brain, even after the drug has been eliminated from the other organs (Koppanyi and Dille, 1934). Differences in the activity of oxidative liver enzymes also may lead to increased or decreased metabolic rates of barbiturates in species in which these drugs have a short (or long) duration of action. After injecting a dose of 100 mg/kg of hexobarbital in mice, rabbits, and rats, and of 50 mg/kg in dogs, the order of sleeping-time was: mice, 12 min; rabbits, 49 min; rats, 95 min; dogs, 260 min. The half-life of the drug was shown to parallel the sleeping-time and to be correlated with enzyme activity (Quinn et al., 1954).

In contrast with the findings on differences in rate of drug metabolism resulting from variations in enzyme activity, the effects of drugs which inhibit enzyme activity, such as cholinesterase (ChE) or monoamino oxidase (MAO) inhibitors, may depend on the amount of substrate present in a given species. For example, because large species variations exist in the rate of synthesis of catecholamines, the effects of MAO inhibitors differ from species to species. Iproniazid or pargyline do not increase the levels of brain catecholamines in dogs and cats, although 5-hydroxytryptophane (5-HTP) levels are elevated. By contrast, the same drugs enhance the levels of noradrenaline and 5-hydroxytryptamine (5-HT) in the brains of rabbits or rats (Gillette, 1965). Therefore, it seems clear that differences in plasma levels are not the sole cause of differences in drug responses. Not only quantitative variations in the amounts of a particular chemical, but also variations in the ratio between two mediators may be responsible for interspecies or intraspecies variations in drug effects. The ratio of noradrenaline to adrenaline, for example, which is quite constant in most animal species, may vary between individual cats from 13 to 91 per cent (Meier, 1963).

A number of biochemical differences between different species or individuals is also responsible for the variety of effects observed after the use of many psychopharmacological agents and for drug-induced abnormalities in the central nervous system. This is the case for lysergic acid diethylamide (Piala et al., 1959), phenothiazines (Goldenberg and Fishman, 1961), or 6-azauracil (Shnider et al., 1960). While in some instances, as is the case for phenothiazines (Myrian-Thopoulos et al., 1962) or imipramine (Burnett et al., 1964), there is evidence for a clear-cut genetic control which might even depend on a major gene effect, in other instances a complex of factors is involved.

The examples of primaquine sensitivity, isoniazid inactivation, or reactivity to phenothiazine and imipramine show very clearly that not only interspecific but also intraspecific differences may produce variation in drug response, and that the principles of quantitative genetics may be applied to the analysis of the effects of drugs. In addition, pharmacogenetics may represent a useful experimental approach for assessing the nature of individual differences at the CNS level.

In recent years, the large number of genetic experiments on the laboratory mouse has resulted in many easily available lines and inbred strains (Staats, 1972) and has provided much information on the neurological and behavioural maturation of this species which resembles, in its initial patterns, that of man (Fox, 1964). In addition, clear intraspecific phenotypic variations at the brain level have been established, indicating large individual differences for the weights and sizes of some cerebral structures (Wimer et al., 1971), the levels of cholinergic and adrenergic substances (Ebel et al., 1973; Eleftheriou, 1971), or the electroencephalographic patterns (Valatx et al., 1972). The existence of these differences, the availability of about 300 single-gene mutants, and the identification of its karyotype (Green, 1972) make the laboratory mouse a very useful tool for pharmacogenetic analyses.

This review deals with some genetic factors which control drug effects on the behaviour or random-bred, inbred, and recombinant inbred strains of mice.

15.1. Heterogeneous stocks

One of the problems which the psychopharmacologist encounters in the study of psychotropic drugs is the large behavioural variability evident in random-bred strains of mice or in the so-called "common" laboratory Swiss-Webster mice. For example the results of a study in which three different tests assessing exploratory behaviour were used indicate that the performance of random-bred animals varied from day to day. Not only large fluctuations were observed in the mean performance of the group, but also the individual scores were scattered widely around the mean (Fontenay et al., 1970). The variable performance of random-bred mice in different behavioural tests (Bovet et al., 1969; Oliverio, 1971) is one of the factors connected with differences in sensitivity or with different modalities of reaction to the same psychotropic agent. In fact, a number of findings indicate that stimulating agents such as amphetamine may enhance response rates in rats which are normally characterized by low avoidance levels (Rech, 1966) and may be ineffective in rats characterized by high rates of responding. Similarly, nicotine was found to exert a stimulating effect on the activity of rats

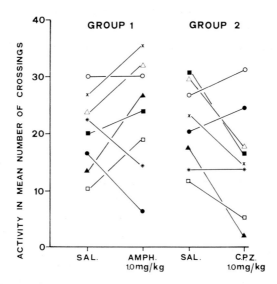

Fig. 15.1. Exploratory behaviour in a tilt-floor box by two groups of random-bred mice under control conditions (saline) and 15 min after injection of 1 mg/kg of amphetamine (AMPH) or 1 mg/kg of chlorpromazine (CPZ). Each group included 8 mice.

during their inactive phase (day-time) and a sedative effect during their active phase (at night) (Bovet et al., 1966). However the level of activity (or of responding) observed under control conditions does not seem to be the only key to predicting the type of behavioural effect produced by a drug. An example of large individual differences in drug reactivity, unrelated to previous performance levels in the absence of drug, is given in Fig. 15.1. Two groups of random-bred mice belonging to an 8-way cross derived from 8 different inbred strains were subjected to a short test of exploratory activity without drug treatment and, 24 hrs later, to a second session in which the mice were previously injected with amphetamine or chlorpromazine. Evidently, amphetamine exerted a stimulating effect on exploratory activity in some mice (5 out of 8), while the levels of activity remained unaffected or were even depressed in other animals. Using larger groups, we obtained results pointing in the same direction. Similarly, chlorpromazine reduced the levels of exploratory activity in some mice, but it was ineffective in others.

The existence of such differences in reactivity to psychotropic agents is not very surprising. Extensive examination of feral or random-bred populations of mice from several areas of the world has shown that the percentage of polymorphic loci in the species is about 40 % and that these polymorphisms are widely spread and stable (Selander et al., 1969). Findings along this line show that, as far as the frequencies of several loci controlling isozymes in separate populations of mice are concerned, the average animal is heterozygous at 10.3 % of his loci (Roderick et al., 1971). A similar degree of polymorphism is likely to be responsible for the different effects of a drug on brain function within a noninbred mouse population.

15.2. Inbred strains

There are two important reasons for using inbred strains. The first is the extreme behavioural homogeneity of individuals belonging to the same line, and the second is the difference between the behavioural traits characterizing each strain. These strain differences reflect a large allelic variability among independent inbred strains, indicating that for many behavioural traits variability among different strains is as large as that found among the different individuals belonging to a random-bred population. In a study conducted by Roderick et al. (1971), thirty-nine inbred strains were characterized for their alleles for as many as 16 polymorphic loci. Among these strains, variability was at least as large as in any single feral or random-bred population of mice. When a number of behavioural measures are considered, large differences are also evident among inbred strains (McClearn et al., 1970; Lindzey et al., 1971; Wahlsten, 1972). Similarly, drugs acting upon different behavioural responses are apt to be strain-dependent in their effects. The strain-dependent action of some drugs has been correlated to quantitative or qualitative strain differences in specific enzyme content or to the genetic architecture of the strains. While some of the strain differences observed for drug effects are due to altered mechanisms of excretion or liver metabolism, other effects could be ascribed to different modes of action at the brain level and these results have provided an insight into the neurophysiology of behavioural brain mechanisms.

The findings on the sleeping-times of mice after nembutal administration represent one of the first accomplishments in the field of psychopharmacogenetics. Large differences in sleeping-time became evident between strains injected with the same dose of nembutal (80 mg/kg): mice of some strains slept as short as 160 min (BALB/c), while others as long as 270 min (CBA) (Brown, 1959). Differences in the rate of inactivation in the liver account for the observed variability. Since then, more refined and detailed biometric analyses have been conducted: measures of heritability, dominance, and segregation, and estimates of the number of genes involved have been calculated for alcohol preference and morphine susceptibility in inbred mice. Strain differences in a variety of behaviours related to alcohol consumption have been reported (Eriksson, 1969; Fuller, 1964; McClearn and Rodgers, 1959). Animals of the C57BL/6 strain consumed about two-thirds of their total daily liquid intake from a 10 % ethanol solution. Several other strains displayed intermediate levels of preference and others, like the DBA/2 strain, avoided the alcohol solution almost completely. Comparisons of high- and low-preferring strains by a number of investigators provided evidence that the activities of the liver enzymes alcohol dehydrogenase (ADH) and aldehyde dehydrogenase (ALDH) are related to spontaneous preference. These behavioural and biochemical findings have been confirmed by a selective breeding programme for high and low alcohol intake (Eriksson, 1969). However, estimates of segregation in the F_2 generation and measures of correlations between alcohol dehydrogenase levels and ethanol consumption in heterogeneous strains showed that only a small fraction to the observed variability in alcohol preference was related to individual differences in respect to the liver dehydrogenase activity.

A more thorough and refined biometric analysis has been conducted by Eriksson and Kiianmaa (1971) for morphine addiction. Susceptibility to morphine addiction was analyzed genetically in two inbred strains of mice, CBA/Ca and C57BL. Morphine consumption was higher in the C57BL mice than in the CBA/Ca animals. Heritability estimates obtained by application of biometrical methods to the parental, F_1, F_2, and backcross generations were in the range of 0.80–0.90, indicating a very large genetic component and polygenic determination. The existence of positive strain correlations for alcohol consumption and morphine addiction seems to suggest a common biological basis. In fact, a number of findings seems to prove that inherited differences in morphine addiction may be determined by differences in ADH and ALDH activities.

Let us now consider a more sophisticated behavioural pattern such as active avoidance learning. While the examples given above reveal quantitative differences in drug sensitivity between strains, other findings demonstrate that one and the same psychotropic agent may affect performance in opposite directions, depending on the genetic background of the strain or on the behavioural patterns present under control conditions.

Clear differences in strain reactivity with regard to the effects of a psychoactive agent on learning have been found by Bovet et al. (1969). The effects of different doses of arecoline, a typical cholinergic agent, were compared in two inbred strains subjected to the same avoidance schedule in a shuttle-box. A decrement of performance was evident after treatment with various dosages of this drug, the magnitude of this impairment being strain-dependent. This is shown in Fig. 15.2 in which the effects of arecoline are

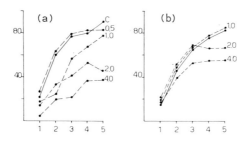

Fig. 15.2. Effects of (i.p.) arecoline on avoidance learning of BALB/c (a) and DBA/2J (b) mice. The points represent the mean of 8 mice over five consecutive daily sessions of 100 trials (= controls). The numbers indicate the dose of the drug. C = control groups.

compared in DBA/2J and BALB/c mice. In this case, the magnitude of the impairment was independent of the levels of performance as displayed by the two strains under control conditions, since here both exhibited comparable levels of responding.

On the other hand, the dose-response curves of a given drug may vary depending on the levels of "kinetic drive" present in the strains. Fuller (1966, 1970) tested four inbred strains for an active and a passive avoidance task in a two-compartment cage. The strain differences in avoidance learning were related to variations in strength of kinetic drive. While no association was found between the effectiveness of chlorpromazine and genotype, the dose-response curves with chlordiazepoxide (CDE) did vary with respect to mean levels of activity in the strains. For CDE the values of the regression

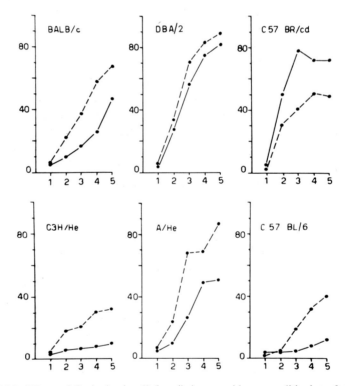

Fig. 15.3. Effects of (i.p.) nicotine (0.5 mg/kg) on avoidance conditioning of six inbred-strains of mice. Dotted lines, nicotine-injected groups; solid lines, control groups. Abscissae: sessions; ordinates: avoidances.

coefficient for the strains ranged from +26.5 in C57BL/6J mice to −133.7 in DBA/2J mice. In general, CDE appeared to enhance responding in the least active strain and decrease it in the more active strains.

Strain-dependent effects of CDE on avoidance behaviour of mice were also evident in some findings of Sansone and Messeri (1973). The dose which induced a 50% impairment of performance in a previously acquired avoidance task was about two times larger in the SEC/1ReJ strain than in BALB/c mice.

Findings essentially similar in their import are those dealing with the effects of nicotine on avoidance learning (Bovet et al., 1966). The effects of the same dose of nicotine (0.5 mg/kg) were compared in nine inbred strains and in a random-bred "Swiss" stock subjected to identical avoidance schedules in a shuttle-box. The results show that under similar conditions the various strains attain quite different levels of performance and that there are important differences between the effects of nicotine in the various strains. Nicotine exerted a facilitating effect in six out of nine inbred strains: the level of avoidance responding increased to 35 per cent above the level of the control mice (C3H/He). The same dose had a smaller effect in DBA/2J mice (15 per cent) and a reducing effect on the performance of another strain (Fig. 15.3). The facilitating effect of nicotine was generally higher in the strains characterized by low performance levels, a significant positive correlation being evident between the avoidance levels in the absence of drug and those in the presence of nicotine (Fig. 15.4). In general, these findings suggest that nicotine improves avoidance learning by acting on memory consolidation processes. More recent experiments support this hypothesis that nicotine facilitates acquisition and retention processes (Oliverio, 1968; Evangelista et al., 1970).

An interesting possibility in psychopharmacogenetic studies derives from the use of crosses between inbred strains. Our attention has recently been focused on three different strains of mice for the purpose of conducting a genetic, biochemical, and pharmacological analysis of learning. A genetic analysis of avoidance, maze learning, and wheel-running activity has been carried out with mice from three inbred strains and their F_1, F_2, and F_3 progenies. Whereas two strains (SEC1/ReJ and DBA/2J) are characterized by high levels of avoidance learning but low levels of running activity, a third strain (C57BL/6J) showed poor avoidance and maze learning but it was very active (Oliverio et al., 1972). The two main findings emerging from the biometric analysis were:

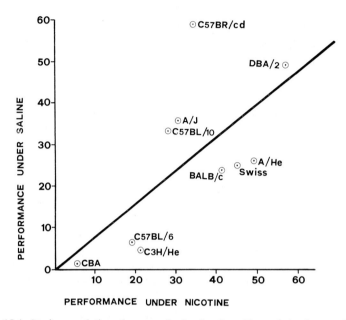

Fig. 15.4. Strain correlations between the levels of avoidance behaviour under control conditions (saline) and following treatment with nicotine. Each point refers to the performance of two groups of mice trained for five avoidance sessions, in the absence or presence of nicotine, respectively. Each group incuded 18 mice. The numbers indicate the mean percentage of avoidances over 5 sessions.

1. The mode of inheritance of a given behavioural measure was shown to depend on the crosses considered. Crossing the C57 strain with the high avoiding-low running SEC mice resulted in SEC-like progeny, while crossing with the high avoiding-low running DBA mice yielded an offspring similar to the C57 phenotype. A similar dominance pattern was evident for maze learning.

2. Genetic correlations were assessed in F_3 mice and their F_2 parents. The estimates of the correlations between avoidance and activity were negative and large enough to be attributable to a pleiotropic effect. The significant positive genetic correlation found between avoidance and maze learning suggested that these behaviours were also influenced by many of the same genes, although this effect might also be ascribed to linkage. Thus, the two C57 × SEC and C57 × DBA F_1 hybrids were very similar for avoidance, maze learning, and activity patterns to their SEC or C57 parents, respectively.

These strains and their hybrids may be used with profit for the study of the biochemical and pharmacogenetic aspects of memory and learning. Much is known already about the behaviour of DBA/2 and C57BL/6 mice; the strains are characterized by low and high activity patterns, respectively (van Abeelen, 1966; DeFries and Hegman, 1970; McClearn et al., 1970). Supporting our findings, Sprott (1971) has reported that DBA/2 and C57BL/6 mice show good and poor passive avoidance learning, respectively, and that their F_1 hybrids are like the C57 parents. Thus it seems that strain differences as observed for the active avoidance task are detectable also for other paradigms such as maze learning or passive avoidance. There is evidence that these strains differ also in brain chemicals. Pryor et al. (1966) have measured brain acetylcholinesterase (AChE) activity in five strains of mice. They found large differences between the strains, and these differences become even more pronounced if a regional analysis is performed: it is particularly interesting to note that the two strains characterized by high-avoidance levels (A and DBA/2) also show a higher AChE activity than the two low-avoiding strains (C3H and C57BL). A more detailed analysis carried out by Ebel et al. (1973) in DBA, C57BL, and their F_1 revealed that DBA mice have higher AChE and cholineacetyltransferase (ChA) activities in the temporal lobe than C57 mice.

As for the adrenergic system, Eleftheriou (1971) has demonstrated that clear differences exist between the regional brain norepinephrine turnover rates of different strains of mice. Of the strains studied, DBA/2J mice exhibited the highest turnover rates while C57BL/6J and C3H/HeJ mice displayed the lowest in all brain regions examined.

The effects of different drugs on the behaviour of the three strains considered above and their F_1 hybrids have been investigated. It seemed particularly interesting to see whether strains differing in brain neurochemistry react in different ways to drugs acting at the CNS level. Previous findings by Schlesinger and Griek (1970) indicated that DBA mice have lower ECS and metrazol seizure thresholds than either C57BL/6 mice or F_1 hybrid mice. Our data (Oliverio and Castellano, 1973) are in agreement with these findings and show that DBA and SEC mice have lower metrazol thresholds than C57 animals. Here, too, the C57 × SEC hybrids resembled the SEC phenotype, and the C57 × DBA mice resembled the C57 inbred strain, thus confirming the dominance order observed for behavioural phenotypes such as avoidance activity or maze learning (Fig. 15.5).

In a second experiment, the central effects of an adrenergic and two

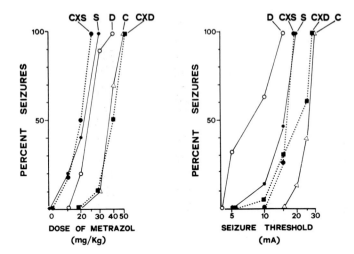

Fig. 15.5. Metrazol and electroconvulsive shock (ECS) seizure thresholds in C57BL/6J(C), DBA/2J(D), and SEC/1ReJ(S) inbred mice and their F_1 hybrids (CxS and CxD).

cholinergic agents on the exploratory activity of the different strains were assessed. The number of crossings occurring in a tilt-box was measured for 30 min. Exploratory activity clearly differed in the three strains, which agrees with previous findings (van Abeelen, 1966). Actually, the number of crossings was about two times higher in the C57 strain than in the other two lines: C57, 122.9±10.1; SEC, 56.9±4.1; DBA, 65.0±3.9). Again, the offspring of the active C57 mice shows a low or a high exploratory activity depending on whether they came from a cross with SEC or with DBA (C57×SEC, 60.2±6.1; C57×DBA, 115.0±11.2). Differences in responses to amphetamine and to scopolamine were also found. Rather surprisingly, both drugs reduced the activity in the C57 strain, while they increased the number of crossings in SEC and DBA mice (Fig. 15.6). The responses of the C57×SEC hybrids to amphetamine and scopolamine were essentially similar to those of the SEC mice. However, when examining the C57×DBA hybrid, the usual pattern of dominance was absent. In fact, amphetamine decreased the number of crossings (like it did in the C57 strain), while scopolamine enhanced the activity of the C57×DBA mice, as was observed in the DBA strain. The latter findings seem to indicate that the C57 genotype is dominant over the DBA genotype as far as the response to amphetamine is concerned, but that it is recessive to the DBA genotype with regard to the effects of scopolamine. Physostygmine diminished exploratory behaviour in all strains.

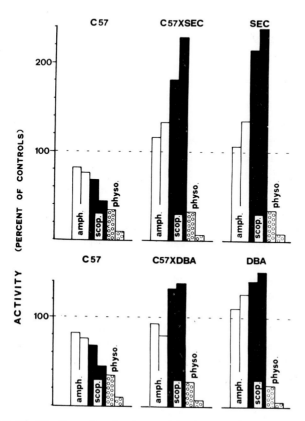

Fig. 15.6. Modifications of exploratory activity (expressed as percent of control level) in three strains of mice and two F_1 hybrids. Amphetamine sulphate (0.5 and 1.0 mg/kg, i.p.), scopolamine hydrobromide (2.5 and 5.0 mg/kg, i.p.), and physostigmine sulphate (0.15 and 0.30 mg/kg, i.p.) were injected 30 min. before the test. The lower and the higher dose of each drug are represented by the left and right columns, respectively. Each column represents the mean of crossings in a tilt-box for 15 mice.

These results are consistent with other evidence showing that C57 and DBA mice differ in activity and may respond to the same centrally active drugs in opposite directions (van Abeelen et al., 1969, 1971, and 1972). In addition it is evident that, although amphetamine or scopolamine produce opposite effects, depending on the strain considered, a second cholinergic agent, physostigmine, exerts a reducing effect in all genotypes. Since the levels of adrenergic and cholinergic mediators differ in the brains of mice of these strains, it is possible that their different reactivities to various adrenergic and cholinergic agents is related to this neuro-chemical diversity. A further study of the possible connections with learning abilities seems also worthwhile.

15.3. Recombinant inbred strains

Recently, application of more refined genetic methods has permitted us to analyze the exploratory activity of mice and its modification by scopolamine and amphetamine in greater detail. The search for the gene(s) responsible for the differences in basal exploratory activity and in the reactivity to scopolamine and amphetamine was founded on a genetic analytic method called the strain distribution pattern (SDP) method (Bailey, 1971). This type of genetic analysis was developed in order to determine quickly the probable location of a new gene by matching its pattern of distribution (SDP) with that of other genes among a set of inbred strains. Used were seven recombinant inbred (RI) strains, derived from an initial cross of the BALB/cBy strain with the C57BL/6By strain (Bailey, 1971). RI strains are bred from an F_2 cross between two unrelated but highly inbred progenitor strains and then maintained with strict inbreeding. This procedure will fix the chance recombination of genes, although in ever-decreasing amounts, in each successive generation after the F_1. The resulting group of strains can be looked upon, in a sense, as a replicable recombinant population. The SDP's of distinctive loci (histocompatibility, isozymes, hemoglobin, etc.) have been determined in the seven inbred strains. The SDP method, therefore, is useful for locating genes, since identical or similar SDP's for any pair of traits indicate possible pleiotropism or linkage of controlling genes.

In addition, a group of congenic lines (which are not to be confused with the RI strains) was developed from an initial cross between BALB/cBy(C) and C57BL/6By (B6) using a system of skin-graft testing, and backcrossing to B6 for at least 12 generations. This procedure resulted in congenic B6 lines, each of which differs from B6 itself by only one introduced chromosomal segment containing a C strain allele at a distinctive histocompatibility (H) locus. Each of these congenic lines was tested against the RI strains by means of skin-grafts to see which of the RI strains carry the C strain allele and which the B6 allele at a particular H locus, thereby establishing their strain distribution patterns.

The pattern the RI strains exhibited for low or high activity and the pattern of its modifications by scopolamine were compared with the strain distribution patterns (SDP's) already determined for the H-loci. Matching SDP's indicate identity or close linkage of two genes. However, two SDP's may match merely by chance $[P = (\frac{1}{2})^7]$.

A short-cut method of verifying the linkage suggested by the SDP of the new gene is using the congenic lines that carry H-genes with matching SDP's. If linkage had caused the matching SDP's, it is likely that the gene(s) responsible for activity and its modification by scopolamine or amphetamine were introduced into the congenic lines together with the H-gene lying on the same chromosomal segment. The effects of alleles at the individual loci of interest were then tested directly by comparing the activity – or the effects of scopolamine upon the activity – of the appropriate congenic lines with those of their background strain, C57BL/6By.

Seven RI strains (CxBD, CxBE, CxBG, CxBH, CxBI, CxBJ, and CxBK), derived from a cross between strains BALB/cBy (C) and C57BL/6By (B6) followed by more than 40 generations of full-sib mating, were used for this analysis (Oliverio et al., 1973).

Statistical analyses of the data obtained in the testing of all strains for basal exploratory activity revealed that these data are distinctly divided into two major categories. Strains CxBD, CxBI, and CxBJ and the progenitor strain BALB/cBy exhibited low basal exploratory activity, while all other RI strains and the two reciprocal F_1 hybrids were high in basal exploratory activity and similar to the other progenitor strain, C57BL/6By.

For the purpose of classifying all strains with regard to their exploratory activity after injection of the drug, two major response categories are considered. The first category groups all strains according to the occurrence of a significant increase or decrease in the activity following the lowest (2 mg/kg) scopolamine dose. It appears that strains CxBI, CxBJ, CxBD, CxBH, and progenitor strain BALB/cBy showed a trend toward an increase of basal exploratory activity, while all other strains tended to decline (Fig. 15.7).

The second classification groups all strains according to their overall responses to scopolamine. It is considered whether a particular strain persisted in its increase or decrease of exploratory activity over all three doses of scopolamine. Strains CxBE, CxBG, and CxBK, both reciprocal hybrids $B6CF_1$ and $CB6F_1$, and progenitor strain C57BL/6By showed a persistent decline in exploratory activity with increasing dose-levels of scopolamine. Strains CxBI, CxBJ, and progenitor strain BALB/cBy, on the other hand,

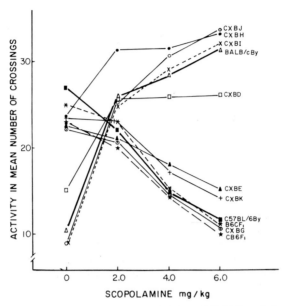

Fig. 15.7. Mean activity exhibited by BALB/cBy, C57BL/6By, their reciprocal F_1 hybrids $B6CF_1$ and $CB6F_1$, and the seven recombinant inbred strains after i.p. injection of saline or scopolamine in doses of 2, 4, or 6 mg/kg/body weight.

went up. However, strains CxBD and CxBH did not significantly change with increasing doses of scopolamine. In short, the latter two strains exhibited their maximum scopolamine-induced modification of exploratory activity with the lowest dose of the drug and, after this, remained at a plateau (Fig. 15.7).

Our statistical approach (based on analysis of variance) of the strains and crosses used has permitted us to arrive at a genetic model for basal exploratory activity based on the following considerations: (1) the RI strains were either high or low, like the progenitor strains C57BL/6By and BALB/cBy; intermediate groups were absent; (2) the reciprocal F_1 hybrids $B6CF_1$ and $CB6F_1$ were similar and, moreover, resembled the C57BL/6By progenitor strain, indicating dominance of the allele for high activity (or of the C57BL/6B genotype); (3) there was no evidence of maternal effects judging from the reciprocal F_1 hybrids; (4) there is evidence that the controlling gene is located in chromosome 4, since the congenic line H(w26) was BALB/cBy-like for exploratory activity; (5) we obtained data from a backcross, $B6CF_1 \times$ BALB/cBy, which consisted of two distinct groups. This is in agreement with previous results demonstrating dominance of the C57BL/6J genotype over the BALB/cJ genotype for some other measures of activity (DeFries and Hegmann, 1970; Oliverio and Messeri, 1973).

Different single genes have been found to contribute to the genetic variance of this type of exploratory activity of mice, as measured in longer sessions (25 min). However, none of these single-gene mutants differed from normal mice for initial (10 min) exploratory behaviour. The present results suggest that a single major gene may be responsible for the initial short-term exploratory activity of mice, apart from the presence of other genes that may contribute to the genetic variance of this trait (DeFries and Hegmann, 1970; Oliverio and Messeri, 1973; Oliverio et al., 1972).

The genetic model which best explains the data obtained for the sustained scopolamine-response postulates the existence of a second locus at which one allele controls the activity changes related to dose. These conclusions are based on: (1) the existence of a plateau effect in strains CXBD and CXBH, although the two strains exhibited quite dissimilar basal exploratory activities; (2) the lack of a pharmacological "ceiling effect" in the CXBD strain; (3) the confirmation, by the use of congenic lines H(w19) and H(w41), that this effect is not a pleiotropic effect of the H-2 locus, although the relevant allele is probably closely linked to it.

The symbol *Exa* is suggested to designate the locus, linked with H(w26) in chromosome 4, linkage group VIII (unpublished data), for basal short-term exploratory activity, with Exa^h to designate the allele determining high level and Exa^l the allele determining low level of activity. The symbol *Sco* is proposed for the second locus, linked with *H-2* in chromosome 17, linkage group IX, one allele controlling the sustained reactivity to scopolamine once the initial response has occurred, the other allele permitting a reversal of the exploratory activity level determined by *Exa*. It is suggested that Sco^a be used to designate the allele determining the absence of activity changes related to dose of scopolamine, and Sco^p for the allele controlling dose-dependent activity changes (Oliverio et al., 1973).

A further analysis of the exact nature of the genic control might be performed by extensive research into the acetylcholine levels, the enzymes involved in its synthesis and inactivation, and the turnover rates of acetylcholine in different brain regions.

The search for the gene(s) regulating the effects of amphetamine on exploratory behaviour indicated a highly varied response of the RI strains to amphetamine injection. This followed from the separation of the strains into six groups which differed from one another, suggesting that the amphetamine-response is under the influence of several genes which may control one or more critical events in the pharmacological modulation of this drug's activity.

Application of a genetic analysis based on recombinant inbred strains seems to be a new and promising approach in the field of psychopharmacogenetics. It has allowed us to establish that the effects of chlorpromazine on shuttle-box avoidance behaviour are determined by two genes. In experiments still in progress, it has been possible to obtain proof of locus identity for one of these two genes (Castellano et al., 1974).

Genetic segregation and recombination have provided the very foundation of classical genetics by revealing the nature of the gene and the modalities of linkage. On the other hand, inbreeding, which is the counterpart of segregation and recombination, has shown great utility as a tool for generating replicable genotypes. The advantages of these opposing processes can be exploited simultaneously by the use of RI strains, which appear to be very promising materials in behavioural genetics. In general, we may conclude by asserting that psychopharmacogenetics has developed from previous findings, concerned with individual differences in drug responses, to a promising branch in which the study of genetic factors in the control of psychoactive drugs will offer new means of analyzing the brain mechanisms underlying behavioural individuality.

References

Abeelen, J. H. F. van (1966): Effects of genotype on mouse behaviour. *Anim. Behav.*, **14**, 218.
Abeelen, J. H. F. van and Strijbosch, H. (1969): Genotype-dependent effects of scopolamine and eserine on exploratory behaviour in mice. *Psychopharmacologia (Berl.)*, **16**, 81.
Abeelen, J. H. F. van, Smits, A. J. M. and Raaijmakers, W. G. M. (1971): Central location of a genotype-dependent cholinergic mechanism controlling exploratory behaviour in mice. *Psychopharmacologia (Berl.)*, **19**, 324.
Abeelen, J. H. F. van, Gilissen, L., Hanssen, Th., and Lenders, A. (1972): Effects of intrahippocampal injections with methylscopolamine and neostigmine upon exploratory behaviour in two inbred mouse strains. *Psychopharmacologia (Berl.)*, **24**, 270.
Bailey, D. W. (1971): Recombinant inbred strains. *Transplantation*, **11**, 325.
Bovet, D., Bovet-Nitti, F. and Oliverio, A. (1966): Action of nicotine on spontaneous and acquired behavior in rats and mice. *Ann. N.Y. Acad. Sci.*, **142**, 261.

Bovet, D., Bovet-Nitti, F. and Oliverio, A. (1969): Genetic aspects of learning and memory in mice. *Science*, **163**, 139.

Brown, A. M. (1959): The investigation of specific responses in laboratory animals. Symp. Laboratory Animals Centre, Royal Veterinary College, London, 9–16.

Burnett, C. H., Dent, C. R., Harper, C. and Warland, B. J. (1964): *Amer. J. Med.*, **36**, 222.

Castellano, C., Eleftheriou, B. E., Bailey, D. W. and Oliverio, A. (1974): Chlorpromazine and avoidance behavior: A genetic analysis. *Psychopharmacologia*, **34**, 309

DeFries, J. C. and Hegmann, J. P. (1970): Genetic analysis of open-field behavior. In: Contributions to Behavior-Genetic Analysis: The Mouse as a Prototype. G. Lindzey and D. D. Thiessen, eds. Appleton-Century-Crofts, New York, p. 23.

Ebel, A., Hermetet, J. C. and Mandel, P. (1973): Comparative study of acetylcholinesterase and choline acetyltransferase enzyme activity in brain of DBA and C57 mice. *Nature New Biol.*, **242**, 56.

Eleftheriou, B. E. (1971): Regional brain norepinephrine turnover rates in four strains of mice. *Neuroendocrinology*, **7**, 329.

Eriksson, K. (1969): Factors affecting voluntary alcohol consumption in the albino rat. *Ann. Zool. Fennici*, **6**, 227.

Eriksson, K., and Kiianmaa, K. (1971): Genetic analysis of susceptibility to morphine addiction in inbred mice. *Ann. Med Exper. Biol. Fenn.*, **49**, 73.

Evangelista, A. M., Gattoni, R. C. and Izquierdo, I. (1970): Effects of amphetamine, nicotine and hexamethonium on performance of a conditioned response during acquisition and retention trials. *Pharmacology*, **3**, 91–96.

Evangelista, A. M. and Izquierdo, I. (1972): Effects of atropine on avoidance condition: interaction with nicotine and comparison with N-methyl-atropine. *Psychopharmacologia (Berl.)*, **27**, 241.

Fontenay, M., LeCornec, J., Zaczinska, M., Debarle, M. C., Simon, P. and Boissier, J. R. (1970): Problèmes posès par l'utilisation de trois tests de comportement du rat pour l'étude des médicaments psychotropes. *J. Pharmacol. (Paris)*, **1**, 243.

Fox, M. W. (1964): A phylogenetic analysis of behavioral neuro-ontogeny in precocial and non-precocial mammals. *Can. J. comp. Med.* **28**, 197.

Fuller, J. L. (1964): Measurement of alcohol preference in genetic experiments. *J. comp. physiol. Psychol.*, **57**, 85.

Fuller, J. L. (1966): Variation of effects of chlorpromazine in three strains of mice. *Psychopharmacologia (Berl.)* **8**, 408.

Fuller, J. L. (1970): Strain differences in the effects of chlorpromazine and chlordiazepoxide upon active and passive avoidance in mice. *Psychopharmacologia (Berl.)*, **16**, 261.

Gillette, J. R. (1965): Drug toxicity as a result of interference with physiological mechanisms. *Ann. N.Y. Acad. Sci.*, **123**, 42–53.

Goldenberg, H. and Fishman, V. (1961): Species dependence of chloropromazine. *Proc. Soc. exp. Biol. Med.*, **108**, 178.

Green, M. C. (1972): (For the Committee on Standardized Genetic Nomenclature for Mice). Standard karyotype of the mouse. *Mus musculus. J. Hered.*, **63**, 69.

Kalow, K. (1966): Genetic aspects of drug safety. *Appl. Ther.*, **8**, 44.

Koppanyi, T. and Dille, J. M. (1934): Studies on barbiturates. VII. Experimental analysis of barbital action. *J. Pharmacol. exp. Therap.*, **52**, 91.

Koppanyi, T. and Avry, M. A. (1966): Species differences and the clinical trial of new drugs: a review. *Clin. Pharmacol. Therap.* **7**, 250.

Lindzey, G., Loehlin, J., Manosevitz, M. and Thiessen, D. D. (1971): Behavioral genetics. *Ann. Rev. Psychol.*, **22**, 39.

McClearn, G. E. and Rodgers, D. A. (1959): Differences in alcohol preference among inbred strains of mice. *Quart. J. Stud. Alcohol*, **20**, 691.

McClearn, G. E., Wilson, J. R. and Meredith, W. (1970): The use of isogenic and heterogenic mouse stocks in behavioral research. In: Contributions to Behavior-genetic Analysis: The Mouse as a Prototype. Lindzey, G. and Thiessen, D. D., eds. Appleton-Century-Crofts, New York, p. 5.

Meier, H. (1963): Factors influencing drug metabolism. In: Experimental Pharmacogenetics. Academic Press, New York, pp. 9–75.

Myrianthopoulos, N. C., Kurland, A. A. and Kurland, L. T. (1962): Hereditary predisposition in drug-induced parkinsonism. *Arch. Neurol.*, **6**, 5.

Oliverio, A. (1968): Neurohumoral systems and learning. In: D. H. Efron, ed. Psychopharmacology, a review of progress, 1957–1967, (Public Health Service Publication, No. 1936, Washington, D.C.) pp. 867–868.

Oliverio, A. (1971): Genetic variations and heritability in a measure of avoidance learning in mice. *J. comp. physiol. Psychol.*, **74**, 390.

Oliverio, A. and Castellano, C. (1973): Pharmacogenetic aspects of learning and memory. In: Proceedings Fifth International Congress of Pharmacology, San Francisco, 1972. G. H. Acheson, F. E. Bloom, J. Cochin, T. A. Loomis, R. A. Maxwell and G. T. Okita, eds. (Karger, Basel).

Oliverio, A., Castellano, C. and Messeri, P. (1972): A genetic analysis of avoidance, maze and wheel running behavior in the mouse. *J. comp. physiol. Psychol.*, **79**, 459.

Oliverio, A. and Messeri, P. (1973): An analysis of single-gene effects on avoidance, maze and wheel-running and exploratory behavior in the mouse. *Behav. Biol.*, **8**, 771.

Oliverio, A., Eleftheriou, B. E., and Bailey, D. W. (1973): Exploratory activity: Genetic analysis of its modification by scopolamine and amphetamine. *Physiol. Behav.* **10**, 893.

Piala, J. J., High, J. P., Hassert, G. L., Burke, J. C. and Craver, B. N. (1959): Pharmacological and acute toxicological comparisons of triflupromazine and chlorpromazine. *J. pharmacol. exp. Therap.*, **127**, 55.

Pryor, G. T., Schlesinger, K. and Calhoum, W. H. (1966): Differences in brain enzymes among five inbred strains of mice. *Life Sci.*, **5**, 2105.

Quinn, G., Axelrod, J. and Brodie, B. B. (1954): Species and sex differences in metabolism and duration of hexobarbital. *Fed. Proc.*, **13**, 395.

Rech, R. H. (1966): Amphetamine effects on poor performance of rats in a shuttle-box. *Psychopharmacologia (Berl.)*, **9**, 110.

Roderick, T. H., Ruddle, F. H., Chapman, V. M. and Shows, T. B. (1971): Biochemical polymorphisms in feral and inbred mice (*Mus musculus*). *Biochem. Genet.*, **5**, 457.

Sansone, M. and Messeri, P. (1973): Strain differences in the effects of chlordiazepoxide and chlorpromazine on avoidance behavior of mice. *Pharmacol. Res. Comm.* (in press).

Schlesinger, K. and Griek, B. J. (1970): The genetics and biochemistry of audiogenic seizures. In: Contributions to Behavior-genetic Analysis: The Mouse as a Prototype. G. Lindzey and D. D. Thiessen, eds. Appleton-Century-Crofts, New York, p. 219.

Selander R. K., Kunt, W. G. and Yang, S. Y. (1969): Protein polymorphism and genic heterozygosity in two European subspecies of the house mouse. *Evolution*, **23**, 379.

Shnider, B. I., Frei, E., Tuohi, J., Gorman, J., Freireich, E., Brindley, C. O. and Clements J. (1960): Clinical studies on 6-azauracil. *Cancer Res.*, **20**, 28.

Spinks, A. (1965): Justification of clinical trial of new drugs. In: Evaluation of new drugs in man. Zaimis, E., ed. Pergamon Press, London, pp. 1–19.

Sprott, R. L. (1971): Inheritance of avoidance learning. The Jackson Laboratory's Annual Report, **42**, 78.

Staats, J. (1972): Standard nomenclature for inbred strains of mice: Fifth listing. *Cancer Res.*, **32**, 1609.

Valatx, J. L., Bugat, R. and Jouvet, M. (1972): Genetic studies of sleep in mice. *Nature*, **238**, 226.

Vesell, E. (1972): Introduction: genetic and environmental factors affecting drug response in man. *Fed. Proc.*, **31**, 1253.

Wahlsten, D. (1972): Genetic experiments with animal learning: a critical review. *Behav. Biol.*, **7**, 143.

Williams, R. T. (1962): Altered drug metabolism. Ciba Foundation Symposium on Enzymes and Drug Action. Little, Brown & Co., Boston, pp. 239–244.

Wimer, C. C., Wimer, R. E. and Roderick, T. H. (1971): Some behavioral differences associated with relative size of hippocampus in the mouse. *J. comp. physiol. Psychol.*, **76**, 57

CHAPTER 16

Neurochemical correlates of behaviour in inbred strains of mice

P. Mandel, A. Ebel, G. Mack and E. Kempf

16.1. Introduction

The recent discoveries in molecular biology may provide new perspectives for the genetics of behaviour and for the study of hereditary diseases. However, it seems probable that in higher species very complex regulatory mechanisms operate in addition to the basic systems described in the genetics of bacteria and fungi. Thus the establishment of relationships between genetic variation and behavioural complexity appears to be a difficult task. Human geneticists are concerned with the genetic analysis of intelligence, mental deficiency, and psychiatric disorders which offer an opportunity to study the genetic control of behaviour. It is very worthwhile to trace the intermediate pathways between genes and behavioural traits in humans: the fear of an incurable inherited disease may be alleviated because its expression involves a molecular alteration and one may hope – as already holds for some somatic hereditary diseases, e.g. phenylketonuria or Refsum's disease – to be able to correct such metabolic defects. It is necessary, however, to point out the limitations of investigations into genetically determined behaviour. Humans represent a very heterogeneous population and can hardly offer a model in obtaining conclusive evidence of gene-behaviour relationships.

The species that can be used as models for such research should satisfy the following criteria: they must be prolific and easy to breed and they must have a relatively small number of chromosomes. The fruit fly, *Drosophila melanogaster*, meets these requirements, as do certain other species low on the phylogenetic scale. Effort has been expended upon sexual (Sturtevant, 1915; Merrell, 1949; Tebb and Thoday, 1956), adaptive (Pittendrigh, 1958), locomotory (Connolly, 1966, 1967), and visually elicited (Hotta and Benzer, 1970) behaviour in *Drosophila*. However, these animals are unsuited for studying the genetic control of complex behaviour; the obvious feasibility of genetic techniques is counteracted by the difficulties in extrapolating the behavioural data to those of humans. An animal fit for such studies is the mouse, *Mus musculus*, especially convenient because of the ease of genetic manipulation and the existence of many neurological mutants. The "waltzing mouse" is one of the first mutants recorded and its history in China goes back to 80 B.C. At present, about 100 neurological mutants of mice are known. Inbred mouse strains, studied with respect to behaviour or selected for special characteristics, provide excellent material for physiological-behavioural analysis in intact organisms. Moreover, the contribution of studies on inherited factors might be superior to those on surgery, electrical stimulation, drug administration, or other techniques. In this chapter we present some examples of neurochemical correlates possibly involved in genetically determined behavioural differences between inbred strains of mice, focusing on the complexity of such analyses.

16.2. Behavioural characteristics of inbred strains of mice

Striking differences between the behaviours of various mouse strains have been established. Differences have been found for locomotory activity, emotionality, learning, aggressiveness, susceptibility to audiogenic seizures, response to stressors, alcohol preference, and a number of other behavioural measures (Ginsburg and Allee, 1942; Scott, 1942; Lindzey, 1951; Thompson, 1953; Fuller and Thompson, 1960; Royce and Covington, 1960; Thiessen, 1961; McClearn, 1965, Fuller, 1967, Fuller and Sjursen, 1967; Al-Ani et al., 1970; Lindzey and Thiessen, 1970). Thus, any correlations between behaviour and neurophysiology, on the one hand, and neurochemical patterns, on the other, can be tested. In addition, by altering behavioural parameters by means of drugs or cross-breeding and investigating concomitant changes at the biochemical level, one may hope to characterize the neurochemical basis of the behavioural reactivity of the strains. It is important to measure many aspects of the behavioural phenotype when studying genetically different groups.

A survey of the literature on strains DBA/2, C57BL/6, and their F_1 hybrid, shows for several characteristics either similarities of the F_1 with one of the parental strains (McGill and Blight, 1963; Mordkoff et al., 1964; Bovet et al., 1969; Oliverio et al., 1972) or intermediate performances (Fuller et al., 1950; Schlesinger and Mordkoff, 1963). Royce and Covington (1960) have shown that C57BL/6J mice condition very poorly but that DBA/2J mice can be conditioned well in an electric shock avoidance experiment. Van Abeelen (1966) analyzed stereotyped behaviour patterns, general activity, and a simple learning response in these strains. He found that strain DBA/2J is mainly characterized by low score for various locomotory and exploratory activities and strain C57BL/6J by low values for feeding behaviour, defecation, grooming, and aggresiveness. For some frequencies the behaviour of C57BL/6J resembled that of the F_1 although the latter is much more aggressive.

16.3. Serotonin, norepinephrine, gamma-aminobutyric acid, and behaviour in mouse strains

Determinations of levels of serotonin (5-HT) and norepinephrine (NE) have been made by Maas (1962), using brain parts comprising the diencephalon, mesencephalon, and pons taken from two strains of mice, C57BL/10J and BALB/cJ. The C57BL/10 mice exhibit more exploratory activity and less emotionality, as measured in an open-field defecation test, and superiority in fighting when comparing them with the BALB strain. In these portions of the brain, the amount of 5-HT per gram of tissue was significantly higher in the BALB strain, about 30 % while the levels of NE did not differ (Maas, 1962). Significantly higher NE levels were observed in the hippocampus and pyriform cortex portions of C57BL/10 mice but in the brain stem the content of NE was higher in BALB (Sudak and Maas, 1964). In another report by Maas (1963), dealing with the whole brain and with the part of the brain that remains after removal of the diencephalon, mesencephalon, and pons, the 5-HT contents appeared to be similar in the BALB and C57BL/10 strains. The strain difference mentioned earlier became smaller when monoamine oxidase (MAO) was inhibited by injecting a MAOI compound. The C57BL/10 strain seems to show a higher amount and specific activity of MAO; if MAO is completely inhibited, a slight increase in 5-HT results. Large differences in behaviour between the two strains were observed following the drug administration: the BALB mice were periodically hyperactive and hypersensitive to stimuli during the periods of quietness, but the

C57BL/10 mice showed only a moderate sensitivity to stimuli and remained in the corner of the cage (Maas, 1963). When measuring the 5-HT and NE levels in "emotional" strains, C57BL/10, BALB, and C3H/J, and in "non-emotional" strains, A/J and AKR/J, it was found that the latter group showed lower 5-HT levels than the former (Sudak and Maas, 1964). According to these authors, neither the hippocampus, nor the pyriform cortex revealed differences between strains C57BL/10 and BALB with regard to 5-HT, only in the stem portions a significant difference was present.

Seiden and Peterson (1968) have shown that the suppression of a conditioned avoidance response by reserpine was identical for DBA/1 and C57BL/10 mice. However, the time course of the reversal of the conditioned avoidance response after treatment with L-dopa differed for the two strains, the DBA/1 mice maintaining the response for a longer period than the C57BL/10 mice. Reserpine-induced depletion of catecholamine was similar in DBA/1 and C57BL/10. The endogenous level of dopamine in whole brain was slightly higher in the DBA/1 strain. Moreover, after dopa injection, the level of dopamine (μg/g) turned out to be higher in the DBA/1 strain than in the C57BL/10 strain, but to be similar if taken per total brain. No significant strain differences emerged for the endogenous levels of brain NE; reserpine caused a similar depletion.

Two strains of mice (C57BL and A2G), which differ for a number of behavioural traits, have been used in investigations on gamma-aminobutyric acid (GABA) and glutamate synthesis, acetylcholinesterase (AChE) activity, and the levels of biogenic amines in the midbrain and brain stem. C57BL is characterized by a high spontaneous activity, a low reactivity to environmental stimulation, and low seizure susceptibility. The more reactive A2G mice showed a lower rate of GABA production in the cerebral cortex and a higher AChE activity as compared to the C57BL strain. Brain dopamine levels were significantly higher in the C57BL and could be related to activity or reactivity differences between the strains. Brain 5-HT and NE levels were similar in these strains (Al-Ani et al., 1970).

Several studies indicate a role of brain 5-HT and NE in short-term emotional arousal (Carlson et al., 1967), sympathetic activation due to environmental changes (Vogt, 1960), and learning ability (Woolley and Van der Hoeven, 1963). Eleftheriou (1971) examined the disappearance of ^{14}C-NE from the hypothalamus, hippocampus, frontal cortex, and amygdala in four strains of mice: DBA/2J, C57BL/6J, C3H/HeJ, and SJL/J. After intracisternal injection of ^{14}C-NE, uptake and disappearance of the amine was exhibited in the following descending order: hippocampus > hypothalamus > frontal cortex > amygdala. DBA/2 and SJL showed the highest rate of disappearance, while C3H exhibited the lowest. This finding might be related with the high emotionality of DBA/2 mice and with the high audiogenic seizure susceptibility of DBA/2 and SJL animals. No analysis of bound and free NE was carried out.

16.4. The cholinergic system and behaviour in mouse strains

It has been postulated that a cholinergic system in the brain acts to suppress non-reinforced behavioural responses in rodents placed in learning situations (Carlton, 1963) and also that a cholinergic system is involved in the control of exploration (Vossen, 1966). There is sufficiently convincing evidence that cholinergic mechanisms play a role in habituation (Carlton, 1969) and in limbic activating systems (Domino et al., 1967), that cholinertic receptors may control operant response strength (Margules and Margules, 1973), and that damage to portions of the limbic system results in a variety of changes in behaviour (Grossman, 1972). Warburton (1969a, 1969b) and Warburton and Russell (1969) provided evidence for a role of the hippocampus in the cholinergic control of behaviour in rats. These experiments neither exclude the possibility that other brain regions, too, are involved in the regulation of exploratory behaviour, nor do they exclude that alternative non-cholinergic systems play a role as well. As regards the functions of the limbic system, it should be pointed out that the effects of limbic lesions are not in all respects identical to those of administered cholinergic blockers; lesions in the limbic system also disrupt neural processes which are not mediated by cholinergic mechanisms. One cannot hope to isolate the role of the cholinergic components of the limbic system without paying closer attention to the functions of non-cholinergic mechanisms (Grossman, 1972).

There is also evidence that anticholinergics like scopolamine may disrupt learning performance and that they may interfere with an animal's ability to habituate to novelty and thus prolong and intensify exploratory activity in rats and mice (Carlton, 1963; Meyers, 1965; Parkes, 1965; Longo, 1966; Calhoun and Smith, 1968; Carlton, 1968; Daly, 1968; Leaton, 1968; Meyers, 1968; Oliverio, 1968a, 1968b, 1968c). A study on two inbred mouse strains and their F_1 hybrid (Van Abeelen, 1966) showed that DBA/2 mice are characterized by low scores for various exploratory activities, as compared to high-scoring C57BL/6 animals; the F_1 rated like C57BL/6. From their experiments with anticholinergic and anticholinesterase drugs, Van Abeelen et al. (1969, 1971, 1972) conclude that in the central nervous system of mice a genetically controlled cholinergic mechanism exists which facilitates exploratory behaviour.

16.5. Multiple neurochemical correlates of behaviour

16.5.1. Choline acetyltransferase and acetylcholinesterase

Since several observations suggest that cholinergic systems as well as biogenic amines play a role in behavioural phenomena (Bennett et al., 1964; Rosenzweig and Bennett, 1969; Igic et al., 1790; Deutsch, 1971; Vogel and

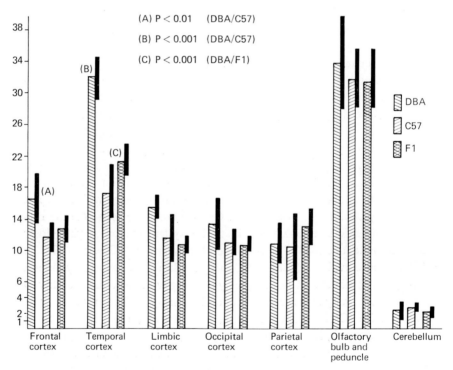

Fig. 16.1. Choline acetyltransferase activity in different brain regions of DBA/2J and C57BL/6J mice and their F_1 offspring. Values are expressed as μmole/g proteins^{-1} H^{-1}.

Leaf, 1972), either genetically determined, or resulting from environmental effects, or both, we have started investigations on cholinergic and aminergic mechanisms in the central nervous system of mice from three strains – C57BL/6J ("C57BL"), DBA/2J ("DBA"), and SECRe/1J ("SEC") – and their F_1 hybrids. These groups differ in wheel-running activity, avoidance behaviour, and maze learning. As shown by Bovet et al. (1968) and Oliverio et al. (1972), the C57BL strain is characterized by high spontaneous activity and low levels of avoidance, while the other two strains, DBA and SEC, attain high avoidance levels and few maze errors but show a much lower activity. Crossing the C57BL with the SEC mice resulted in SEC-like progeny with regard to activity, avoidance, and maze learning, while crossing the C57BL with the DBA strain yielded offspring similar to the C57BL phenotype for activity, avoidance, and maze learning (Oliverio et al., 1972).

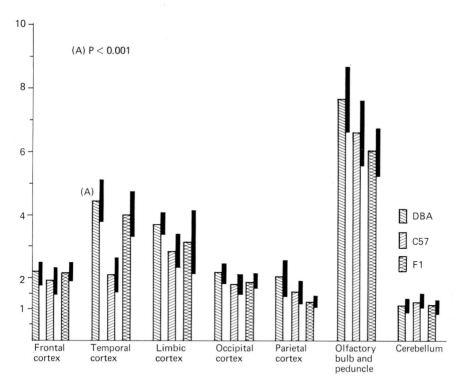

Fig. 16.2. Acetylcholinesterase activity in different brain regions of DBA/2J and C57BL/6J mice and their F_1 offspring. Values are expressed as μmole/g proteins^{-1} H^{-1}.

Cholinergic mechanisms were studied by testing the two main enzymes involved in acetylcholine synthesis and degradation, choline acetyltransferase (ChA; EC 2.3.1.6) and acetylcholinesterase (AChE; EC 3.1.1.7), in several parts of the central nervous system (Ebel, et al., 1973; Mandel et al., 1973, 1974). Our results showed that in the occipital, parietal, and limbic cortical regions, the olfactory bulbs, peduncles, and cerebellum no significant differences were present between the three strains in respect of AChE or ChA activities (Figs. 16.1–4). Slight, but significant differences were observed for ChA in the frontal cortex when comparing DBA to C57BL and SEC, the values being higher in DBA. However, in the temporal cortex the activities of ChA and AChE were much higher in DBA and SEC as compared to C57BL. These differences were highly significant (Student-Fisher test). The temporal ChA activity in the hybrids was close to that of the C57BL strain in the DBA × C57BL cross and close to the SEC strain in the SEC ×

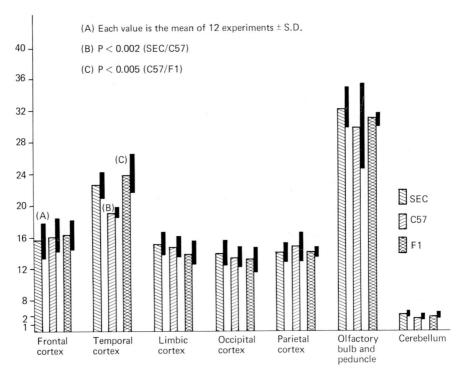

Fig. 16.3. Choline acetyltransferase activity in different brain regions of SECRe/1J and C57BL/6J mice and their F_1 offspring. Values are expressed as μmole/g proteins^{-1} H^{-1}.

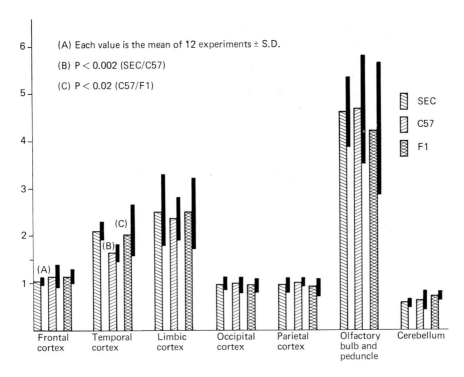

Fig. 16.4. Acetylcholinesterase activity in different brain regions of SECRe/1J and C57BL/6J mice and their F_1 offspring. Values are expressed as μmole/g proteins^{-1} H^{-1}.

C57BL cross, paralleling the learning ability. The AChE activity in the hybrids was close to that of one of the parental strains in both crosses. Our data indicate a relationship between ChA activity in the temporal lobe and learning ability in the three parental strains and their hybrids and suggest a role of the cholinergic system of the temporal lobe in memory processes. These suggestions can be related to the observations mentioned earlier which indicate that the cholinergic system is involved not only in different overt behaviours, but also in learning and memory storage; see also Carlton (1969), Stein (1969), and Deutsch (1971).

The idea that the temporal lobe may play a specific role in learning and discrimination processes is also supported by other observations. This is the case for the Klüver-Bucy syndrome characterized by "psychic blindness", with an inability to distinguish food from other objects (Klüver and Bucy, 1937). Moreover, Penfield and Milner (1958) and Penfield and Perot (1962) attributed to the temporal cortex an important function in the process of memory formation. More recently, Mishkin (1966) produced deficits in form discrimination learning by occipital-temporal disconnection. Horel and Keating (1969) observed that after disconnection of cortical pathways between occipital and temporal lobes, monkeys displayed an inability to distinguish between food and inedible objects, although they were able to locate the objects. Marked changes in escape behaviour were also produced in these animals. Finally, Keating and Horel (1971) have shown that in monkeys parietal-temporal disconnection produces a loss of their ability to distinguish a steel bolt from a piece of food when only using tactile cues. The data suggested that the temporal lobe is involved in the organization of the somato-sensory system, analogous to its function in discrimination processes related to visual stimuli.

16.5.2. Norepinephrine levels and turnover

The NE level is significantly higher in whole brains of C57BL mice as compared to DBA and SEC (Table 16.1). The turnovers measured after α-methyltyrosine injection are rather close in the three strains, although the turnover rate seems to be somewhat higher in C57BL mice since the amount of NE is also higher (Mack et al., 1973; Kempf et al., 1974). When separate brain areas were examined (Table 16.2), a significantly higher amount of NE was found in the pons and medulla in the C57BL strain as compared to DBA and SEC. In contrast, lower values were observed in the hypothalamus of C57BL mice. The turnover time is also shorter in the pons and medulla of the C57BL mice, compared to DBA and SEC. In the F_1 generation of the cross DBA × C57BL, the values for NE in the pons and medulla were closer to C57BL, while those of the SEC × C57BL cross were closer to SEC. For the

TABLE 16.2

Norepinephrine levels ±S.D. in various brain areas of mice from three strains and two F_1 hybrids.

Brain areas	DBA/2J	F_1	C57BL/6J	F_1	SECRe/1J
Pons and medulla	7.0±0.4	7.8±0.4	8.0±0.4	7.3±0.4	7.4±0.6
Hypothalamus	16.5±1.8	14.7±1	14.7±0.2	14.5±1.4	16.3±1.0
Cortex	5.5±0.7	4.7±0.2	5.3±0.4	3.8±0.4	3.8±0.7

TABLE 16.1
Brain serotonin, dopamine, and norepinephrine levels and turnover in three strains of mice.

Strains	Serotonin ng/brain ± S.D. N	turnover rate (ng/g/hr)	Dopamine ng/brain ± S.D. N	turnover rate (ng/g/hr)	Norepinephrine ng/brain ± S.D. N	turnover rate (ng/g/hr)
C57BL/6J	472 ± 45 (22)	192	703 ± 91 (29)	348	266 ± 19 (29)	104.5
DBA/2J	397 ± 38* (22)	185	687 ± 82 (28)	313	219 ± 17* (29)	93.2
SECRe/1J	431 ± 26* (18)	179	581 ± 83* (25)	427	228 ± 17* (25)	98.5

* $P < 0.001$ (Student-Fisher test)

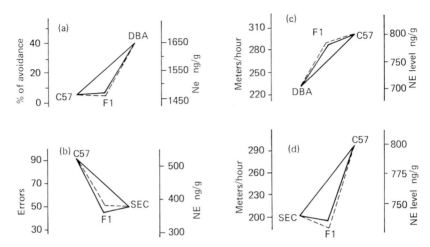

Fig. 16.5. Correlations between performances and norepinephrine (NE) levels in the parental strains and their F_1 hybrids. A: correlation between avoidance and norepinephrine level in the hypothalamus. B: correlation between maze learning and norepinephrine level in the cortex. C and D: correlation between activity and norepinephrine level in the pontine medulla.

Unbroken lines: performances (Oliverio et al., 1972).
Broken lines: norepinephrine level in the different brain areas.

hypothalamus the NE levels of both hybrid types resembled those of C57BL.

The higher levels of NE in the C57BL pons and medulla – where the reticular formation is located – may be related with the higher activity of these mice; a similar relationship seems to be present when considering the two different hybrid types (Fig. 16.5). Since the hypothalamus is involved in many aspects of behaviour, one can speculate on the higher values for the hypothalamic NE observed in DBA and SEC, as compared to C57BL. A correlation between hypothalamic NE levels and avoidance behaviour in the parental strains may be proposed. In the F_1 (C57BL × DBA) the correlation still exists, but the hypothalamic NE levels of the F_1 (C57BL × SEC) were close to the C57BL values.

16.5.3. Dopamine levels and turnover

The dopamine levels per whole brain, unlike NE, are equally high for DBA, but significantly lower in SEC, as compared to C57BL. The turnover times after α-methyltyrosine injection differ between the three strains: low in

SEC and high in DBA compared to C57BL (Kempf et al., 1974). Dopamine, which is almost exclusively found in the extrapyramidal system, may play a role in differences in psychomotor sensitivity and/or locomotory activity (Van Rossum and Hurkmans, 1964). The sensitivity to chlorpromazine of the SEC mice may also be connected with their dopamine level and turnover (Oliverio et al., 1973; Castellano et al., 1974).

16.5.4. Serotonin levels and turnover

The amount of 5-HT per gram of fresh weight was the same in all three strains. Since brain weight differs between the three strains, high total values can be expected for animals having high brain weight. Thus, the 5-HT level of whole brain seems to be more convenient for evaluating the situation. In fact, whole brain content is lower in DBA and SEC when compared to C57BL. Furthermore, the turnover time after pargyline injection is similar for the three strains (Kempf et al., 1974). It would be interesting to investigate different brain areas and to try to correlate them with behaviours which are, possibly, controlled by serotoninergic mechanisms.

16.6. Conclusions

Attempts to elucidate how transmitters are related to behaviour have been made by the induction of metabolic changes during and after training by means of drugs or anatomical lesions. A more natural approach is available: the study of various inbred strains of mice which present clear-cut differences in behaviour. The relevance of genetics to the analysis of behaviour and its neurophysiological and neurochemical bases is beyond doubt (Van Abeelen, 1966; Pryor et al., 1966; Bovet et al., 1969; Van Abeelen et al., 1969, 1971, 1972; Elias, 1970; Eleftheriou and Bailey, 1972; Oliverio et al., 1972, 1973; Oliverio and Messeri, 1973). In view of the complexity of behavioural traits, a large number of neurochemical correlates have to be investigated simultaneously in the parental strains and their hybrids. Not only the absolute levels of transmitter substances have to be investigated in different specific anatomical and functional brain areas, but also the rates of their synthesis, release, and degradation.

In our experiments, a plausible relationship could be established between ChA activity in the temporal limbic lobe and NE content of the hypothalamus, on the one hand, and learning ability on the other (see the survey in

TABLE 16.3

Behavioural characteristic of the C57BL/6J, DBA/2J, and SECRe/1J strains of mice and of two F_1 hybrid groups, and some neurochemical correlates.

	C57BL/6J	DBA/2J	F_1(C57×DBA)	SECRe/1J	F_1(C57×SEC)
Maze learning ability	low*	high**	low*	high**	high**
Avoidance behaviour performance	low	high	low	high	high
Wheel-running activity	high	low	high	low	low
Acetylcholinesterase in temporal lobe	low	high	high	high	high
Choline acetyltransferase in temporal lobe	low	high	low	high	high
Norepinephrine in total brain	high	low	intermediate	low	intermediate
Norepinephrine in pontine medulla	high	low	high	low	low
Norepinephrine in hypothalamus	low	high	low	high	low
Serotonin in total brain	high	low	n.d.	low	n.d.
Dopamine in total brain	high	high	n.d.	low	n.d.

* compared to DBA/2J and SECRe/1J
** compared to C57BL/6J
n.d.: not determined

Table 16.3). A good correlation could also be established between NE content and turnover rate in the pons and medulla and locomotory activity. However, many other differences exist between the investigated strains with regard to dopamine and 5-HT. The significance of these findings is not clear at present. Differences in free amino acids and ATPase activity (to be published later) were also found between the three strains.

In conclusion, the genetically determined behavioural variation paralleling the neurochemical variation requires extensive investigations. It would be dangerous to centre the neurochemical approach on a single transmitter or macromolecular metabolic pathway. In addition to genetically determined neurochemical correlates, secondary alterations in different pathways may be found as we did in neurological mutants (Kempf et al., 1973).

References

Abeelen, J. H. F. van (1966): Effects of genotype on mouse behaviour *Anim. Behav.* **14**, 218,

Abeelen, J. van, L. Gilissen, Th. Hanssen and A. Lenders (1972): Effects of intrahippocampal injections with methylscopolamine and neostigmine upon exploratory behaviour in two inbred mouse strains. *Psychopharmacologia (Berl.)*, **24**, 470.

Abeelen, J. H. F. van, A. J. M. Smits and W. G. M. Raaijmakers (1971): Central location of a genotype-dependent cholinergic mechanism controlling esploratory behaviour in mice. *Psychopharmacologia (Berl.)*, **19**, 324.

Abeelen, J. H. F. van and H. Strijbosch (1969): Genotype-dependent effects of scopolamine and eserine on exploratory behaviour in mice. *Psychopharmacologia (Berl.)*, **16**, 81.

Al-Ani, A. T., G. Tunnicliff, G. Rick and G. A. Kerkut (1970): GABA production, acetylcholinesterase activity and biogenic amine levels in brain for mouse strains differing in spontaneous activity and reactivity. *Life Sci.*, **9**, 21.

Bennett, E. L., M. C. Diamond, D. Krech and M. R. Rosenzweig (1964): Chemical and anatomical plasticity of brain. *Science*, **146**, 610.

Bovet, D., F. Bovet-Nitti and A. Oliverio (1968): Memory and consolidation mechanisms in avoidance learning of inbred mice. *Brain Res.* **10**, 168.

Bovet, D., F. Bovet-Nitti and A. Oliverio (1969): Genetic aspects of learning and memory in mice. *Science*, **163**, 139.

Calhoun, W. H. and A. A. Smith (1968): Effects of scopolamine on acquisition of passive avoidance. *Psychopharmacologia (Berl.)*, **13**, 201.

Carlson, L. A., L. Hevi and L. Orö (1967): Plasma lipids and urinary excretion of catecholamines during acute emotional stress in man and their modification by nicotinic acid. *Försvars medicin*, **3**, suppl. 2, 129.

Carlton, P. L. (1963): Cholinergic mechanisms in the control of behavior by the brain. *Psychol. Rev.*, **70**, 19.

Carlton, P. L. (1968): Brain acetylcholine and habituation. In: Anticholinergic drugs and Brain Functions in Animals and Man, Bradley, P. B. and M. Fink, eds. (Elsevier, Amsterdam) pp. 48–60.

Carlton, P. L. (1969): Brain-acetylcholine and inhibition. In: Reinforcement and Behavior, Tapp, J. T., ed. (Acad. Press, New York) pp. 286–327.

Castellano, C., B. E. Eleftheriou, D. W. Bailey and A. Oliverio (1974): Chlorpromazine and avoidance: a genetic analysis. *Psychopharmacologia (Berl.)*, **34**, 309.

Connolly, K. (1966): Locomotor activity in *Drosophila*. II. Selection for active and inactive strains. *Anim. Behav.*, **14**, 444.

Connolly, K. (1967): Locomotor activity in *Drosophila*. III. A distinction between activity and reactivity. *Anim. Behav.*, **15**, 149.

Daly, H. B. (1968): Disruptive effects of scopolamine on fear conditioning and on instrumental escape learning. *J. comp. phsyiol. Psychol.*, **66**, 579.

Deutsch, J. A. (1971): The cholinergic synapse and the site of memory. *Science* **174**, 788.

Domino, E. F., A. T. Dren and K. I. Yamamoto (1967): Pharmacologic evidence for cholinergic mechanisms in neocortical and limbic activating systems. In: Structure and Function of the Limbic System, Adey, W. R. and T. Tokizane, eds. (Elsevier, Amsterdam) pp. 337–364.

Ebel, A., J. C. Hermetet and P. Mandel (1973): Comparison of acetylcholinesterase and choline acetyltransferase in temporal cortex of DBA and C57 mice. *Nature New Biol.*, **242**, 56.

Eleftheriou, B. E. (1971): Regional brain norepinephrine turnover rates in four strains of mice. *Neuroendocrinology*, **7**, 329.

Eleftheriou, B. E. and D. W. Bailey (1972): A gene controlling plasma serotonin levels in mice. *J. Endocrinol.*, **55**, 225.

Elias, M. F. (1970): Differences in reversal learning between two inbred mouse strains. *Psychon. Sci.*, **20**, 179.

Fuller, J. L. (1967): Effect of drinking schedule upon alcohol preference in mice. *Quart. J. Stud. Alcohol*, **28**, 22.

Fuller, J. L., C. Easler and M. E. Smith (1950): Inheritance of audiogenic seizure susceptibility in the mouse. *Genetics*, **35**, 622.

Fuller, J. L. and F. H. Sjursen (1967): Audiogenic seizures in eleven mouse strains. *J. Hered.*, **58**, 135.

Fuller, J. L. and W. R. Thompson (1960): Behavior Genetics (Wiley, New York) 396 pp.

Ginsburg, B. and W. C. Allee (1942): Some effects of conditioning and social dominance and subordination in inbred strains of mice. *Physiol. Zool.*, **15**, 485.

Grossman, S. P. (1972): Cholinergic synapses in the limbic system and behavioral inhibition. *Neurotransmitters*, **50**, 315.

Horel, J. A. and E. G. Keating (1969): Partial Klüver-Bucy syndrome produced by cortical disconnection. *Brain Res.*, **16**, 281.

Hotta, Y. and S. Benzer (1970): Genetic dissection of the *Drosophila* nervous system by means of mosaics. *Proc. Natl. Acad. Sci. US*, **67**, 1156.

Igic, R., P. Stern and E. Basagic (1970): Changes in emotional behaviour after application of cholinesterase inhibitor in the septal and amygdala region. *Neuropharmacology*, **9**, 73.

Keating, E. G. and Horel, J. A. (1971): Somatosensory deficit produced by parietal-temporal cortical disconnection in the monkey. *Exp. Neurology*, **33**, 547.

Kempf, E., J. Greilsamer, G. Mack and P. Mandel (1973): Turnover of brain adrenergic transmitters in quaking mice. *J. Neurochem.*, **20**, 1269.

Kempf, E., J. Greilsamer, G. Mack and P. Mandel (1974): Correlation of behavioural differences in three strains of mice with differences in brain amines. *Nature New Biol.*, **247**, 483.

Klüver, H. and P. C. Bucy (1937): 'Psychic blindness' and other symptoms following bilateral temporal lobectomy in rhesus monkeys. *Amer. J. Physiol.*, **119**, 352.

Leaton, R. N. (1968): Effects of scopolamine on exploratory motivated behavior, *J. comp. physiol. Psychol.*, **66**, 524.

Lindzey, G. (1951): Emotionality and audiogenic seizure susceptibility in 5 inbred strains of mice. *J. comp. physiol. Psychol.*, **44**, 389.

Lindzey, G. and D. D. Thiessen (1970): Contribution to Behavior-Genetic Analysis. The Mouse as a Prototype. Century Psychology Series (Appleton-Century-Crofts, New York).

Longo, V. G. (1966): Behavioral and electroencephalographic effects of atropine and related compounds. *Pharmacol. Rev.*, **18**, 965.

Maas, J. W. (1962): Neurochemical differences between two strains of mice. *Science* **137**, 621.

Maas, J. W. (1963): Neurochemical differences between two strains of mice. *Nature*, **197**, 255.

Mack, G., J. Greilsamer, E. Kempf and P. Mandel (1973): Neurochemical correlates of genetically determined differences in performance levels in mice. 4th Intern. Meeting Neurochemistry, Tokyo, August, Abstr. p. 419, n° 531.

Mandel, P., G. Ayad, J. C. Hermetet and A. Ebel (1974): Correlation between choline acetyltransferase activity and learning ability in different mice strains and their offsprings. *Brain Res.*, **72**, 65.

Mandel, P., A. Ebel, J. C. Hermetet, D. Bovet and A. Oliverio (1973): Etude des enzymes du système cholinergique chez les hybrides F_1 de souris se distinguant par leur aptitude au conditionnement. *C. R. Acad. Sci. Paris, Série D*, **276**, 395.

Margules, D. L. and A. Margules (1973): Cholinergic receptors in the ventromedial hypothalamus govern operant response strength. *Amer. J. Physiol.*, **221**, 1475.

McClearn, G. E. (1965): Genotype and mouse behaviour. In: Genetics Today, Proc. XI Intern. Congr. Genet., vol. 3, Geerts, S. J., ed. (Pergamon Press, Oxford) pp. 795–805.

McGill, T. E. and W. C. Blight (1963): The sexual behavior of hybrid male mice compared with the sexual behavior of the inbred parent strains. *Anim. Behav.*, 11, 480.

Merrell, D. J. (1949): Selective mating in *Drosophila* melanogaster. *Genetics*, **34**, 370.

Meyers, B. (1965): Some effects of scopolamine on a passive avoidance response in rats. *Psychopharmacologia (Berl.)*, **8**, 111.

Meyers, B. (1968): Comment to effects of scopolamine on avoidance conditioning and habituation of mice. *Psychopharmacologia (Berl.)*, **13**, 354.

Mishkin, M. (1966): Visual mechanisms beyond the striate cortex. In: Frontiers in Physiological Psychology, Russel, R. W., ed. (Academic Press, New York) pp. 93–119.

Mordkoff, A. M., K. Schlesinger and R. A. Lavine (1964): Developmental homeostasis in behavior of mice: locomotor activity and grooming. *J. Hered.*, **55**, 84.

Oliverio, A. (1968a): Effects of scopolamine on avoidance conditioning and habituation of mice. *Psychopharmacologia (Berl.)*, **12**, 214.

Oliverio, A, (1968b): Some additional data on the effects of scopolamine on habituation and conditioning. *Psychopharmacologia (Berl.)*, **13**, 356.

Oliverio, A. (1968c): Neurohumoral systems and learning. Adrenergic and cholinergic facilitatory and inhibitory mechanisms. In: Psychopharmacology. A Review of Progress, Publ. n⁰ 1836, Efron, D. H., ed., (National Institute of Mental Health, Chevy Chase, Md., USA) pp. 867–878.

Oliverio, A., C. Castellano and P. Messeri (1972): Genetic analysis of avoidance, maze and wheel-running behaviours in the mouse. *J. comp. physiol., Psychol.*, **79**, 459.

Oliverio, A., C. Castellano, P. Renzi and M. Sansone (1973): Decreased sensitivity of septal mice to impairment of two-way avoidance by chlorpromazine. *Psychopharmacologia (Berl.)*, **29**, 12.

Oliverio, A., B. E. Eleftheriou and D. W. Bailey (1973): Exploratory activity: genetic analysis of its modification by scopolamine and amphetamine. *Physiol. Behav.*, **10**, 893.

Oliverio, A. and P. Messeri (1973): An analysis of single-gene effects on avoidance, maze wheel-running and exploratory behavior in the mouse. *Behav. Biol.*, **8**, 771.

Parkes, M. W. (1965): An examination of central actions characteristic of scopolamine: comparison of central and peripheral activity in scopolamine, atropine and some synthetic basic esters. *Psychopharmacologia (Berl.)*, **7**, 1.

Penfield, W. and B. Milner (1958): Memory deficit produced by bilateral lesions in the hippocampal zone. *Arch. Neur. Psychiat. (Chicago)*, **79**, 475.

Penfield, W. and P. Perot (1962): The brain's record of auditory and visual experience. A final summary and discussion, *Brain*, **86**, 595.

Pittendrigh, C. S. (1958): Adaptation, natural selection and behavior. In: Behavior and Evolution Simpson, G. G. and A. Roe, eds., (Yale University Press, New Haven) pp. 390–416.

Pryor, G. T., K. Schlesinger and W. H. Calhoun (1966): Differences in brain enzymes among live inbred strains of mice. *Life Sci.*, **5**, 2105.

Rosenzweig, M. R. and E. L. Bennett (1969): Effects of differential environments on brain weights and enzyme activities in gerbils, rats and mice. *Develop. Psychobiol.*, **2**, 87.

Rossum, J. M. van, and J. A. Th. M. Hurkmans, (1964): Mechanism of action of psychomotor stimulant drugs, Int. J. Neuropharmacol. **3**, 227–239.

Royce, J. R. and M. Covington (1960): Genetic differences in the avoidance conditioning of mice. *J. comp. physiol. Psychol.*, 53, 197.

Schlesinger, K. and A. M. Mordkoff (1963): Locomotor activity and oxygen consumption. Variability in two inbred strains of mice and their F_1 hybrids. *J. Hered.*, **54**, 177.

Scott, J. P. (1942): Genetic differences in the social behavior of inbred strains of mice. *J. Hered.*, **33**, 11.

Seiden, L. S. and D. D. Peterson (1968): Reversal of the reserpine-induced suppression of the conditioned avoidance response by L-dopa: correlation of behavioral and biochemical differences in two strains of mice. *J. Pharmacol. Exp. Therapeutics*, **159**, 422

Stein, L. (1969): Chemistry of purposive behavior. In: Reinforcement and Behavior Tapp, J. T., ed., (Adademic Press, New York) pp. 329–355.

Sturtevant, A. H. (1915): Experiments on sex recognition and the problem of sexual selection in Drosophila. *J. Anim. Behav.*, 5, 351.

Sudak, H. S. and J. W. Maas (1964): Central nervous system serotonin and norepinephrine localization in emotional and non-emotional strains in mice. *Nature*, **203**, 1254.

Tebb, G. and J. M. Thoday (1956): Reversal of mating preferences by crossing strains of Drosophila melanogaster. *Nature*, **177**, 707.

Thiessen, D. D. (1961): Mouse exploratory behavior and body weight. *Psychol. Rec.* **11**, 299.

Thompson, W. R. (1953): The inheritance of behaviour: behavioural differences in fifteen mouse strains. *Canad. J. Psychol.*, **7**, 145.

Vogel, J. R. and R. C. Leaf (1972): Initiation of mouse killing in non-killer rats by repeated pilocarpine treatment. *Physiol. Behav.*, **9**, 421.

Vogt, M. (1960): Hormones in Human Plasma: Nature and Transport (Little Brown, and Co, Boston).

Vossen, J. M. H. (1966): Exploratief gedrag en leergedrag bij de rat (with summary in English), Ph. D. Diss., Nijmegen University.

Warburton, D. M. (1969a): Effects of atropine sulfate on single alteration in hippocampectomized rats. *Physiol. Behav.*, **4**, 641.

Warburton, D. M. (1969b): Behavioral effects of central and peripheral changes in acetylcholine systems. *J. comp. physiol. Psychol.*, **68**, 56.

Warburton, D. M. and R. W. Russell (1969): Some behavioral effects of cholinergic stimulation in the hippocampus. *Life Sci.*, **8**, 617.

Woolley, D. W. and Th. van der Hoeven (1963): Alteration in learning ability caused by changes in cerebral serotonin and catecholamines. *Science*, **139**, 610.

CHAPTER 17

Behavioural effects of psychoactive drugs influencing the metabolism of brain monoamines in mice of different strains

I. P. Lapin

17.1. Introduction

An important practical purpose of studying different strains of laboratory animals in psychoneuropharmacological experiments is that one may find subjects which are suitable for extrapolating data obtained in animals to man and thus for predicting psychotropic effects in human subjects. For example, administration of a combination of desmethylimipramine and reserpine-like benzoquinolizine Ro 4-1284 causes a marked hyperactivity, a phenomenon used for the screening of imipramine-like antidepressants, in the Sprague-Dawley and Holzman strains of rats but not in the Osborne-Mendel and NIH Black strains (Bickel, 1968). Because a number of psychotropic ("emotiotropic") drugs, particularly antidepressants, neuroleptics, and stimulants, markedly affect the metabolism of monoamines and because this influence seems to be the basis of their psychotherapeutic action, our attention is focused on the possibility of detecting reliable models and subjects for predictive psychopharmacology among different strains of rodents, taking into consideration their patterns of emotional behaviour as well as their responses to drugs affecting the metabolism of monoamines. A recent paper illustrates this point (Maengwyn-Davies et al., 1973). In this study, three inbred strains of mice (BALB/c or NIH, C57BR/cdJ, and A/HeJ) were compared with regard to the influences of isolation, fighting, and various psychotropic drugs upon the activities of adrenal tyrosine hydroxylase and phenylethanolamine-N-methyltransferase. Strain differences have been reported for emotional behaviour (Maas, 1962; Bourgault et al., 1963; Sudak and Maas, 1964), levels of brain monoamines (Maas, 1962; Sudak and Maas, 1964), and for responses to drugs such as amphetamine (Weaver and Kerley, 1962; Lapin, 1966; Karczmar and Scudder, 1967), chlorpromazine and chlordiazepoxide (Fuller, 1970), and hypnotics (Mackintosh, 1962; Catz and Jaffe, 1967). Low levels of brain monoamines, associated with low levels of emotionality and excitability of the brain, have been reported for mice of the C57BL/10J strain (Sudak and Maas, 1964).

In our studies on antidepressants (Lapin, 1962–1971), we have also employed different strains of mice in order to detect reliable subjects for predictive purposes. Aggressiveness and fighting were used as indices of emotional behaviour. Since a remarkable parallelism seems to exist between many kinds of aggressiveness and the general level of nervous reactivity (Welch, 1969) and since, moreover, a relationship between the effects of the psychotropic drugs on animal behaviour and this reactivity level has been found (see Lapin, 1971), we carried out a series of experiments on spontaneous motor activity (response to a novel environment) and aggressive behaviour in various strains of mice. Among the drugs used, amphetamine and reserpine would seem to be of special interest, since interactions with these drugs are generally used for screening. Anti-depressants are screened in tests for synergism with amphetamine and in tests for antagonism to reserpine (or its derivatives). Neuroleptics counteract amphetamine. Anti cholinergics were included in our studies because their pharmacological profiles are very similar to those of antidepressants, which requires special procedures to differentiate between them (Lapin, 1967, 1970a). Some of the data obtained in our investigations are reported and discussed in the present chapter.

17.2. Methods

Male mice from different strains, weighing 18–22 g, were supplied by Rappolovo Farm (near Leningrad). They were housed in animal rooms in groups of 20–30 mice per cage ($30 \times 27 \times 28$ cm). Standard food and cow whole milk were given ad libitum. Prior to testing, the mice were kept in the laboratory in groups of 6–10 in metal cages of $20 \times 15 \times 10$ cm without food or milk. For comparison, random-bred male albino mice (called "common") obtained from the same farm were used. The rearing frequency of single mice was counted during sessions of 2 or 3 min, using plexiglass boxes ($10 \times 5 \times 20$ cm).

The total motor activity of grouped or isolated mice was measured by means of an electronic integrator (Zilberman and Lapin, 1965) in sessions of 5–90 min. The principle of the apparatus is that vibrations of the floor are transformed into electric signals which are integrated and then added up by the counter. With tuned sensitivity of the apparatus, practically only locomotion is registered as long as abnormal movements like stereotypies or seizures do not occur. The intensity of emotional excitement, accompanying amphetamine group toxicity, was measured by recording the loudness of squealing of fighting mice and other excited individuals in the group. For this purpose, a sensitive microphone was connected to one channel (and its counter) of the integrator.

Drugs were dissolved in distilled water and this solvent was also injected into the controls. All solutions and distilled water were injected intraperitoneally (i.p.) in a volume of 10 ml/kg of body weight. Dose-levels and times of injection are given in the Tables. Statistical evaluation of the data was performed by using Student's t-test or the χ^2 test.

17.3. Results

17.3.1. Rearing and locomotion

Levels of spontaneous rearing and locomotion. Comparisons were made between mice that were placed into the box for the first time. All came from

TABLE 17.1a

Rearing and locomotion in four different mouse strains.

Strain	Rearing* Mean±S.E. (N = 30)	Locomotion** Mean±S.E.	
		Single mice (N = 12)	Groups of 3 mice (N = 5)
C57BR	20.13±1.33	351±26	761±46
BALB	12.77±1.56	345±23	623±44
C57BL	9.63±1.33	259±31	428±59
Common	21.51±1.48	–	–

* Observed in sessions of 3 min.
** Values counted in sessions of 10 min. The comparison between strains for one particular test (rearing or locomotion was) carried out on the same day.

TABLE 17.1b

Significance levels of differences between strains (Student's t-test).

Strain	Rearing				Locomotion					
					Single mice			Grouped mice		
	C57BR	BALB	C57BL	Common	C57BR	BALB	C57BL	C57BR	BALB	C57BL
C57BR	—	0.01	0.001	n.s.	—	n.s.	0.05	—	n.s.	0.01
BALB	0.01	—	n.s.	0.01	n.s.	—	0.05	n.s.	—	0.05
C57BL	0.001	n.s.	—	0.001	0.05	0.05	—	0.01	0.05	—
Common	n.s.	0.01	0.001	—	—	—	—	—	—	—

TABLE 17.2

Extinction of rearing in three different mouse strains.

Strain	N	Mean±S.E. Ist test (initial)	2nd test (1 hr later)	P
BALB	6	16.2±0.95	11.8±0.8	< 0.01
C57BR	8	16.7±0.42	5.5±0.9	< 0.001
Common	7	18.7±1.8	11.4±2.4	< 0.05

The results of this experiment turned out to be reproducible in replicated experiments. C57BL mice were not tested in this experiment, but in a similar experiment they showed an extinction of rearing resembling that of BALB and common mice.

the animal farm on the same day and were housed in the same room. The data are shown in Table 17.1a and b. The C57BL mice displayed the lowest exploratory and motor activity. Mice of the C57BR strain reared more frequently than BALB or C57BL animals. The score for motor activity (locomotion) was highest in the C57BR mice and lowest in the C57BL mice. The decreasing order of strains was for both forms of motor activity: C57BR – BALB – C57BL.

Extinction of rearing. Because drug effects on motor activity are usually measured by comparing scores obtained after and before the injection, i.e. in repetitive trials and against a background of spontaneous decline in the activity, it is important to compare the rates of extinction of rearing in control mice of different strains.

On the basis of the data presented in Table 17.2, one can expect that with the procedure mentioned above, an inhibitory action of a drug (e.g. a sedative or tranquillizer) would be less marked in the C57BR mice than in the other strains. By contrast, a stimulant action of a drug would be more pronounced in the C57BR mice.

Spontaneous and experimentally-induced aggressiveness. As reported elsewhere (Lapin, 1966, 1970a), the C57BL mice differ from other strains in their lack of both spontaneous and shock-induced aggressive behaviour. The C57BR and the BALB animals are more aggressive than the common male mice. Our observations are in agreement with data on the low level of "emotionality" in C57BL mice (Maas, 1962; Sudak and Maas, 1964).

17.3.2. Effects of d,l-amphetamine sulfate

Toxicity. In single mice placed in plexiglass boxes of $6 \times 6 \times 9$ cm, similar signs of amphetamine intoxication were found in all three strains tested, namely in respect of excitement, tremor, clonic seizures, and hyperthermia. The LD50 of amphetamine was about 10 times lower in the single BALBs than in mice of the C57BR and C57BL strains (Table 17.3). About the same difference was observed between the BALB and common mice.

TABLE 17.3

Toxicity of subcutaneous injections with amphetamine in single and in grouped mice of three different strains.

Strain	Aggregation	LD50* mg/kg		Ratio single/grouped
BALB	Single	19.5	20.8	about 2
	Groups of 10	10.5	11.0	
C57BL	Single	152.2	200.3	about 3
	Groups of 10	48.0	64.0	
C57BR	Single	200.4	206.2	about 24
	Groups of 10	8.3	8.6	

* LD50 was determined in two colonies of mice in two seasons. Grouping was made in plexiglass boxes of $20 \times 10 \times 20$ cm under a room temperature of 19–20 °C. Lethality was measured during a 24 hr period after injection. In common male mice, the LD50 for amphetamine varied in several colonies and in different seasons: single mice, 90 to 150 mg/kg; grouped mice, 12 to 20 mg/kg.

Grouping lowered the resistance to amphetamine in BALB and C57BL mice by a factor about 2 or 3, and in C57BR mice by a factor 24 (Table 17.3). In male mice of other strains (Swiss, Swiss-Webster, GFF) this ratio lies between 5 and 10 (Weaver and Kerley, 1962; Consolo et al., 1965); therefore the latter strains can be placed between the C57BR and the C57BL strain.

Marked strain differences in behaviour were observed in grouped mice. Equitoxic doses of amphetamine produced strong emotional excitement (rage, squealing, fighting) in the BALBs, moderate excitement in the C57BRs and only slight emotional activation in the C57BLs (Table 4). In the grouped C57BL mice, the LD50 of amphetamine was 4–7 times higher than in the other strains. Differences in locomotor stimulation were less conspicuous (Table 17.4). Mice of the BALB strain showed the largest degree of hyper-

TABLE 17.4

Effects of equitoxic doses of amphetamine on total activity and squealing in grouped mice of three different strains.

Strain	Lethality 0/10		Lethality 5/10		Lethality 8/10	
	Activity	Squealing	Activity	Squealing	Activity	Squealing
BALB	+55%	599	+306%	506	+70%	562
C57BR	+11%	60	+160%	228	+52%	205
C57BL	+20%*	38	+114%	97	+52%	40

These data were collected in three experiments.
Activity (in periods of 10 min) was measured for 20 min before and 90 min after the injection. Of these nine 10 min intervals, those four were selected during which excitement was maximal. The values represent the differences between the mean scores for 10 min periods after and before injection. Squealing is represented as total values counted by the electronic integrator for four 10 min intervals during which squealing and excitement were maximal.
* This value represents mainly licking and gnawing of the platform of the box, whereas in the other two strains the counts almost exclusively represent locomotion.

activity. Similar differences in the behaviour of amphetamine-intoxicated mice of different strains were earlier reported (Lapin, 1966).

Hyperactivity. The data on amphetamine-induced motor excitement of grouped mice are also presented in Table 4. Amphetamine (10 mg/kg) enhanced rearing in isolated common and C57BL mice almost to the same extent (Lapin et al., 1972). As mentioned above, these two populations differ very substantially from each other in respect of spontaneous rearing. This enhancing effect was prevented by the antiserotonin drug BOL-148 (diethylamide of 2-bromolysergic acid) but increased by the precursor of serotonin, 5-hydroxytryptophan. These data suggest an involvement of brain serotonin in the stimulant action of amphetamine.

Hyperthermia. Different strains show different degrees of hyperthermia after injection of amphetamine (Lapin, 1966). In the grouped C57BLs, the maximum body temperature was significantly higher in animals which died later (38.4±0.9; max. $\Delta t = 2.8 \pm 0.3$) than in surviving animals (36.9±0.3; max. $\Delta t = 1.4 \pm 0.2$). A similar difference was observed in male mice of the C3HA strain (Lapin and Samsonova, 1964), but in mice of other strains and in common males this difference was absent. However, common female mice did show it (Lapin and Samsonova, 1964). The latter observation confirms data reported by Askew (1962).

Stereotypies. Amphetamine caused qualitatively different stereotyped behaviours in grouped as well as in single mice (Table 17.5). These differences remained unchanged over a wide range of doses of amphetamine. The amphetamine-induced stereotypies in the BALB strain are very similar to those described in ordinary rats and mice by Schelkunov (1964). The stereotypies found in the C57BL mice resemble the stereotypies produced in common mice by apomorphine, a drug which directly stimulates dopaminergic receptors in extrapyramidal structures of the brain (Ernst, 1967). It remains unknown in what way the strain differences for the amphetamine stereotypies are related to the dopamine metabolism in extrapyramidal systems, which is responsible for this phenomenon (Randrup and Munkvad, 1970).

TABLE 17.5

Stereotyped behaviour induced by amphetamine in three different mouse strains.

Strain	Predominant manifestation of behavioural stereotypy
BALB	Sniffing and cleaning in motionless animals
C57BR	Frequent up-and-down movements*
C57BL	Permanent gnawing of walls and floor

* In these stereotypies, the mice did not come down to a normal position with all four legs resting on the floor.

Thus, for studying amphetamine group toxicity as a test for experimentally produced emotional excitement, males of the C57BL and the C3HA strain and common females seem to be less suitable material because they hardly exhibit aggressive behaviour. The BALB strain is the one to be preferred for the study of aggressive behaviour produced by amphetamine. Moreover, in amphetamine-intoxicated BALB mice the lethality and hyperthermia are increased by pretreatment with the antidepressant imipramine (0.5 and 1 mg/kg) much more distinctly than they are in other strains (Lapin, 1970b). It is reported elsewhere (Lapin, 1962 and 1964; Lapin and Schelkunov, 1965) that potentiation of amphetamine group toxicity by small doses (0.5–2 mg/kg) of imipramine and its derivatives can be used as a test for screening antidepressants.

17.3.3. Effects of anticholinergics

The stimulating effects of anticholinergics (in doses higher than 0.02–0.03 mM/kg or about 5–7 mg/kg) on rearing and locomotion were practically identical in mice of the three strains tested (Lapin, 1967). Activation of locomotion in mice and rats by anticholinergics is a well-known phenomenon (Harris, 1961; Frances and Jacob, 1971), but stimulation of rearing has not been described before. The stimulating effect of anticholinergics is offered as a means for differentiating pharmacologically this class of drugs from antidepressants. In screening tests, these two classes of drugs have many characteristics in common.

Quaternary analogues of the anticholinergics did not affect rearing and locomotion (Lapin, 1967). Because these compounds penetrate into the brain only with difficulty, a central origin of the stimulation produced by tertiary anticholinergics is indicated.

It has previously been reported (Lapin et al., 1972) that the stimulating effect of anticholinergics (benactyzine, atropine, etc.) is associated with a lowering of the concentration of brain serotonin and noradrenaline, and that it can be prevented by depletors of brain serotonin (p-chlorphenylalanine) or brain noradrenaline (alpha-methyltyrosine). This suggests involvement of the two brain monoamines in the mechanism of stimulation by anticholinergics. The stimulation of rearing, produced by amphetamine or anticholinergics, is more dependent on brain serotonin than is the stimulation of locomotion. Strain differences between the Sprague-Dawley and Fisher strains of rats were found with regard to rearing and brain serotonin functioning (Rosecrans and Schecter, 1972) which suggested a correlation between the two.

The similarities between anticholinergics and amphetamine regarding the neurochemical nature of their stimulating action have led us to test whether an "anticholinergics group toxicity" does exist. In doses of 10–100 mg/kg, benactyzine, atropine, scopolamine, and parpanit failed to produce emotional group excitement as it is typical for amphetamine. No differences in LD50 values were found for these four drugs between single and grouped mice.

Starting from the idea that brain acetylcholine may be a universal trigger of aggressiveness and other emotional reactions (Allikmets, 1971), we sought an answer to the question whether an excess of brain acetylcholine can produce emotional excitement and fighting in grouped mice. Intraperitoneal injections of eserine (1 and 2 mg/kg) or proserine (0.5 and 1 mg/kg) into both control mice and mice pretreated with atropine methyliodide (20 mg/kg) or merpanit (20 mg/kg) failed to produce the phenomenon of "group toxicity" (the latter two quaternary compounds, which penetrate into the brain with difficulty, were used to check for peripheral toxic effects). Doses of eserine 10 and 20 times lower than those used in our experiments significantly increased the concentration of acetylcholine in the mouse brain (Frances and Jacob, 1971).

17.3.4. Imipramine-like antidepressants

Sedative action. The sedative action of large doses of imipramine (14–28 mg/kg or 0.05–1.00 mM/kg) on rearing manifested itself very drastically in the C57BR strain (Lapin, 1967). Therefore, this strain was used to trace the sedative effects of lower doses (Lapin, 1969). In testing motor activity, however, the sedative effect of imipramine turned out to be stronger in the BALB strain (Lapin, 1967), although mice of this genotype were more resistant to the sedative effect of the drug on rearing.

Reportedly (Campbell and Richter, 1967; Bickel, 1968; Lapin, 1964a and 1971), tricyclic antidepressants do not produce any specific effects when injected alone, and their action is tested mainly in combination with other drugs (reserpine, amphetamine, etc.). However, some of these nonspecific effects of antidepressants are of predictive value, e.g. their sedative action (Lapin, 1964b; Stille, 1968) and their effect on emotional behaviour (Allikmets, 1970; Lapin, 1971).

Emotionality. Table 17.6 shows that a single injection of imipramine tended to increase emotionality. Imipramine lowered the thresholds of squealing in mice of all three strains studied. There was also a tendency for an increase of aggressiveness in the C57BR mice. These findings are in agreement with data on rats reported by Allikmets (1964) and with a brilliant generalization by Welch (1969). Foot-shock failed to induce fighting in both the control and the imipramine-treated C57BL animals.

A further series of five experiments with each mouse strain showed that the lowering of the threshold of squealing is most pronounced in the C57BR strain after treatment with imipramine. A dose of 25 mg/kg of other typical antidepressants, desmethyl-imipramine and amitryptiline, did not alter the squealing threshold (or fighting), which means that this phenomenon cannot be used as a reliable test for this class of drugs. This conclusion is a tentative

TABLE 17.6

Effect of imipramine on threshold of squealing and on fighting in mice of three different strains.

Strain	Treatment	Threshold (volts) of squealing		Number of fighting pairs	
		before treatment	1 hr after injection	before treatment	1 hr after injection
C57BR	Dist. water	22.5±1.3	22.6±1.7	0/7	0/6
	Imipramine	25.6±1.5	20.4±1.9*	0/6	3/6
BALB	Dist. water	22.7±2.2	25.0±2.7	1/5	2/5
	Imipramine	25.0±1.2	22.0±0.6*	2/6	4/6
C57BL	Dist. water	26.0±1.8	33.0±2.1	0/6	0/6
	Imipramine	33.0±1.5	20.2±1.2*	0/6	0/6

* $P < 0.05$ (as compared with the original threshold, *i.e.* before treatment). Threshold of squealing was measured by means of the foot-shock technique. The number of fighting pairs is shown in the numerator, the number of pairs of mice tested in the denominator. Imipramine was given i. p. in a dose of 25 mg/kg. All mice were deprived of food and milk for one day before testing. A separate series of experiments has shown that starvation is a necessary condition for producing the effects of imipramine.

one because the antidepressants mentioned were not used in a wide dose-range.

Other effects. Chronic injection of imipramine (25 mg/kg) during 9 days did not change locomotion, thresholds of the flexor reflex, squealing, and fighting under foot-shock, nor did it change the effects of amphetamine and reserpine on the three strains (Lapin, 1970a). Negative findings were also obtained in common mice treated during 9, 10, and 16 days with imipramine, desmethylimipramine, and amitryptiline (Lapin, 1970a). These observations prove that in mice chronic injections are not advantageous as compared to single injections in screening antidepressants (Lapin, 1970a, 1971).

17.3.5. Antagonism to reserpine-like benzquinolizine Ro 4-1284

Considering ptosis and hypothermia, the strongest antagonist of Ro 4-1284 which is known among the antidepressants, desmethylimipramine (Bickel, 1968), is almost equipotent in all three strains of mice (Lapin,

1970b, 1971). Desmethylimipramine, in a dose of 10 mg/kg, was injected i.p. 30 min. prior to Ro 4-1284 (10 and 20 mg/kg; i.p.). Ptosis and hypothermia were registered during 3 hr after the injection of Ro 4–1284. In the BALB strain, the antagonism of desmethylimipramine to Ro 4-1284 was most stable.

An antagonism of imipramine-like antidepressants to the sedative effects of reserpine or its derivatives is difficult to detect in mice of various strains (Lapin, 1964c, 1970b, 1971). This is not in keeping with some other observations made in mice (see Sigg, 1968). In rats there is the well-known phenomenon of motor excitement produced by a combination of antidepressants and Ro 4-1284 (Sulser et al., 1962; Bickel, 1968).

17.3.6. Effects of reserpine and its derivatives

Reserpine in i.p. doses of 1 and 2 mg/kg produced the typical syndrome in all strains of mice studied, i.e. sedation, the characteristic posture ("hunched back"), ptosis, hypothermia, and diarrhoea. In the BALB strain, the value for hypothermia was significantly lower than in the other strains (Table 17.7). The BALB strain did not differ from the other strains with regard to its hypothermic response to Ro 4-1284 (10 and 20 mg/kg, i.p.).

TABLE 17.7

Hypothermia produced by intraperitoneal reserpine in three different mouse strains.

Reserpine	Strain	Hypothermia Cumulative indices* over 4 hr	
		Mean±S.E. (N = 6)	P
1 mg/kg	1. BALB	3.74±0.9	1–2 < 0.02
	2. C57BL	6.88±0.7	2–3 n.s.
	3. C57BR	7.63±1.3	3–1 < 0.05
2 mg/kg	1. BALB	7.91±0.9	1–2 < 0.002
	2. C57BL	16.42±1.7	2–3 n.s.
	3. C57BR	17.41±2.5	3–1 < 0.01

* Cumulative index (Garattini and Jori, 1967) is the sum of $\Delta t°$ for each measurement, as compared with the initial temperature. The temperature was measured 1, 2, 3, and 4 hr after injection. No strain difference was found for the scores of ptosis.

17.4. Conclusion

The data presented show that our mouse strains differ markedly in their behaviour and their responses to psychotropic drugs influencing brain monoamines (Table 17.8).

TABLE 17.8

A summary of the strain differences found for behaviour and for the responses to drugs influencing brain monoamines in mice.

Variable	Strain			Common mice
	BALB	C57BR	C57BL	
Behaviour				
Rearing rate	+	+++	(+)	+++
Total motor activity	+	+	(+)	*
Aggressiveness	+++	+++	(+)	+++
Amphetamine				
Resistance of single mice	(+)	+++	+++	++
Resistance of grouped mice	+	(+)	+++	+
Ratio of LD50 in single to LD50 in grouped mice	2	24	3	7–10
Difference in maximum body temperature between dying and surviving males	−	−	+	−
Aggression in grouped mice	++	+	(+)	++
Motor excitement in a group	++	+	(+)	*
Stereotypies	sniffing, cleaning	up-and-down movements	licking, gnawing	all kinds
Anticholinergics				
Stimulation of locomotion and rearing	identical			*
Antidepressants (imipramine)				
Inhibition of rearing	+	++	+	*
Inhibition of locomotion	++	+	+	*
Lowering of thresholds of squealing under foot-shock	identical			*
Antagonism of desmethyl-imipramine to Ro 4-1284	identical			*
Reserpine				
Hypothermia	(+)	++	++	*
Ptosis, sedation	identical			*

* No comparison in the same experiment. Key: +++ = very strong (high); ++ = strong; + = moderate; (+) = low; − = absent.

The BALB strain is very fit for studying amphetamine group toxicity. Mice of the C57BR strain can also be successfully used for this purpose. The C57BR strain seems to yield the best results in measuring the sedative effect of imipramine (and probably of other antidepressants) when testing inhibition of rearings, and the BALB strain when testing inhibition of locomotion. For studying the antagonism of antidepressants to reserpine and its derivatives, the strains turned out to be equally fit; the results in the BALB strain seemed to be somewhat more stable, though.

Further studies are necessary to discover among the strains the material most suitable for predictive psychopharmacology. The choices must be based on further systematic comparisons, particularly as far as emotional behaviour is concerned. Because tests for the screening of imipramine-like antidepressants are apt to detect effects on adrenergic processes and because there is also evidence (Lapin and Oxenkrug, 1969) that enhancement of central serotoninergic processes may be a determinant of the thymoleptic effect, it seems profitable to use a number of different strains of mice when searching for predictive tests with regard to the activation of central serotoninergic processes.

References

Allikmets, L. H. (1964): Effects of Psychotropic Drugs on Aggressiveness of Rats With Lesioned Septum or Amygdala. Transactions of Tartu University, Section "Medicine", **9**, 123–127 (In Russian).

Allikmets, L. H. (1970): Neurophysiological Analysis of the Action of Psychotropic Drugs on Limbic and Related Structures of the Brain, Vilnius (In Russian).

Allikmets, L. H. (1971): Is the Hypothalamic Trigger Mechanism of Aggression Cholinergic? In: Structural, Functional and Neurochemical Organization of Emotions, Papers of the All-Union Symposium (Leningrad, "Nauka"), 144–147 (In Russian).

Askew, B. M. (1962): Hyperpyrexia as a contributory factor in the toxicity of amphetamine in aggregated mice. *Brit. J. Pharmacol.*, **19**, 245.

Bickel, M. H. (1968): Untersuchungen zur Biochemie und Pharmakologie der Thymoleptika. In: Progress in Drug Research (Birkhäuser Verlag, Basel and Stuttgart) 121–225.

Bourgault, P. C., Karczmar, A. G. and Scudder, C. L. (1963): Contrasting behavioral, pharmacological, neurophysiological and biochemical profiles of C57BL/6 and SC-I strains of mice. *Life Sci.*, **8**, 533.

Campbell, D. E. S. and Richter, W. (1967): An observational method estimating toxicity and drug actions in mice applied to 68 reference drugs. *Acta pharmacol. toxicol.*, **25**, 345.

Catz, C. and Jaffe S. (1967): Strain and age variations in hexobarbital response. *J. Pharmacol. exp. Therap.*, **155**, 152.

Consolo, S., Garattini, S. and Valzelli, L. (1965): Amphetamine toxicity in aggressive mice. *J. Pharm. Pharmacol.*, **17**, 53.

Ernst, A. M. (1967): Mode of action of apomorphine and dexamphetamine on gnawing compulsion in rats. *Psychopharmacologia (Berl.)*, **10**, 316.

Frances, H. and Jacob, J. (1971): Comparaison des effets de substances cholinergiques et anticholinergiques sur les taux cerebraux d'acetylcholine et sur la motilité chez la souris. *Psychopharmacologia (Berl.)*, **21**, 338.

Fuller, J. L. (1970): Strain differences in the effects of chlorpromazine and chlordiazepoxide upon active and passive avoidance in mice. *Psychopharmacologia (Berl.)*, **16**, 261.

Garattini, S., and Jori, A. (1967): Interactions between imipramine-like drugs and reserpine on body temperature. In: Antidepressant Drugs, Garattini, S. and Dukes, M. N. G. eds. (Excerpta Medica, Amsterdam) pp. 179–193.

Harris, L. S. (1961): The effect of various anti-cholinergics on the spontaneous activity in mice. *Fed. Proc.*, **20**, 395 (abstract).

Karzcmar, A. G. and Scudder, C. L. (1967): Behavioral responses to drugs and brain catecholamine level in mice of different strains and genera. *Fed. Proc.*, **26**. 1186 (abstract).

Lapin, I. P. (1962): Qualitative and quantitative relationships between the effects of imipramine and chlorpromazine on amphetamine group toxicity. *Psychopharmacologia (Berl.)*, **3**, 413.

Lapin, I. P. (1964a): Characteristiques pharmacologiques de l'Imipramine. *Therapie (Paris)* **19**, 1107.

Lapin, I. P. (1964b): Comparative pharmacological data relating to the use of chloracizine and Tofranil in psychiatric practice. *Z. Nevropatol. i Psychiatr. (Moscow)*, **64** 281 (In Russian).

Lapin, I. P. (1964c): Biochemical Pharmacology of Nonhydrazine Antidepressants and Their Use for the Treatment of Diseases of the Nervous System. *Z Vsesojuzn. Khim. Ob. (Moscow)*, **9**, 438 (In Russian).

Lapin, I. P. (1966): Influence of grouping on the behavioral effects of amphetamine in various strains of mice. In: Psychopharmacology and Regulation of Behavior, Symposium of the XVIII International Congress of Psychology, Moscow, (Nauka, Moscow) pp. 85–86.

Lapin, I. P., Simple Pharmacological Procedures to Differentiate Antidepressants and Cholinolytics in Mice and Rats, 1967, Psychopharmacologia (Berl.), **11**, 79–87.

Lapin, I. P. (1969): Pharmacological Activity of Quaternary Derivatives of Imipramine and Diethylaminopropionyl-iminodibenzyl. *Pharmakopsychiat.-Neuropsychopharmakol.*, **2**, 14.

Lapin, I. P. (1970a): A Pharmacological study of Antidepressants of the Imipramine Group. Abstract of the D. Med. Sci. Thesis, Leningrad (In Russian).

Lapin, I. P. (1970b): A Pharmacological Study of Antidepressants of the Imipramine Group. D. Med. Sci. Thesis, Leningrad (In Russian).

Lapin, I. P. (1971): Antidepressants. In: Problems of Pharmacology, Advances of Science Series, part: Pharmacology, Chemotherapy Drugs, Toxicology (All-Union Institute of Scientific and Technical Information, Moscow) pp. 7–44 (In Russian).

Lapin, I. P. and Oxenkrug, G. F. (1969): Intensification of the central serotoninergic processes as a possible determinant of the thymoleptic effect. *Lancet*, **1**, 132.

Lapin, I. P., Oxenkrug, G. F. and Azbekyan S. G. (1972): Involvement of brain serotonin in the stimulant action of amphetamine and of cholinolytics. *Arch. int. Pharmacodyn.*, **197**, 350.

Lapin, I. P. and Samsonova, M. L. (1964): Role of hyperthermia in amphetamine group toxicity. *Farmakol. i Toxikol. (Moscow)*, **27**, 379. (In Russian)

Lapin, I. P. and Schelkunov, E. L. (1965): Amphetamine-induced changes in Behaviour of small laboratory animals as simple tests for evaluation of central effects of new drugs. In: Pharmacology of Conditioning, Learning and Retention, Proc. 2nd Int. Pharmacol. Meeting, Prague (Pergamon Press, Prague) pp. 205–215.

Maas, J. W. (1962): Neurochemical differences between two strains of mice. *Science*, **137**, 621.

Mackintosh, J. H. (1962): Effect of strain and group size on the response of mice to seconal anaesthesia. *Nature*, **194**, 1304.

Maengwyn-Davies, G. D., Johnson, D. G., Thoa, N. B., Weise, V. K. and Kopin, I. J. (1973): Influence of isolation and of fighting on adrenal tyrosine hydroxylase and phenylethanolamine-N-methyltransferase activities in three strains of mice. *Psychopharmacologia (Berl.)*, **28**, 339.

Randrup, A. and Munkvad, I. (1970): Biochemical, anatomical and psychological investigations of stereotyped behavior induced by amphetamines. In: Amphetamine and Related Drugs, E. Costa and S. Garattini, eds. (Raven Press, New York) pp. 695–713.

Rosecrans, J. A. and Schechter, M. D. (1972): Brain 5-Hydroxytryptamine correlates of behavior in rats: strain and sex variability. *Physiol. Behav.*, **8**, 503.

Schelkunov, E. L. (1964): The method of "amphetamine stereotypy" for testing the effects of Drugs on central adrenergic processes. *Farmakol. i Toxikol. (Moscow)*, **27**, 628 (In Russian).

Sigg, E. B. (1968): Tricyclic thymoleptic agents and some newer antidepressants. In: Psychopharmacology. A review of Progress, 1957–1967, D. H. Efron, ed. (Public Health Service, Publ. No. 1836, Washington) pp. 655–669.

Stille, G. (1968) Pharmacological investigation of antidepressant compounds. *Pharmakopsychiatr.-Neuropsychopharmakol.*, **1**, 92.

Sudak, H. S. and Maas, J. M. (1964): Central nervous system serotonin and norepinephrine localization in emotional and nonemotional strains in mice. *Nature*, **203**, 1254.

Sulser, F., Watts, J. and Brodie, B. B. (1962) On the mechanism of antidepressant action of imipramine-like drugs. *Ann. N.Y. Acad. Sci.*, **96**, 279.

Weaver, L. C., and T. L. Kerley, (1962): Strain Difference in Response of Mice to D-amphetamine. *J. Pharmacol. exp. Ther.*, **135**, 240.

Welch, B. L. (1969): Symposium Summary. In: Aggressive Behavior, Proc. Symposium on the Biology of Aggressive Behavior, Milan (Excerpta Medica, Amsterdam) pp. 363–369.

Zilberman, N. E. and Lapin, I. P. (1965): An electronic integrator for the quantitative testing of effects of drugs on motor activity of a group of small laboratory animals. *Farmakol. i Toxikol. (Moscow)*, **28**, 495 (In Russian).

Subject index

A, mouse strain, 267f, 337, 385
Acetylcholine (ACh), 333, 360, 368, 370–372, 391, 404, 425f
Acetylcholinesterase (AChE), 333f, 352, 357f, 360, 367f, 370–372, 385, 400, 402ff
Acheta
  rubens, 124
  veletis, 124
Acquiescent posture, in mice, 326
A/Crgl, mouse strain, 14
Active avoidance learning
  in mice, 381ff, 399ff
  in rats, 307–309, 313
Activity
  in *Drosophila*, 205–252
  in *Drosophila*, larval, 206ff
  in *Drosophila*, spontaneous, 51
  in mice, 14, 73, 331, 348, 383ff, 399ff, 418
  in mice, horizontal, 347, 350, 357, 367
  in mice, vertical, 347, 350, 357, 367
  in rats, 378
  see also Locomotor activity
Adaptation, 181, 185–187
  phylogenetic, 126
  social, 278
Adaptive, 3, 34, 47, 49, 51–53, 58, 67, 80f, 88, 126, 210, 294, 397
Addiction, 60, 381
Additive
  components of variation, 11, 15, 23, 33, 35, 46f, 66
  environmental variation, 19
  genetic effects, 5, 8, 27, 36f, 66, 70ff, 91, 217
  variation, 8, 33, 46, 48f, 99, 105, 108, 113
Additivity–dominance model, 15, 20f, 29, 33, 65f, 70, 76, 83
Adenosine triphosphatase (ATPase), 268, 411
Adipsia, 302
Adrenal weight, 333f
Adrenaline, 334, 377
Adrenocorticotropic hormone, 278
Aedes, 233
Affective
  mouse-killing behaviour, 302ff
  states, 293
A2G, mouse strain, 333, 400
Agapornis, 126f
  fischeri, 123
  roseicollis, 123
Aggression–flight balance, 186
Aggressive behaviour
  in birds, 124, 127
  in humans, 274
  in mice, 185–199, 295–303, 321–346
  in poeciliid fish, 125
  in rats, 293ff
Aggressiveness
  in cichlids, 120
  in mice, 119, 193, 294, 321ff, 398f, 418,

421, 424–426, 429
  in rats, 294ff, 339
A/He, mouse strain, 326, 382, 384, 417
A/J, mouse strain, 326, 339, 384, 400
AKR, mouse strain, 326, 400
Alarm substance, 129
Alcohol
  dehydrogenase, 380f
  preference, 380f, 398
Aldehyde dehydrogenase, 380
Alizarin staining, 148, 265
Alkaline phosphatase, 268
Ambidirectional dominance, 6, 18, 47, 50–52, 61, 67, 73, 88, 91, 108
Ambulation
  in mice, 331, 347f, 350
  in rats, 17f, 22, 25f, 51f, 348
1 (2) *Amd* mutant, 242
Amenorrhea, 275
American
  blacks and whites, 96
  foulbrood disease, 136
p–Aminobenzoic acid, 376
Amitryptiline, 426f
Amphetamine, 269, 311, 327, 334, 379, 386–389, 391f, 418, 422–427, 429
  group toxicity, 419, 422–424, 430
  toxicity, 334
Amygdala, 298, 300f, 307f, 312–314, 335, 400f
Anas
  acuta, 123, 126
  platyrhynchos, 123, 126
Androgens, 32, 294, 299f, 306, 309, 335–337
Anger, 293, 299
Anolis
  aeneus, 122
  trinitatis, 122
Anopheles gambiae, 182
Anoptichthys, 129f
Anosmia, 306f
Antennaless, 229
Anticholinergic, 349, 353, 358, 360, 363, 368, 371, 401f, 418, 425, 429
  group toxicity, 425f

Anticholinesterase, 349, 353, 358, 360, 401f
Anticoagulants, 375
Antidepressant, 375, 417f, 424–430
Antipyrine, 375
Aphagia, 302
Apis mellifera, 4
Apodemus sylvaticus, 186
Apomorphine, 424
Appetitively-motivated attack, in killer-rats, 296ff
Arecoline, 381
Aristaless(*al*), 228f
Arithmetic test, 108
Artemia, 147
Artificial selection, 48f, 119f
Arylamines, 376
Assortative mating, 97, 99, 105f, 109–111, 113, 222, 250
Astyanax mexicanus, 129f
Atavism, 125f
Atropine, 425
Attack latency, in mice, 195–198, 339
Attention mechanisms, 349
Audiogenic seizures, 259, 267–270, 398, 400f
Auditory stimuli, in courtship of *Drosophila*, 225ff
Aversively-motivated defence, in killer-rats, 296ff
6–Azauracil, 377

Backing from behind, in *Xiphophorus*, 131f
Bacteria, 397
BALB, mouse strain, 186
BALB/c, mouse strain, 69ff, 325–327, 333, 337, 339, 380–384, 388–391, 399f, 417, 419–430
BALB/Sc, mouse strain, 324, 326, 337, 342
Barbiturate, 375f
Bar locus, 172
Basenjis, 328
Beagles, 328
Behaviour
  biological clusters in, 88
  evolution, 3, 39, 45–61, 119–140, 169, 185, 202, 248

Subject index

genetics, 43, 121, 129, 134, 136, 183, 202f, 273f, 287, 296, 329, 347, 392, 397
hygienic, in honey bees, 4
human, 91–117
inheritance, 1, 3f, 127
mammalian, 51
ontogenesis (development), 134f, 137, 164, 292–295, 305, 312, 336ff
sex–associated, 59
Benactyzine, 425
Benzoquinolizine Ro 4–1284, 417, 427–429
Between–family variation, 7, 20, 92ff
Biometrical
analyses, 65–89, 380f, 383
approach, 1–41, 43–61, 91–117, 202, 236, 348
model, 5–41
Birds, 168f, 322, 331
Biting
in mice, 323, 341
in rats, 324
Blood–brain barrier, 349, 352
Blood group, 53
Boar, 167
Bobbing sequence, in lizards, 123
Body
–image, 276
–schema, 277
–size, in mice, 55
Bony plates, 145ff
Bowing display, in doves, 124f
Brain
stem, 264, 333–335, 400
weight, 357, 368, 409
Breeding
capacity, 148
commercial, 34
cycle, in sticklebacks, 141
design, 23, 52, 60
mouth–, 123–126
plant, 44
programme, 38
random, 378–380, 383, 418
selective, 10, 119f, 135, 294f, 321ff, *see also* Selection
size, effective, 187

structure, 4
studies, 19
substrate–, 123–126
value, 70, 80, 83, 329
Bristle (chaeta) number in *Drosophila*, 49f, 252
2–Bromolysergic acid diethylamide (BOL–148), 423
Brood–care, in *Tilapia*, 123
Brown
(*b*), 365
(*bw*), 224
Buffering
mechanism, 27
to stress, 2
Bufo
americanus, 122
americanus americanus, 122
woodhousei, 122
woodhousei fowleri, 122

Calliphora, 229
Castration, 268, 306, 309, 325, 332, 336
Cat, 299, 304, 310, 376f
Cattanach's translocation, 263
CBA, mouse strain, 69ff, 326, 380f, 384
C57BL, mouse strain, 186
C57BL/6, mouse strain, 69ff, 267, 326, 337, 339, 354–367, 380–391, 398–410
C57BL/10, mouse strain, 324–327, 331, 333, 337, 339f, 342, 384, 399f, 418–429
C57BL/Crgl, mouse strain, 14
C57BL/LiA/Gro mouse strain, 191f
C57BR, mouse strain, 326, 382, 384, 417, 419–430
Cerebellum, 264, 402ff
CF–1, mouse strain, 326
CFW, mouse strain, 326
C3H, mouse strain, 68ff, 324, 326, 339, 382–385, 400f, 423f
Chaeta (bristle) number in *Drosophila*, 49f, 252
Chickens, 321f, 329, 331, 376
Chimpanzees, 293
Chironomus, 147

Chlordiazepoxide, 348, 382f, 418
p-Chlorophenylalanine, 310, 335f, 425
Chlorpromazine, 297, 348, 378f, 382, 392, 409, 418
Choline acetyltransferase (ChA), 308, 313, 372, 385, 402ff
Cholinesterase (ChE), 376
  inhibitor, 313
Chorthippus
  biguttulus, 125
  brunneus, 124
Cichlasoma nigrofasciatum, 120
Cinnabar (*cn*), 172ff
Circling, in mice, 259ff
C57L, mouse strain, 326
Climbing, in mice, 354f
Cocker–spaniels, 328
Cognitive development, 275ff
Coisogenic stocks, 177, 180
Colour pattern, in sticklebacks, 146
Commensals, 186
Concordance, 112f, 115
Conditioning
  avoidance, in *Drosophila*, 203
  avoidance, in mice, 5, 69, 74–76, 348, 381ff, 399ff
  avoidance, in rats, 307–309, 313
  classical (or Pavlovian), 56
  instrumental (or operant), 56, 203, 297
Congenic lines, of mice, 388–391, see also Inbred strains of mice
Consummatory behaviour patterns, 292, 300
Contact behaviour, in cichlids, 125, 127, 129
Convergent evolution, 3
Convulsions, in mice, 267
Copulation, in *Drosophila*, 213ff
Corpus
  allatum, 231ff
  cardiacum, 231ff
Correlated
  environments, 8
  genetic and environmental influences, 99, 105, 108
Correlates, neurochemical, of behaviour, 397–415
Correlation
  coefficient, intraclass, 92
  genetic, 81, 92ff, 384
  marital, 97, 102, 107, 114
  parent–offspring, 29, 97, 103, 105, 112
  phenotypic, 99
  real and foster parents, 100
  rearing behaviour and locomotor activity, 356–357, 367
  tetrachoric, 111f
Cortex
  cerebral, 400
  frontal, 335, 400ff
  limbic, 402ff
  occipital, 402ff
  parietal, 402ff
  pyriform, 333, 399f
  temporal, 358, 385, 402ff
Coumarin, 375
Coupling phase of linkage, 37
Courtship
  in birds, 123–127
  in *Drosophila*, 170f, 174, 178, 210–247
  in poeciliid fish, 130ff
  in sticklebacks, 142
Covariance
  family means, 17, 66ff
  g_2 and e_2 effects, 94f
  sibling, 92
CPB–s/Gro, mouse strain, 190–192, 195–197
Creeping through
  in sticklebacks, 142–164
  refractory period, 144, 162
Criminals, 274
Critical periods in cognitive development, 287
Cross–fostering, 337f
Crowing, in birds, 124f
Cubitus valgus, 275
Cultural norms, 288
Cytochrome oxidase, 268
Cytogenetics, of mice, 347

Dancer (*Dc*), 263f
Dark preference, in mice, 74, 76

DBA/1, mouse strain, 326, 339, 400
DBA/2, mouse strain, 267f, 326, 339, 354–366, 371, 380–387, 398–410
Deaf (v^{df}), 266
Deafness, 259, 261f, 266
  (dn), 266, 270
Decreaser gene, 6, 21, 24, 35–37, 47f, 54, 96
Defecation
  in mice, 325, 331, 348, 399
  in rats, 17f, 22, 32, 52, 307f, 325, 348
Defense posture, in mice, 326
Deleterious genes, 108, 208
Deme, 187, 190, 193
Desmethylimipramine, 417, 426–429
Diallel cross, 11, 17–22, 26, 28f, 51f, 57, 65–89, 217
Diencephalon, 399
Dihybrid ratio, 161f
α–Dimethyl tyrosine, 241f
Dimorphism, 47
  sexual, 59, 167f
Diphenolase (tyrosinase A_2), 240
Diptera, 233
Direction sense, 277
Directional
  dominance, 35, 48, 50–52, 54, 57, 61, 66f, 72ff, 97ff, 106, 217
  selection, 67, 108, 252
Discordant MZ twins, for schizophrenia, 115
Discrimination, left-right, 276
Dispersion of genes, 35, 37, 54, 83
Displacement activities, 126
Disruptive selection, 47, 59, 252
Dizygotic twins, *see* Twins
Dogs, 321, 328, 330, 376
Domesticated
  animals, 50, 135, 330
  form, 127
  mice, 270
Dominance
  absence of, 6
  ambidirectional, 6, 18, 47, 50–52, 61, 67, 73, 88, 91, 108
  complete, 6, 67, 81, 86

components of variation, 11, 15, 19f, 23, 33, 46, 48, 53, 66
  directional, 35, 48, 50–52, 54, 57, 61, 66f, 72ff, 97ff, 106, 217
  for long copulation duration in *Drosophila*, 236
  genetic effects, 8, 36, 66, 70ff
  level of, 91
  order, 68, 70ff, 385
  over-, 6, 35, 54, 67ff
  partial, 67
  properties of genes, 23
  ratio, 19, 47
  relationships, 6, 329
  unidirectional, 6, 49, 54, 91
  unsystematic, 70
  variation, 8, 46, 48, 99, 103, 105f, 110, 113f
Dominant status in mice, 187, 190–197, 322ff, 339f
Dopa (3,4-dihydroxyphenylalanine), 239f, 336, 400
  decarboxylase, 242
  quinone, 239f
Dopamine (DA), 210, 239, 241f, 269, 333, 336, 400, 407–410, 424
  β-hydroxylase, 302
Dose-response curves, 382
Double creeping through, in sticklebacks, 144–164
Dreher (dr), 264
Drive
  -energy, 280
  internal state, 292
Drosophila
  activity, 205–252
  americana americana, 183
  bristle (chaeta) number, 49f, 252
  copulation, 213ff
  courtship, 170f, 174, 178, 210–247
  feeding, in larvae, 206–209
  generation span, 141
  geotactic behaviour, 119, 134f, 207, 252
  licking, 213ff
  locomotor activity, 205–252, 398
  mating behaviour, 170, 210–252, 397f

mating speed, 119, 181, 216ff, 243–248
melanogaster, 38, 46, 49, 51, 119f, 168ff, 201–258, 397f
mounting, 213ff
optomotor reaction, 119
orientation, 213ff
paulistorum, 182
pavani, 178
persimilis, 120, 127, 178, 181, 227, 250
phototactic behaviour, 119, 252
picicornis, 215
populations, 167–183
pseudoobscura, 120, 127, 177, 181f, 217, 227, 250, 252
reactivity, 51, 206
rejection responses, 213ff
robusta, 178
scissoring, 215
simulans, 119, 237, 247
subobscura, 178, 225, 228, 230f, 234
viability, 49–51
visually-elicited behaviour, 398
wing vibration, 213ff
Dwarfism in children, 279f
Dyscalculia, 276, 278
Dyslexia, 276, 278

Ebony (e), 173, 179, 242
Educational attainments, 108–111
Electroconvulsive shock, 385f
Electroencephalographic patterns, 377
Electroretinogram, 242
Emotional
  elimination, in rats. 26, 32, 51f
  reactivity, in rats, 17, 209
  responses, in mice, 348, 400
  responsiveness, 52, 292ff, 331
Emotionality
  in mice, 348, 398f, 401, 417f, 421, 426
  in rats, 18, 22, 325, 348
Environmental
  components of variation, 11, 16, 20, 29, 46
  manipulation, 4
  variance, 7, 93ff
  variation, 2, 330

Epistasis, 23, 48, 51, 54, 67ff
Erotic development, 286
Eroticism, 285
Escape
  -avoidance conditioning, in rats, 18, 26, 52, 56–58, 304
  behaviour, in monkeys, 406
  from water, in rats, 52
  learning, in mice, 69, 76–81
Eserine, 425f, see also Physostigmine
Estrogen, 32f, 275, 284f
Ethanol, 375, 380
Ethology, 45, 314
Euscelis
  lineolatus, 125
  plebejus, 125
Evans blue, 352f
Evolution
  convergent, 3
  history, 108
  of behaviour, 3, 39, 45–61, 119–140, 169, 185, 202, 248
Experience, 22, 134, 291ff, 325f, 336, 340–343
Exploratory
  activity (Exa), 391
  behaviour, in mice, 65, 69–74, 261, 347–374, 378f, 386–391, 399, 401
  behaviour, in rats, 302
Expressivity of shaker-waltzer mutants, 260, 265
Extrachromosomal components of variation, 8, 11
Eye colour, 53

Factorial analysis, 323
Fecundity, in Drosophila, 182f, 231
Feeding, in Drosophila larvae, 206–209
Feral mice, 186f, 191, 195–197, 270, 379f
Fertility
  in Drosophila, 230–232
  in mice, 196
Fidget (fi), 262f
Fisher strain of rats, 425
Fitness, 48–51, 60, 108, 185, 216, 225, 235, 250

Fixable components of variation, 8, 46, 49
Fixation, 48, 365
Fixed action patterns, 122, 124, 134, 203
Fleeing, adaptive, 58
Follicle stimulating hormone (FSH), 335
Forebrain, 334f
Forked (*f*), 173, 177, 179, 182
Fostering, 95f, 99, 337f, 350
Fraying, in mice, 197f
Freezing, 56, 58, 81, 307, 340, 352
Fright reaction, in fish, 129f
Frontal position, in *Xiphophorus*, 132f
Frustration, 293
Full siblings
  reared apart, 93ff
  reared together, 93ff
Fungi, 397

Gallus sonnerati, 124f, 127
Gamma-aminobutyric acid (GABA), 209, 333, 399f
Gamma-hydroxybutyric acid, 210
Gasterosteus aculeatus, 141–165
Gender
  identity, 281ff
  role, 281ff
Gene(s), loci
  activation, 337
  association, 36f
  balanced effect, 5
  -blocks, 135, 178, 198
  decreaser, 6, 21, 24, 35–37, 47f, 54, 96
  deleterious, 108, 208
  dispersion, 35, 37, 54, 83
  increaser, 6, 21, 24, 35–37, 47f, 54, 96
  linked, 135, 206, 328
  location, 388, 390
  major, 4, 10, 38, 164, 222, 236, 377, 391
  minor supplementary effect, 4
  modifier, 133
  number of, 36f, 67ff, 98, 106, 114f, 132, 329, 372, 380
  number of linked, 37
  pool, 120, 169, 181, 183, 194, 206, 235, 248, 252

  single, 5, 27, 103, 115, 136f, 179, 202, 204, 236f, 377, 391
  super-, 178
  unequal frequencies, 66
Genetic(al)
  additive effects, 5, 8, 27, 36f, 66, 70ff, 91, 217
  architecture, 3, 24, 29, 43–60, 67, 91, 108, 217, 227, 330, 380
  background, 53, 60, 164, 180, 186, 237, 239, 260f, 264f, 270, 343, 365, 381
  components of variation, 2, 46, 66, 120, 134
  correlations, 81, 92ff, 384
  determination, degree of (DGD), 67ff, 329
  dominance effects, 8, 36, 66, 70ff
  drift, 169, 365
  fundamental equation, 1, 44
  interrelations, 65
  mosaics, 204, 227, 275f
  non-additive effects, 67, 73, 83, 91
  substrate, 5
  variation, 2, 6f, 216, 239, 329, 341, 367, 391, 397
Genetics
  Bacteria, 397
  behaviour(al), 43, 121, 129, 134, 136, 183, 202f, 273f, 287, 296, 329, 347, 392, 397
  Fungi, 397
  Mendelian, 3, 347
  plant, 24, 46
Genotype-environment interaction, *see* Interaction
Geotactic behaviour, in *Drosophila*, 119, 134f, 207, 252
GFF, mouse strain, 422
Glucose-6-phosphate dehydrogenase, 375
Glucuronic transferase, 376
Glutamate, 209, 400
Glutamic acid, 268
Gnashing the teeth, in rats, 324
Gonadal hormones, 31, *see also* Sex-hormones
Gonadotropic hormone, 278

Grooming, in mice, 69–73, 190, 196, 326, 333, 354f, 399
Groszfamilien, mice, 186
Gryllus
  bimaculatus, 128
  campestris, 128
Guarding behaviour, in baboons, 123
Guinea-pig, 59

Habenular nucleus, medial, 310
Habitat, 185, 248
Habituation, 22, 313, 347, 401
Half-siblings, 17
Heritability, 2, 10, 19, 34, 120, 206, 380f
  broad, 20, 33, 49, 53, 67ff, 99, 103, 114, 329
  narrow, 19f, 22, 34, 49, 61, 67ff, 99, 102f, 107, 114, 217, 329
  realisable, 102f, 110
  realised, 208
Hermaphroditism, 274, 283
Herpes simplex virus, 336
Heterogametic sex, 27f
Heterosis, 24, 34–36, 54f, 163, 217, 354
Hexobarbital, 376
Hippocampus, 333, 349, 353, 363f, 371, 399–401
Histocompatibility (*H*) locus, 388–391
Holzman strain of rats, 417
Homogametic sex, 27
Honey bees, 4, 136
Hormonal substitution treatment, 283, 285f, 288
House mice, 185ff, 347
HS, mouse strain, 326, 339
Human behavioural development, 273f
Hybrid
  between species, 121–137
  vigour, 80, *see also* Heterosis
Hybridisation, 10, 122, 137, 177, 180, 250, 252
6-Hydroxydopamine, 302
5-Hydroxyindoleacetic acid, 310
5-Hydroxytryptamine (5-HT, serotonin), 210, 310f, 333–336, 376, 399f, 407, 409f, 423, 425

5-Hydroxytryptophan (5-HTP), 310, 376, 423
Hygienic behaviour, of honey bees, 4
Hyperactivity, 259, 261f, 270, 307, 311, 417, 422f
Hyperphagia, 304, 309
Hyperthermia, 422–424
Hypnotics, 418
Hypoactivity, 302
Hypo-cerebral ganglion, 231
Hypogonadism, 275, 285
Hypopituitarism, 278, 280
Hypothalamus, 169, 334f, 400f, 406–410
  lateral, 298–302, 304
  medial, 303f
  ventromedial, 306, 309f
Hypothermia, 427–429

Imipramine, 377, 417, 424, 426–430
Impulse-level, general, 280
Inbred strains of mice
  A, 267f, 337, 385
  A/Crgl, 14
  A2G, 333, 400
  A/He, 326, 382, 384, 417
  A/J, 326, 339, 384, 400
  AKR, 326, 400
  BALB, 186
  BALB/c, 69ff, 325–327, 333, 337, 340, 380–384, 388–391, 399f, 417, 419–430
  BALB/Sc, 324, 326, 337, 342
  CBA, 69ff, 326, 380f, 384
  C57BL, 186
  C57BL/6, 69ff, 267, 326, 337, 339, 354–367, 380–391, 398–410
  C57BL/10, 324–327, 331, 333, 337, 339f, 342, 384, 399f, 418–429
  C57BL/Crgl, 14
  C57BL/LiA/Gro, 191f
  C57BR, 326, 382, 384, 417, 419–430
  CF-1, 326
  CFW, 326
  C3H, 68ff, 324, 326, 339, 382–385, 400f, 423f
  C57L, 326
  CPB-s/Gro, 190–192, 195–197

Subject index

DBA/1, 326, 339, 400
DBA/2, 267f, 326, 339, 354–366, 371, 380–387, 398–410
HS, 326, 339
MO, 326
recombinant (RI), 377, 388ff
RF, 326f
SC-1, 326
SEC/1, 383–387, 403–410
SJL, 326f, 400f
ST, 326
SWR, 326
Inbreeding, 6, 97f, 151, 163, 174, 176, 187, 329, 365, 388, 392
    coefficient, 98
    depression, 97
Incipient speciation, 47, 182
Increaser gene, 6, 21, 24, 35–37, 47f, 54, 96
Individuals
    isogenic, 95
    unrelated, 28
    unrelated, reared together, 93ff
Inertia of emotional arousal, 279f, 288
Infantilism, 275, 278–280
Inheritance
    behaviour, 1, 3f, 127
    chromosomal, 2
    Mendelian, 164
    mono-(uni-)factorial, 128f, 132, 134f, 156
    particulate, 4
    polyfactorial, 130, 136
Inner ear, of the mouse, 259–267
Insects, 4, 18, 168, 202, 231
Instinctive responses, 1
Insulin, 268
Intellectual
    abilities, 96
    functioning, 276
    skills, 94
Intelligence, 98–108, 275f, 348, 397
Intention movement, 214
Interaction
    between gene loci, 11, 137
    between genes, complementary, 23f, 36
    between genes, duplicate, 24, 48–51, 54, 61

between genetic and experiential factors, 294, 296, 327
between offspring genotype and maternal genotype, 25, 28, 338
dominance, see Dominance
epistatic, see Epistasis
genotype-environment, 2f, 9–11, 16–18, 21f, 26, 29–33, 53, 55, 57, 59, 91, 95ff, 100, 164, 202, 273f, 291, 321f, 336ff, 340, 342
non-allelic, 8, 10, 12–18, 21, 23f, 27f, 31, 34–36, 48, 51
parameters, 13, 15
sex × genotype, 66
Intersex, 168
Intraclass correlation coefficient, 92
Inversion heterozygote, 177f
Iproniazid, 376
IQ, 94f, 99–108, 111ff, 276, 283
Isogenic stock of *Drosophila*, 237
Isogenicity, 175f
Isolation
    ethological, 120ff, 222, 248ff
    geographical, 181, 187, 250
    joint index, 250
    mechanism, 169, 202
    sexual, 47, 120, 169, 181f, 211, 216, 226f, 248–252
Isoniazid, 375, 377

Japanese
    children and IQ, 106
    waltzers, 260
Jerker (*je*), 262
Jerking movements of the head, 259, 261
Joint isolation index, 250
Joint scaling test, 13–15, 19
Jumping
    in mice, 69–73
    in rats, 324
Juvenile hormone, 231

Karyotype, 177f, 377
Killer–rat, 294–314, 335
Kinky (Fu^{ki}), 262, 264, 266
Klinefelter's (XXY) syndrome, 274, 276f

442 Subject index

Klüver-Bucy syndrome, 405
Kreisler (*kr*), 263
Kynurenine, 224

Lactosamines, 241
Lactose synthetase, 241
Leaning, in mice, 69–73, 347, 350ff
Learning
  autonomic responses, 342
  curve, 81
  discrimination, 406
  in mice, 5, 65, 69, 74–76, 81–86, 331, 348, 381ff, 398ff
  in rats, 119f, 307–309, 313
  instrumental, 56, 203, 297
  strict sense, 2
  theory, 56–58
Leiura, forma, 145ff, 157ff
Lek, 216
Lesion experiments, 296ff
Licking, in *Drosophila*, 213ff
Limbic activating system, 401
Linkage, 37, 54f, 384, 388f, 391f, *see also* Sex-linkage
  coupling phase, 37
  repulsion phase, 37, 47f, 61
Lion, 167
Litters
  difference between, 7
  difference within, 7
Locomotor activity
  in cichlids, 120
  in *Drosophila*, 205–252, 398
  in mice, 15, 70–73, 350, 354–357, 366f, 369–371, 399ff, 419ff
  in rats, 300, *see also* Ambulation
Loco–motor anomalies, 136
Locus coeruleus, 302
Loriculus galgulus, 126
Lozenge (*lz*), 240
Lysergic acid diethylamide (LSD), 377

Macro-environmental effects, 7f, 55–57, 59
Major gene, 4, 10, 38, 164, 222, 236, 377, 391
Malic dehydrogenase, 268

Maltese dilution (*d*), 365
Mammals, 7, 12f, 27, 168, 201, 270, 292f, 321
Maniac, sticklebacks, 148ff
MAO inhibitor, 376, 399
Marital correlation, 97, 102, 107, 114
Maternal
  components of variation, 11
  effects, 7f, 19, 24, 26, 28f, 66, 70ff, 337f, 390
  motivations, 261
  stress, 26
Mating
  ability, 224, 226, 228
  assortative, 97, 99, 105f, 109–111, 113, 222, 250
  behaviour, in *Drosophila*, 170, 210–252, 397f
  behaviour, in poeciliid fish, 123
  -calls, in anurans, 122
  non-random, 171
  random, 224, 249
  refusal, 233–235
  sib-, 11, 27, 175, 356, 365, 389
  site (lek), 216
  speed, in *Drosophila*, 119, 181, 216ff, 243–248
  success, 180, 223f, 236, 239, 249
  system, 46
Maudsley
  emotionally non-reactive (MNR) rats, 26, 31f
  emotionally reactive (MR) rats, 26, 31f
Medulla, 406–411
Meiosis, 2
Melanin, 240
Memory, 383, 385, 405f
Mendelian
  classical analyses, 4
  genetics, 3, 347
  inheritance, 164
  models, 5
  ratio, 153, 156, 163
Mental deficiency, 397
Mercapturic acid, 376
Merpanit, 426

Mergus merganser, 123, 125
Mesencephalic central grey, 303f, 306, 311
Mesencephalon, 399
  ventral, 302
Methylatropine, 426
α-Methyldopa, 242
Methylscopolamine, 352f, 360ff
α-Methyltyrosine, 406, 408, 425
Metrazol seizure, 385f
Mice
  acquiescent posture, 326
  active avoidance learning, 381ff, 399ff
  activity, 14, 73, 331, 348, 383ff, 399ff, 418, *see also* Locomotor activity
  aggressive behaviour, 185–199, 295–303, 321–346
  aggressiveness, 119, 193, 294, 321ff, 398f, 418, 421, 424–426, 429
  ambulation, 331, 347f, 350
  biting, 323, 341
  body size, 55
  circling, 259ff
  climbing, 354f
  common, 418ff
  conditioning, avoidance, 5, 69, 74–76, 348, 381ff, 399ff
  defecation, 325, 331, 348, 399
  defence posture, 326
  domesticated, 270
  emotionality, 348, 398f, 401, 417f, 421, 426
  escape learning, 69, 76–81
  exploratory behaviour, 65, 69–74, 261, 347–374, 378f, 386–391, 399, 401
  feral, 186f, 191, 195–197, 270, 379f
  field, 190, 193
  fraying, 197f
  grooming, 69–73, 190, 196, 326, 333, 354f, 399
  house-, 185ff, 347
  inner ear, 259–267
  jumping, 69–73
  leaning, 69–73, 347, 350ff
  learning, 5, 65, 69, 74–76, 81–86, 331, 348, 381ff, 398ff
  locomotor activity, 15, 70–73, 350, 354–357, 366f, 369–371, 399ff, 419ff, *see also* Ambulation
  mounting, 332
  Nanking, 260
  neophila, 347
  neurological mutants, 259, 398, 411
  nosing, 196, 323
  passive avoidance, 69, 74–76, 381ff
  plague, 193f
  reared by rats, 338
  rearing behaviour, 69–73, 347, 350ff, 418ff
  resting, 69–73
  righting reflexes, 261
  running, 323
  sitting, 69–73
  sniffing, 69–73, 347, 350ff
  squealing, 419, 423, 426f, 429
  tail-rattling, 323, 325f, 341
  tolerant, 190–197
  turning, 69–73
  vibrating forepaws, 69–73
  walking, 69–73
  washing, 69–73
Micro–environmental effects, 5, 7f, 53
Micromys minutus, 186
Microphyla
  carolinensis, 122
  olivacea, 122
Microtine rodents, 185
Microtus, 326
  arvalis, 186
Mid–parental
  score, 107
  value, 6, 115
Migration, 187, 193, 195
MO, mouse strain, 326
Model
  Fisher's, 96f, 102, 104f, 111, 114
  -fitting, 23, 95, 100–105, 109, 111–113
  Mendelian, 5
  two-gene, 35
Modifiers, 133
Monkeys, 59, 406
Monoamino oxidase (MAO), 376, 399
Monogamous, 168

Monophenolase (tyrosinase–A_1), 239–241
Monozygotic twins, see Twins
Morphine susceptibility, 380f
Morphs, behavioural, 197
Mosaic, 204, 227, 275f
Mounting
  in *Drosophila*, 213ff
  in mice, 332
Mouse-killing, in rats, 294–314, 335
Mouse strains, see Inbred strains of mice
Multiple factors, genetic and environmental, 5f
Mus
  musculus, 185, 326, 347, 398
  musculus domesticus, 186ff
Mutations, 10, 120, 135–137, 179, 204, 222, 224, 231, 235-237, 240, 242, 260, 270, 274

Nanking mice, 260
Nature–nurture controversy, 1, 3, 273
Nembutal, 380
Nemobius
  allardi, 124
  tinnulus, 124
Neophilia, in mice, 347
Neophobia, in rats, 347
Neostigmine, 352f, 360ff
Nest material carrying, in parrots, 123–126
Neuroleptic, 417f
Neurological mutants of mice, 259, 398, 411
Nicotine, 348, 378, 382–384
Nightingale, 168
Nigrostriatal dopamine system, 269
NIH Black strain of rats, 417
Non-additive genetic effects, 67, 73, 83, 91
Non-fixable components of variation, 8, 46, 49
Noradrenaline (NA; norepinephrine, NE), 210, 239, 241, 302, 310, 314, 333–335, 376f, 385, 399–401, 406–411, 425
Nosing, in mice, 196, 323
Novelty, 349, 371, 401
Number of genes (loci), 36f, 67ff, 98, 106, 114f, 132, 329, 372, 380,
  of linked genes, 37

Nijmegen waltzer (*nv*), 262ff

Olfactory bulbs, 299, 306–309, 313, 335, 352, 402ff
Ommochrome, 224
Ontogenesis (development) of behaviour, 134f, 137, 164, 292–295, 305, 312, 336ff
Onychomus, 326
Optomotor reaction, in *Drosophila*, 119
Orientation
  in *Drosophila*, 213ff
  in space and/or in time, 276
Osborne-Mendel strain of rats, 417
Overdominance, 6, 35, 54, 67ff

Panmixia, 168, 170, 172
Papio
  anubis, 123
  hamadryas, 123
Paragonial glands, 233–235
Pargyline, 376, 409
Parpanit, 425
Particulate inheritance, 4
Partitioning
  of genetic variation, 66
  of heritable and environmental variation, 10
  of phenotypic variation, 11, 92
Passive avoidance, in mice, 69, 74–76, 381ff
Paternity, 194
Peas, Mendel's, 141
Peduncle, 402ff
Penetrance, shaker-waltzer mutants, 260, 265
Pentylenetetrazol, 269
Perception, space-form, 276f
Periaqueductal grey matter, 311
Peromyscus maniculatus bairdii, 326
Personality
  absorption, 280
  development, 273f, 278ff
Phasianus colchicus, 123f
Phenes, 237
Phenocopy, 224, 241, 342
Phenolase, 241
Phenol oxidase, 240f

Subject index

Phenothiazine, 377
Phenotypic variation, 8, 11, 15, 33, 53, 94, 227, 329, 332, 377
Phenylbutazone, 375
Phenylethanolamine-N-methyltransferase, 417
Phenylketonuria, 397
Pheromones, 59, 230, 233–235
Philomachus pugnax, 168
Phototactic behaviour, in *Drosophila*, 119, 252
Physostigmine, 352f, 358ff, 386f, *see also* Eserine
Pirouette (*pi*), 262
Pituitary, 335
Placement effects, 96ff
Plant, 4, 18, 23f, 38, 44, 46
Plague, of mice, 193f
Platypoecilus, 130
Pleiotropy, 50, 236, 260, 328, 384, 388, 391
Plumage, 127, 168
Poecilia sphenops, 125
Polder, 187
Polygamous, 168
Polygene, 36, 176, 365
Polygenic, 5, 73, 130, 134f, 137, 267, 329, 348, 381
  interactions, 181
  system, 5ff, 50, 55, 108, 181
Polymorphism, 47, 379f
  balanced, 197f
  chromosomal, 178
  stable, 164
Pons, 399, 406–411
Population(s)
  anadromic, 146
  breeding structure, 4
  cages, 50, 169–172
  cave-, 129
  coastal, sticklebacks, 146f
  degree of inbreeding, 54
  density, 185, 193, 235
  dynamics, 185, 193
  equilibrium within, 47
  feral, 186f, 192
  heterogeneous, 180, 348, 397
  human, 91
  incidence, 113, 115
  inland, sticklebacks, 146f
  invertebrates, 168
  monomorphic, 146
  natural, 51, 172, 174, 177, 181, 183, 187, 194, 197
  non-inbred, 29, 33, 39
  outbred, 97
  polymorphic, 146
  random-bred, 379f
  randomly-mating, 21, 33, 96, 102
  recombinant, 388
  size, 185
  structure, 185–199
  unselected, 49
  wild, 60
Potence ratio, 6, 19, 23, 27
Preoptic area, 309
Primaquine, 375, 377
Primates, 28, 59
Propylthiouracil, 268
Proserine, 425
Psychiatric disorders, 397
Psychogenetics, 4, 10, 28, 37, 43, 45, 54f, 58, 60
Psychology, 43, 45, 55, 59f, 201, 236, 314
Psychopharmacogenetic study, 60, 349, 380, 383, 392
Psychoticism, 113
Pterin, 224
Ptosis, 427–429
Pubertal development, 275

Rabbit, 270, 376f
Races
  ecological, 121, 128, 135
  of dogs, 135
Raphé nuclei, 306, 310f
Rat–reared mice, 338
Rats
  active avoidance learning, 307–309, 313
  activity, 378, *see also* Locomotor activity
  ambulation, 17f, 22, 25f, 51f, 348

aggressive behaviour, 293ff
aggressiveness, 294ff, 339
biting, 324
conditioning, avoidance, 307–309, 313
defecation, 17f, 22, 32, 52, 307f, 325, 348
emotional elimination, 26, 31, 51f
emotional reactivity, 17, 209
emotionality, 18, 22, 325, 348
escape-avoidance conditioning, 18, 26, 52, 56–58
escape from water, 52
exploratory behaviour, 302
Fisher strain, 425
gnashing the teeth, 324
Holzman strain, 417
jumping, 324
learning, 119f, 307–309, 313
locomotor activity, 300, *see also* Ambulation
Maudsley emotionally non-reactive (MNR), 26, 31f
Maudsley emotionally reactive (MR), 26, 31f
mouse–killing, 294–314, 335
neophobia, 347
NIH Black strain, 417
Osborne–Mendel strain, 417
Sprague–Dawley strain, 417, 425
squealing, 324
Wistar strain, 306
Reactivity, in *Drosophila*, 51, 206
Reading test, 108
Rearing behaviour, in mice, 69–73, 347, 350ff, 418ff
Receptivity, sexual, female, 170, 172ff, 216, 226ff, 245–247
Reciprocal crosses, 24f, 27–29, 66, 153, 156, 177, 179f, 336
Recombinant inbred (RI) strains of mice, 377, 388ff
Refractory period, in creeping through, 144, 162
Refsum's disease, 397
Regression
  parent-offspring, 206
  W_r on V_r, 17f, 66ff

Reinforcement
  of agressiveness in mice, 327, 342
  of killing in rats, 297, 305, 313
Rejection responses, in *Drosophila*, 213ff
Repulsion phase of linkage, 37, 47f, 61
Reserpine, 297, 400, 417f, 426–430
Response
  disinhibition, 349, 360, 368
  instinctive or learned, 1
Responsiveness
  emotional, 52, 292ff, 331
  social, 292ff
Resting, in mice, 69–73
Reticular
  activating system, 300
  formation, 408
RF, mouse strain, 326f
Rhabditis, 135
  inermis inermis, 128
  inermis inermoides, 128
Rhesus monkeys, 59
Rhinencephalon, 371
Rhode Island Red chickens, 331
Righting reflexes, in mice, 261
Ring–X chromosome, 204
Robin, 168
Rodents, 270, 297, 321, 324, 333, 349, 401, 417
Romantic interests, 285f
Rotating (*rg*), 263
Running, in mice, 323

Salmon, 167
Sample skewness, 98
Savageness, 322, 324, 330
SC–1, mouse strain, 326
Scale
  of measurement, 6, 10ff, 25
  5–point, 324
  7–point, 325
  rating, 322f
  transformation, 11ff
Scaling
  criteria, 19
  test, joint, 13–15, 19
  tests, 11ff

Schizophrenia, 111–115, 302
Scissoring, in *Drosophila*, 215
Scopolamine, 352f, 358ff, 368ff, 386, 388–391, 401, 425
   -induced modification of exploration (*Sco*), 391
SEC/1, mouse strain, 383–387, 403–410
Selection
   artificial, 48f, 119f
   differential, 34
   directional, 67, 108, 252
   disruptive, 47, 59, 252
   extreme, 36
   for aggressiveness in mice, 193, 294f, 321ff
   for alcohol intake, 380
   for attack latencies in mice, 195, 197
   for audiogenic seizure susceptibility, 267
   for body size in mice, 55
   for brain weight in mice, 327
   for emotional elimination in rats, 26
   for emotional reactivity in rats, 209
   for emotionality in rats, 325
   for larval feeding rate in *Drosophila*, 208
   for locomotor activity in *Drosophila*, 205ff
   for mating speed in *Drosophila*, 217, 243–248
   for rearing frequency in mice, 349, 364–367
   for running performance in mice, 83
   for sexual behaviour in *Drosophila*, 180
   for tolerance in mice, 193
   gain, 181
   in sticklebacks, 163f
   natural, 43–61, 108, 120, 167–183, 216f, 222, 250
   population change, 34
   relaxed, 206
   response to, 34, 55, 72, 205ff, 217, 226
   sexual, 167–183
   stabilising, 47ff, 67, 88
   strains, 24
Self-stimulation, 298, 300–302, 304, 311
Semi-armata, forma, 145ff, 157ff
Semi-circular canals, 263, 265f
Seminal vesicle weight, 333, 335
Sensitivity
   of neural detector-systems, 292
   to drugs, 60, 375
   to the environment, 31
Sepia (*se*), 177, 182
Septal rage syndrome, 308
Septum, 306, 308
Serotonin, see 5-Hydroxytryptamine, 5-HT
Sex
   assignment, 281f
   -associated behaviour, 59
   -chromosomal patterns, aberrant, 274
   -chromosome, 283
   determination, 168
   -hormones, 275, 277f, 280, 284–288, 299, 309, 337
   -linkage, 24, 27–29, 164, 336
   -peptide, 233–235
   pheromone, 230
   ratio, 189f
   reassignment, 281, 283
Sexual
   behaviour in TA and TNA selected mice, 332, 335, 341f
   dimorphism, 59, 167f
   female receptivity, 170, 172ff, 216, 226ff, 245–247
   infantilism, 275
   isolation, 47, 120, 169, 181f, 211, 216, 226f, 248–252
   motivation, 131, 239
   phase, in sticklebacks, 142f
   selection, 167–183
   vigour, 169, 172ff
Shaker
   -1 (*sh*-1), 262f
   -2 (*sh*-2), 262
   -waltzer syndrome in mice, 259ff
   -with-syndactylism (*sy*), 263
S-high selected mouse strain, 365–371
Siblings reared together, 92ff
Sib-mating, 11, 27, 175, 356, 365, 389
Single genes (loci), 5, 27, 103, 115, 136f, 179, 202, 204, 236f, 377, 391
Sitting, in mice, 69–73
SJL, mouse strain, 326f, 400f
Sleeping-time

after hexobarbital, 376
after nembutal, 380
S-low selected mouse strain, 365–372
Snell's waltzer (*sv*), 262
Sniffing, in mice, 69–73, 347, 350ff
Social
  adaptation, 278
  behaviour of baboons, 123
  dependence, 280
  responsiveness, 292ff
Socio-economic status (SES), 100, 107
Song
  in crickets, 124
  in doves, 125
  in *Drosophila*, 127, 225
  in insects, 124
  in nightingale, 168
Spawning movements, in cichlids, 125
Speciation, 248, 252
  incipient, 47, 182
  sympatric, 182
Spotting, white (Mi^{wh}, W^v and s^l), 267
Sprague–Dawley strain of rats, 417, 425
Square–root transformation, 14–16, 19f, 22
Squealing
  in mice, 419, 423, 426f, 429
  in rats, 324
ST, mouse strain, 326
Stabilising selection, 47ff, 67, 88
Stereotyped
  attack in killer-rats, 297
  behavioural act, 2
  mouse behaviour, 399, 419, 424, 429
Stickleback, three-spined, 141–165
Stimulant drug, 417
Strain distribution pattern (SDP) method, 388f
Streptopelia, 124
  roseogrisea var. *risoria*, 125
  senegalensis, 125
Stress
  buffering to, 2
  fighting-, 334
  maternal, 26
  prior to mating, 26

Stressors, response to, 398
Stria medullaris, 310
Stunted growth, 275–280, 284
Subnormality, in children of subnormal parents, 107f
Subnormals, educable, 275
Subordinate status, in mice, 190–194, 340
Succinic dehydrogenase, 268
Sulfadimidine, 376
Sulfonamides, 376
Super
  -genes, 178
  males, 274
Susceptibility, differential, to the environment, 95f, 100
Swiss–Webster mice, 378, 422
Sword bending, in *Xiphophorus*, 124, 131f
SWR, mouse strain, 326
Sympatric speciation, 182

Tadorna tadorna, 123, 125
Tail-rattling, in mice, 323, 325f, 341
t-alleles, 187
Tameness, 322, 330
Tan (*tn*), 242
Tantrum behaviour, 293
TA selected mouse strain, 328, 331ff
Taxis component, 124
Tegmentum, mesencephalic, ventral, 298, 301, 304
Temperament, 287
Terrier, 328
Territory, 141f, 167f, 185–189, 191, 194f, 197
Testis weight, 333f
Testosterone propionate, 325, 337
Thalamus, dorsomedial, 303, 306, 310
Thread (*th*), 228f
Threat behaviour, in lizards, 123
Threshold
  alterations, 136
  of acceptance, 231f
  of arousal, in humans, 287
  of creeping through, in sticklebacks, 162
  of emotional susceptibility, 339
  of liability to disease, 112

of receptivity, 221, 245–247
of response to electric shocks, 269
of sexual response, 218
of squealing, in mice, 427, 429
phenomena, 211
Thyroid hormone, 278
Thyrotropic hormone, 278
Tilapia, 123, 126
   heudeloti macrocephala, 127, 129
   nilotica, 125, 127, 129
   tholloni, 125
Timidity, 324
Tolerant mice, 190–197
TNA selected mouse strain, 328, 331ff
Trachura, forma, 145ff, 157ff
Translocation, Cattanach's, 263
Transsexualism, 281
Tremble-shoving, in parrots, 126
Trembling movements, in crickets, 128
Triiodothyroxine, 268
Triple-test cross, 18, 23, 29, 60
Trotzperiod, 279
Tryptophan, 224
   pyrrolase, 224
Tubifex, 147
Turner's (XO) syndrome, 273–290
Turning, in mice, 69–73
Twins, 91
   differences in metabolism, 375
   dizygotic, reared together, 93ff
   monozygotic, reared apart, 93ff
   monozygotic, reared together, 93ff
Two-factor theory of escape-avoidance conditioning, 56–58
Typological
   approach, 183
   thinking, 169
Tyr-1 locus, 240
Tyrosine, 239f
   hydroxylase, 241, 307, 417

Unidirectional dominance, 6, 49, 54, 91
Unsystematic dominance, 70

Variance
   assortative mating, 110f
   environmental, 7, 93ff, 273
   family means, 17, 66ff
   genetic, 2, 93ff, 273
   within-family, 17f
Variation
   additive, 8, 33, 46, 48f, 99, 105, 108, 113
   additive components of 11, 15, 19, 23, 33, 35, 46f, 66
   additive environmental, 19
   between families, 7, 20, 92ff
   continuous, 4f, 9, 44, 91, 96
   dominance, 8, 46, 48, 99, 103, 105f, 110, 113f
   dominance components of, 11, 15, 19f, 23, 33, 46, 48, 53, 66
   environmental, 2, 330
   environmental components of, 11, 16, 20, 29, 46
   extrachromosomal components of 8, 11
   fixable components of, 8, 46, 49
   free, 48
   genetic, 2, 6f, 216, 239, 329, 341, 367, 391, 397
   genetic components of 2, 46, 66, 120, 134
   genetic, in drug effects, 348ff, 375ff, 398ff, 417ff
   human, 91
   maternal components of, 11
   non-fixable components of, 8, 46, 49
   phenotypic, 8, 11, 16, 33, 53, 94, 227, 329, 332, 377
   potential, 7, 48
   quantal and discontinuous, 5
   quantitative, 4, 45
   random, 8
   total, 5, 46, 92ff
   total genetical, 6
   within families, 7, 20, 92ff
Varitint-waddler (*Va*), 262
Vermillion (*v*), 172ff, 224
Vestigial (*vg*), 173, 179, 226f
Viability, in *Drosophila*, 49–51
Vibrating forepaws, in mice, 69–73
Vigour
   general, 163, 167, 179f, 182
   sexual, 169, 172ff

Visually-elicited behaviour, in *Drosophila*, 398
Visual stimuli, in courtship of *Drosophila*, 222

WAIS test, 283
Walking, in mice, 69–73
Waltzer (*v*), 260, 398
Waltzer-type (*Wt*), 263
Washing, in mice, 69–73
Waving behaviour of larval nematodes, 128
Webbed neck, 275
White (*w*), 173, 179, 223, 236
  -apricot (*w^a*), 223
White Leghorn chicken, 331
Wildness, 270, 322, 324, 330
Wing vibration, in *Drosophila*, 213ff
WISC test, 277
Wistar strain of rats, 306
Within-family
  skewness, 98, 106
  variances, 17f
  variation, 7, 20, 92ff
Word
  -constructions, 277
  -mobilisation, 276f
  -usage, 277

X-chromosome, 127, 180, 222, 227, 240, 275, 285, 287
Xiphophorus, 127, 137
  helleri, 123, 130–132
  montezumae, 132f
  montezumae cortezi, 123, 130
  montezumae montezumae, 131
XO (Turner's) syndrome, 273–290
X-ray irradiation, 120
XYY sex-chromosome configuration, 274

Y-chromosome, 274
Yellow (*y*), 179, 237ff, 252

Zig-zag dance, in sticklebacks, 142